Astrophysics and Space Science Library

Volume 463

The Astrophysics and Space Science Library is a series of high-level monographs and edited volumes covering a broad range of subjects in Astrophysics, Astronomy, Cosmology, and Space Science. The authors are distinguished specialists with international reputations in their fields of expertise. Each title is carefully supervised and aims to provide an in-depth understanding by offering detailed background and the results of state-of-the-art research. The subjects are placed in the broader context of related disciplines such as Engineering, Computer Science, Environmental Science, and Nuclear and Particle Physics. The ASSL series offers a reliable resource for scientific professional researchers and advanced graduate students.

More information about this series at http://www.springer.com/series/5664

Ivan I. Shevchenko

Dynamical Chaos in Planetary Systems

Springer

Ivan I. Shevchenko
Pulkovo Observatory, Saint Petersburg State University
Saint Petersburg, Russia

ISSN 0067-0057 ISSN 2214-7985 (electronic)
Astrophysics and Space Science Library
ISBN 978-3-030-52143-1 ISBN 978-3-030-52144-8 (eBook)
https://doi.org/10.1007/978-3-030-52144-8

Cover illustration: A planet at the border of circumbinary chaos, a sketch. Provided by the author.

This Springer imprint is published by the registered company Springer Nature Switzerland AG.
The registered company address is: Gewerbestrasse 11, 6330 Cham, Switzerland

Preface

We are living in a new age of discovery.

C. Murray, S. Dermott,
Solar System Dynamics (1999)

Dynamical chaos (the nondeterministic dynamical behaviour in the absence of any random perturbations) plays a major, decisive role in the appearance of the observed dynamical architectures of planetary systems, including our own Solar system. Enough to say, the planet definition itself (as adopted by the General Assembly of the International Astronomical Union in Prague in 2006) implies the existence of a planetary chaotic zone cleared from low-mass material.

Therefore, three rapidly developing branches of modern science are exposed in the book, exhibiting their interplay: the resonant and chaotic dynamics of Hamiltonian systems, the dynamics of the Solar system bodies, and the dynamics of exoplanetary systems. The necessary notions, methods, and tools of the dynamical chaos science (such as symplectic maps, Lyapunov exponents and timescales, chaotic diffusion rates, and stability diagrams and charts) are described and then used to show in detail how the observed dynamical architectures arise in the Solar system (and its subsystems) and in exoplanetary systems. Especially, we concentrate on chaotic diffusion and clearing effects.

The dynamical chaos, as a plenipotentiary subject of research in physics and dynamical astronomy, is a relatively recent development. The first milestone was set 60 years ago by Boris Chirikov in his work on the overlap of nonlinear resonances in Hamiltonian dynamics (Chirikov 1959). The work was soon followed by large-scale studies of chaotic dynamics in physical applications, such as physics of particle accelerators and tokamaks. In celestial mechanics, the chaotic dynamics of celestial bodies became a well-defined field of research in the 1980s of the twentieth century. Today, studies of various manifestations and effects of dynamical chaos in systems of celestial bodies form an integral part of modern astrophysics, celestial mechanics, and exoplanetary science.

Indeed, this is a general trend in physical sciences: the theory of dynamical chaos is becoming warranted in more and more applications. This trend is illustrated in Fig. 1: a steady permanent rise is evident in the citing score of the classical

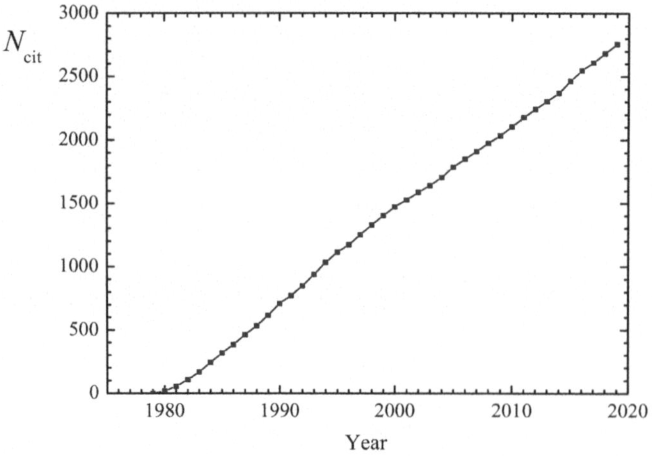

Fig. 1 The integral number of citations of Chirikov's (1979) work, as a function of time, up to year 2019 inclusive, as inferred from NASA ADS citation data

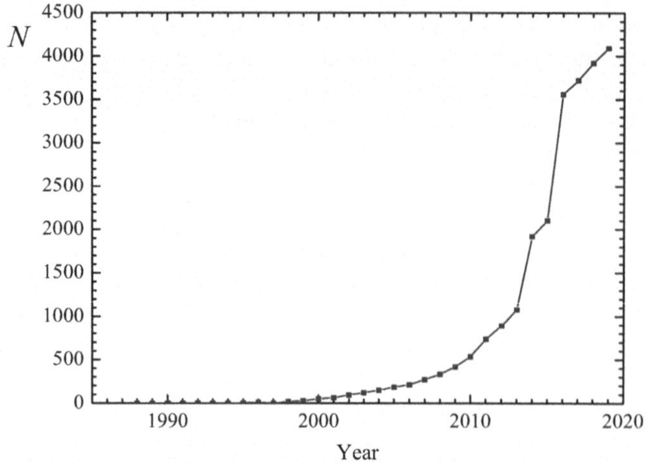

Fig. 2 The total number of confirmed discovered exoplanets, as a function of time, up to year 2019 inclusive, as inferred from Exoplanet.eu data

pioneering review by Chirikov (1979), from 1980s to nowadays. On the other hand, the current rise in the discovery score of exoplanets is burst-like; see Fig. 2. There is no doubt that "we are living in a new age of discovery", as cited in the epigraph to this Preface.

The book consists of three parts.

In Part I, the origins and manifestations of dynamical chaos in Hamiltonian systems are considered. The nonlinear resonances and their interaction and overlap are discussed in detail. The methods for analytical estimating the Lyapunov exponents and timescales, diffusion rates and timescales, widths of chaotic layers, and extents of chaotic domains are described. Paradigmatic symplectic maps, such as the standard map and the separatrix map, are introduced. An introduction to modern numerical tools for characterizing qualitative dynamics is also presented. These techniques are especially useful for building stability charts, for example, by means of massive computations of the Lyapunov exponents on grids of initial data.

In Part II, based on the tools developed in Part I, we analyse resonances and chaos in the Solar system and its subsystems, namely in the motion of large and minor bodies of the Solar system—planets, planetary satellites, asteroids, and comets. We demonstrate how the dynamical chaos affects the observed orbital and spin-orbit properties of ensembles of the Solar system bodies.

In Part III, we focus on the dynamics of exoplanets. Resonances and chaos in multiplanet systems of single stars and in planetary systems of binary stars are considered. A number of spectacular examples of outstanding dynamical behaviours of planets and planetesimal disks in model and real exoplanetary systems are given. We concentrate on estimating the sizes of planetary chaotic zones and the Lyapunov and diffusion timescales and explore how the effects of chaotic clearing and dynamical packing determine the observed architectures of planetary systems.

In the limited volume of the book, it is impossible to provide a complete panorama of such a broad field of studies. Personal preferences in the approach to the general theme and in the choice of particular topics are inevitable. Concerning the adopted style, it is more physical than mathematical, and in this respect, it largely belongs to the physical tradition, founded in this field by Boris Chirikov.

To perceive a more "stereoscopic" and complete view of the subject, an advanced reading of modern outstanding monographs and reviews in celestial mechanics, dynamical chaos theory, and planetary science is recommended.

Especially, in the monograph by Boccaletti and Pucacco (1996, 1999), mathematical foundations of modern celestial mechanics, including dynamical chaos theory and modern numerical tools, are addressed in detail. This monograph is recommended as an advanced reading on the topics covered in the first part of our book.

As different-styled introductions to problems of chaotic behaviour of Solar system bodies, the reviews by Malhotra (1994, 1998), Lissauer (1999), and Davies et al. (2014) are most helpful. In the book by Murray and Dermott (1999), one may find an excellent general introduction to the dynamics of Solar system bodies, including topics relevant to dynamical chaos. The book by Morbidelli (2002) is highly recommended as an advanced course on the resonant and chaotic dynamics of the Solar system and its subsystems.

In 2018, two encyclopedias on exoplanets (Perryman 2018; Deeg and Belmonte, eds. 2018) were published. In these voluminous editions, an interested reader can find a detailed modern astrophysical, astrochemical, and astrobiological material in exoplanetology, with vast bibliographical lists.

Returning to our book, note that it was designed generally to be self-contained: only basic knowledge in mathematics and mechanics is required for understanding the material, if read from the beginning. I hope that the book can be helpful for researchers working in planetary and exoplanetary sciences, astrophysics, celestial mechanics, nonlinear dynamics, at any level (researcher, graduate student, undergraduate student), depending on the interests of the reader.

I am most thankful to Konstantin Kholshevnikov, Leonid Sokolov, and Alexander Melnikov for useful discussions.

Saint Petersburg, Russia
2020

Ivan I. Shevchenko

Contents

Part III Dynamics of Exoplanets

Part I
Origins and Manifestations of Dynamical Chaos

This book is focussed on the Solar system and exoplanetary dynamics. However, an introductory theoretical part is necessary for understanding the further basic material on chaotic dynamics in planetary systems. Analytical tools, addressed in this introductory part, mostly concern an apparatus of paradigmatic symplectic maps, such as standard and separatrix maps. This is needed to provide means for estimating the diffusion and clearing rates, Lyapunov timescales, and extents of chaotic zones in various dynamical applications, exposed further on in the book. We concentrate on basic notions; detailed materials on symplectic maps can be found in the classical review by Chirikov (1979) and in the monographs by Lichtenberg and Lieberman (1992) and Abdullaev (2006). Here we treat the maps in a unified manner, necessary to provide means for the planetary-focussed analysis. Various models of interaction and overlap of non-linear resonances are addressed.

The presented material on numerical tools mostly includes means of construction of charts and diagrams of global dynamics, such as methods based on computations of LCE (Lyapunov characteristic exponents), MEGNO (mean exponential growth numbers), FLI (fast Lyapunov indicators), and other chaos indicators. These numerical techniques have become extremely popular in recent years, in particular in studies of stability of exoplanetary systems. Here we address numerical techniques for constructing global dynamical charts in a detailed and unified manner.

Chapter 1
Chaotic Behaviour

La plus haute perfection . . . se trouve
dans l'union de l'ordre et de l'anarchie.

Pierre-Joseph Proudhon,
Premier Mémoire (1840)

With a bit of imagination and fantasy
one may even conjecture that any macroscopic
event in this World, which formally is
a result of some quantum "measurement,"
would be impossible without chaos.

Boris Chirikov,
Patterns in Chaos (1990)

In this chapter, generic concepts of non-linear dynamics are considered: the non-linear pendulum, resonances and chaos, among them. Various models of resonances, phenomena of interaction and overlap of non-linear resonances are described. Analytical instruments, addressed in this chapter, comprise symplectic maps in general, the standard and separatrix maps, generalized separatrix maps, geometry of chaotic layers. These tools are necessary for estimating the diffusion and clearing rates, Lyapunov timescales, and extents of chaotic zones in various dynamical applications, exposed further on in the book. The presentation partially follows (in Sects. 1.1 and 1.3) the lecture by Shevchenko (2011c). In Sect. 1.6, it partially follows the paper by Shevchenko (1999a), by permission from Springer Nature, © 1999.

1.1 Pendulum, Resonances and Chaos

Resonance represents a generic concept of non-linear dynamics, if not the principal one. According to Chirikov (1982), "resonance is understood as such a situation when some frequencies of a non-perturbed system are close to each other or to frequencies of an external perturbation."

I. I. Shevchenko, *Dynamical Chaos in Planetary Systems*, Astrophysics and Space Science Library 463, https://doi.org/10.1007/978-3-030-52144-8_1

This informal definition may seem rather comprehensive, but how one can be convinced, in practice, in the existence of resonance in a system of any particular celestial bodies? In fact, observed commensurabilities between frequencies in any system may never be absolutely exact, at least due to observational errors. To solve this problem, astronomers use the notion of the *resonant phase* (named also the *resonant* or *critical angle*, or *resonant* or *critical argument*), defined as a linear combination (algebraic sum) of angular variables of the system with integer coefficients, the choice of which determines a resonant relation between the frequencies. If the amplitude of variation of the resonant phase is limited, i.e., this angle librates, similarly to libration of a pendulum, then the system is said to be in resonance; otherwise, if the resonant phase increases or decreases unlimitedly, i.e., rotates, then resonance is absent. Trajectories at the border separating libration and rotation are called the *separatrices*.

The dynamics of a rigid pendulum (Fig. 1.1), therefore, provides a model of resonance. The pendulum model of resonance is ubiquitous, and can even be called universal (Chirikov 1979, 1982), in what concerns the general requirement for libration of the resonant argument and the existence of separatrices. However note that, in mathematical formulations, various models of resonance exist that are different from the mathematical pendulum model. Along with the mathematical pendulum model, they are also considered henceforth in this book.

In celestial mechanics, one deals mostly with *non-linear resonances*, taking place when the frequency of phase oscillations on resonance depends on the amplitude (energy) of the oscillations, as in the pendulum example. Conversely, in linear resonances, the frequency does not depend on the amplitude.

A slightest external push of the rigid pendulum, placed initially at the upper position of equilibrium ($\varphi = \pm\pi = \pm 180°$; the angle φ is defined in Fig. 1.1),

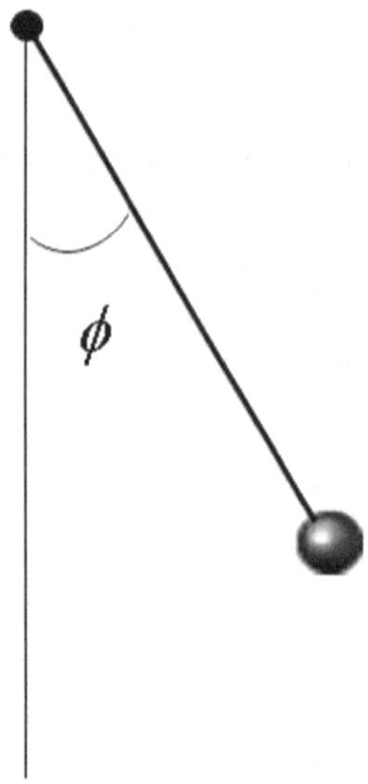

Fig. 1.1 The pendulum

can effect the dynamics *qualitatively*: it can replace oscillation with rotation or vice versa. This phenomenon represents nothing but the so-called *sensitive (essential) dependence on initial conditions.* What would occur, if the pendulum, initially positioned at the separatrices, were subject to a periodic perturbation? It turns out that, generically (for the majority of initial conditions), the motion becomes most unusual. This is just the *chaotic behaviour.* For the first time, this confusing and intricate behaviour was revealed by Poincaré (1899) in a study of the *three-body problem* in celestial mechanics. However, at that time, the character of this intricate motion was not considered in any sense as "random."

However, in the first half of the twentieth century, the near-separatrix dynamics did not attract much attention of researchers. The studies were limited to analysis of special cases using traditional approaches. In 1908, A. Stephenson published an article on the dynamics of the inverted ($\varphi = \pi = 180°$) rigid pendulum with an oscillating point of suspension (Stephenson 1908). He found that the vertical oscillation of the suspension point is capable to stabilize the inverted pendulum, if the oscillation frequency and amplitude are finely adjusted. Later on, at the end of the 1940s of the twentieth century, Petr Kapitsa demonstrated this effect in a real physical experiment (Kapitsa 1951, 1954). Modern opportunities of computer techniques enable one to look at this effect from a novel viewpoint: if one constructs a section of the phase space of motion of such a pendulum, it becomes evident that the domain of stability represents only a small island in an extensive chaotic "sea" formed by trajectories with obviously irregular behaviour. However, up to the middle of the twentieth century this chaotic motion was not a subject of any scientific study, neither the chaotic motion of any other dynamical system was such a subject. One may say that chaotic regions of phase space rested "invisible" for researchers.

In 1959, Boris Chirikov described the dynamical chaos theoretically, as an outcome of *interaction of resonances* (Chirikov 1959). As a criterion for occurrence of large-scale chaos he offered a *resonance overlap* concept. Using the pendulum model, the criterion can be understood as follows. The phase space of a non-perturbed rigid pendulum has two dimensions, defined by two variables: the angle φ, measuring the pendulum's deviation from the vertical direction, and the momentum $p = ml\dot{\varphi}$, where m is the mass of the pendulum, l is its length, $\dot{\varphi}$ is the rate of variation of the angle φ. In the well-known *phase portrait* φ–p of the non-perturbed pendulum (Fig. 1.2), a single domain ("cell") of librations, bounded by the non-perturbed separatrices, is present. Therefore, the non-perturbed pendulum model describes a single resonance. If one "switches on" a periodic perturbation, e.g., the oscillation of the suspension point, then the phase space of our dynamical system is not two-dimensional any more. To compare the perturbed and non-perturbed cases, it is necessary to construct a *phase space section.* It is built as follows: one plots the values of variables not continuously but discretely, at constant time intervals equal to the perturbation period. Then, on the section constructed in this way, one finds out not one but three domains of libration, which are straightforward to regard as three abstract resonances (Fig. 1.5). If the perturbation frequency is large enough, the separation of the resonances in the momentum is significant and they interact

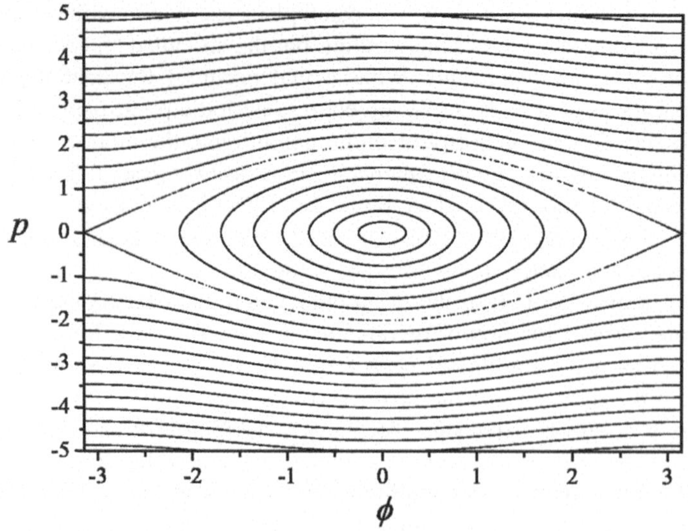

Fig. 1.2 A phase portrait of the non-perturbed pendulum

only weakly. On reducing the perturbation frequency, the resonances approach each other and *chaotic layers* (the bands of obviously irregular behaviour) in the vicinities of their separatrices become visible, as in Fig. 1.5; on reducing the perturbation frequency further on, the layers merge into a single chaotic domain.

1.2 Models of Resonance

The pendulum provides the *first fundamental model* of resonance. It is given by the Hamiltonian

$$\mathcal{H} = Ap^2 + B\cos\varphi, \tag{1.1}$$

where φ is the pendulum angle (the resonance phase angle), p is the momentum, A and B are constant parameters. By a suitable rescaling of the momentum and time, the Hamiltonian (1.1) can be straightforwardly reduced to a form without any parameters:

$$\mathcal{H} = I^2 + \cos\vartheta, \tag{1.2}$$

where I and ϑ are new canonical variables. The phase portrait of the first fundamental model is given in Fig. 1.2.

The pendulum model is often regarded as a "universal" model of non-linear resonance, due to its high usefulness in many applications. However, strictly speaking, it is not universal, because other resonance phenomena exist that do not fit the pendulum model. The next in importance after the first model is the *second fundamental model* of resonance (Henrard and Lemaître 1983; Lemaître 1984); its Hamiltonian is given by

$$\mathcal{H} = I^2 - I^{m/2} \cos m\vartheta. \tag{1.3}$$

The integer m ($m > 0$) is called the *resonance order*. At a given order m, the second fundamental model can be recast to the form in a minimal parametrization by a single parameter. At $m = 1$, this form can be given by

$$\mathcal{H} = I^2 - 3(1+\delta)I - 2(2I)^{1/2} \cos \vartheta, \tag{1.4}$$

where δ is a single parameter (Henrard and Lemaître 1983). Useful formulas, explicitly describing separatrices in the second fundamental model for resonance at various values of order m ($m = 1, 2, 3, 4$), are given in Malhotra (1990).

Bifurcations of contour plots for Hamiltonian (1.4) on increasing δ from negative to positive values are illustrated in Fig. 1.3. Especially note that at high positive δ values the libration (resonance) cell becomes practically identical to that described by the first fundamental model; this property can be readily exploited in applications, because the first model theory (especially in what concerns resonance overlap phenomena) is well developed.

An important particular case of the second fundamental model is represented by the so-called *parametric resonance*. In the parametric resonance model, the dependence on the momentum in the coefficient of the harmonic term in Eq. (1.3) is linear ($m = 2$). The parametric resonance model was extensively explored in the framework of the theory of particle accelerators in the 1950s and 1960s of the twentieth century. Properties of the parametric resonance are considered in detail in Chirikov (1979).

Shinkin (1995) proposed a *third fundamental model* of resonance, described by the parameterized formula

$$\mathcal{H} = \frac{1}{2}p^2 + \alpha \prod_{k=1}^{4} (p + \beta_k)^{j_k/2} \cos m\varphi, \tag{1.5}$$

where the integer numbers $j_1 + j_2 + j_3 + j_4 \le m$; α and β_k ($k = 1, 2, 3, 4$) are arbitrary parameters (see Shinkin 1995; Breiter 2003). An *extended fundamental model of resonance* is presented in Breiter (2003). The third and extended fundamental models allow one to describe resonances in a broad field of dynamical problems; see a particular example concerning the Lidov–Kozai resonance in Shevchenko (2017a).

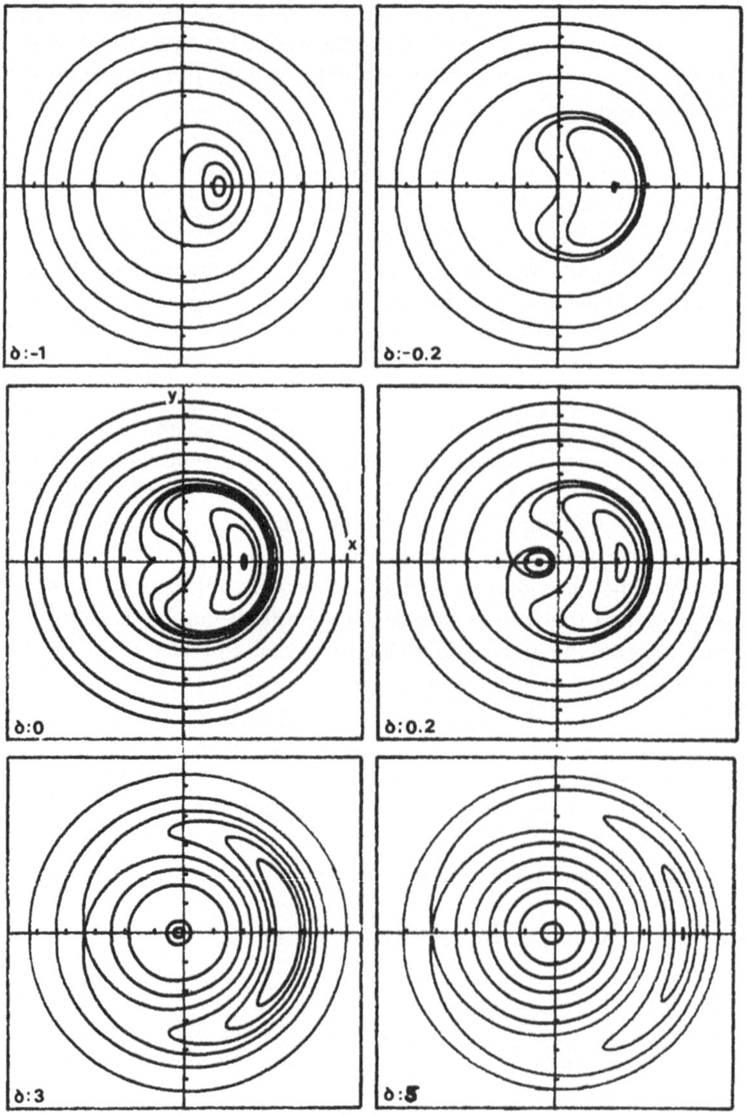

Fig. 1.3 Upper panel: Contour plots for Hamiltonian (1.4) at negative values of δ, in the coordinates $x = (2I)^{1/2}\cos\vartheta$, $y = (2I)^{1/2}\sin\vartheta$. Middle panel: The same at small positive values of δ. Bottom panel: The same at high positive values of δ (Figures 6 and 7 from Henrard and Lemaître (1983), by permission from Springer Nature, © 1983)

1.3 Interaction and Overlap of Resonances

In Chirikov's saying, "...the physicist first of all tries to find out which resonances play role in this or that system and how they interact with each other" (Chirikov 1982). It is just the presence of resonances, often regarded to be an embodiment of order, leads to the unpredictable, chaotic character of the motion. In other words, the presence of resonances in the phase space causes the presence of chaos. However, as we have just seen, for chaos to exist, the presence of not one but two or more resonances in the phase space is required, because their interaction is necessary.

To illustrate this concept, let us take, for the model of perturbed non-linear resonance, the paradigmatic Hamiltonian

$$H = \frac{\mathcal{G}p^2}{2} - \mathcal{F}\cos\varphi + a\cos(k\varphi - \tau) + b\cos(k\varphi + \tau) \qquad (1.6)$$

(see, e.g., Shevchenko 2000a). Its first two terms represent the Hamiltonian H_0 of the unperturbed pendulum with the pendulum angle φ (equivalently, the resonance phase angle) and the momentum p. The last two terms represent periodic perturbations; $\tau = \Omega t + \tau_0$ is the phase angle of perturbation, Ω is the perturbation frequency, τ_0 is the initial perturbation phase. The parameters $\mathcal{F}$, $\mathcal{G}$, a, b, and integer k are constants.

If $k = 1$ and $a = b$, model (1.6) has a well-studied prototype in theoretical mechanics, already mentioned above in Sect. 1.1: it describes a pendulum with the suspension point that vertically oscillates (Bardin and Markeev 1995). The quantity $\varepsilon = a/\mathcal{F} = b/\mathcal{F}$ characterizes the relative perturbation amplitude.

If the perturbations are "switched on" (i.e., $\varepsilon \neq 0$), a section of the phase space of the motion can be constructed. Let us construct the section at $\tau = 0 \bmod 2\pi$, taking the parameters' values as follows: $\Omega = 8$, $\omega_0 = 1$, $k = 1$, $a = b$, $\varepsilon = 0.5$. The resulting section is shown in Fig. 1.4; now not one but three domains of librations, i.e., three resonances, are present.

We define the *adiabaticity parameter* λ to be equal to the ratio of the frequency of perturbation to the frequency of small-amplitude oscillations on resonance. The adiabaticity parameter λ measures the separation of the perturbing and guiding resonances in the units of one quarter of the guiding resonance width. Indeed, $\lambda = |\Omega|/\omega_0$, and the separation of resonances in frequency space is equal to Ω, while the guiding resonance width is equal to $4\omega_0$ (Chirikov 1979). Therefore, the adiabaticity parameter λ can be regarded as a kind of *resonance-overlap parameter*. In the asymptotic limit of the adiabatic perturbation, $\lambda \ll 1$, the resonances in the multiplet strongly overlap, while in the asymptotic limit of the non-adiabatic perturbation, $\lambda \gg 1$, the resonances are separated and do not interact.

If the perturbation frequency is relatively large, as in the case of Fig. 1.4 where $\lambda = 8$, the separation of resonances in the momentum p is also large and they almost do not interact. On reducing the frequency of perturbation, they approach each other, and appreciable chaotic layers emerge in the vicinity of the separatrices

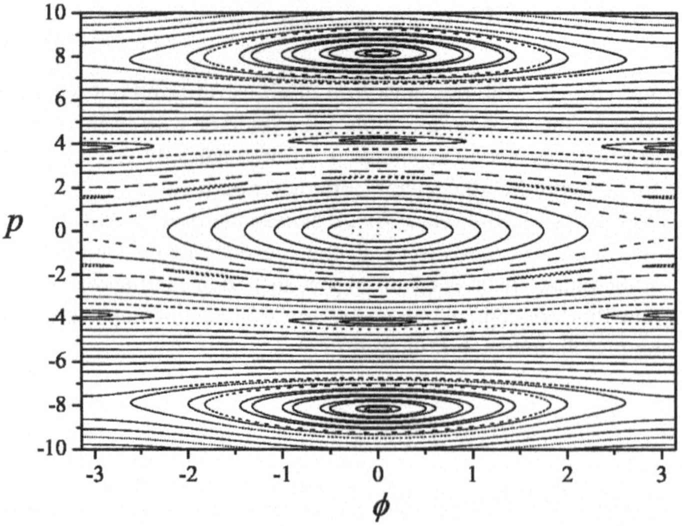

Fig. 1.4 A chaotic resonance triplet. Weak interaction ($\lambda = 8$) (Figure 1 from Shevchenko (2014), with permission from Elsevier)

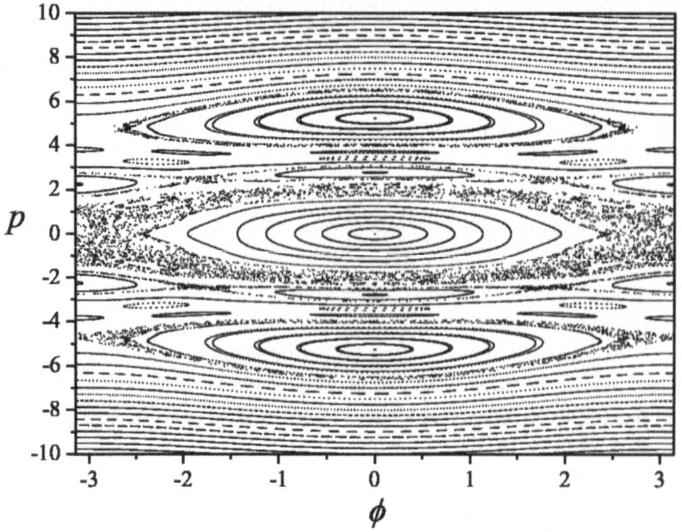

Fig. 1.5 A chaotic resonance triplet. Moderate interaction ($\lambda = 5$) (Figure 2 from Shevchenko (2014), with permission from Elsevier)

(see Fig. 1.5, where $\lambda = 5$ and ε is the same as in Fig. 1.4). As it is well visible in Fig. 1.5, the motion in the vicinity of the separatrices is irregular. On reducing the frequency of perturbation further on, the layers merge into a single chaotic layer, due to the strong resonance overlap (see Fig. 1.6, where $\lambda = 2$). The sequence of phase

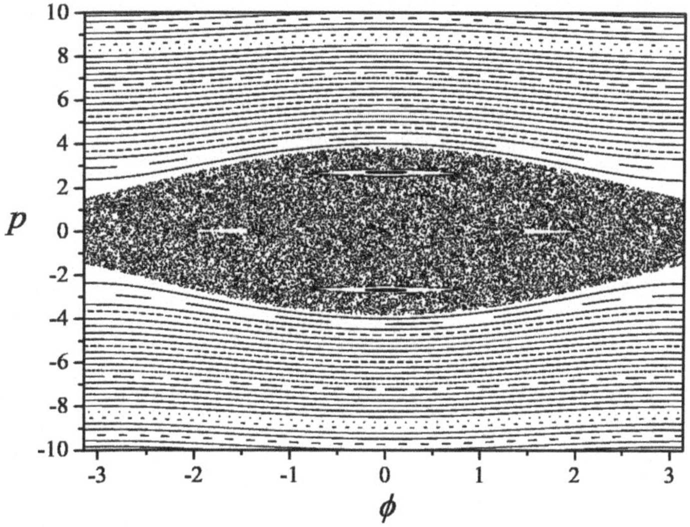

Fig. 1.6 A chaotic resonance triplet. Strong overlap ($\lambda = 2$) (Figure 3 from Shevchenko (2014), with permission from Elsevier)

space sections in Figs. 1.4, 1.5, and 1.6 graphically demonstrates the interaction and overlap of resonances.

1.4 Symplectic Maps in General

Solely basic results (those that are needed henceforth) in the symplectic map theory are presented here, generally following ideas and approaches given in Meiss (1992), Meiss (2007), and Lichtenberg and Lieberman (1992); Abdullaev (2006). For a detailed analysis and complete descriptions, reviews contained in these works are highly recommended.

Dynamical system is defined as a phase space of states plus a set of rules governing the evolution of these states (Meiss 1992). If the rules are given by differential equations, then the evolution is regarded as continuous, and if they are defined by maps, then as discrete. Mappings arise either as outcomes of physical models (as in cases of the paradigmatic *Fermi map* or *Kepler map*, see Chirikov (1979), Chirikov and Vecheslavov (1989), and Lichtenberg and Lieberman (1992)), or as outcomes of purely mathematical analytical methods (as in case of the *separatrix map*, see Chirikov (1979) and Piftankin and Treshchev (2007)), or as outcomes of direct numerical integrations (as in studies of the *Hénon–Heiles problem*, see Hénon and Heiles (1964), Lichtenberg and Lieberman (1992), and Shevchenko and Melnikov (2003)).

We consider solely Hamiltonian systems. Any Hamiltonian flow is given by a scalar function $\mathcal{H}(\mathbf{p}, \mathbf{q}, t)$ that defines the system of differential equations

$$\frac{\mathrm{d}p_k}{\mathrm{d}t} = -\frac{\partial \mathcal{H}}{\partial q_k}, \quad \frac{\mathrm{d}q_k}{\mathrm{d}t} = \frac{\partial \mathcal{H}}{\partial p_k}, \tag{1.7}$$

where vectors $\mathbf{q}$ and $\mathbf{p}$ are conjugate canonical coordinates and momenta, respectively; $k = 1, 2, \ldots N$; N is the number of degrees of freedom.

A *map* is defined as a transformation

$$(\mathbf{p}', \mathbf{q}') = T(\mathbf{p}, \mathbf{q}) \tag{1.8}$$

of canonical variables in the system's phase space: T transforms each point $(\mathbf{p}, \mathbf{q})$ to its image $(\mathbf{p}', \mathbf{q}')$. The map's orbit is given by the sequence of states

$$\{\ldots, (\mathbf{p}_i, \mathbf{q}_i), (\mathbf{p}_{i+1}, \mathbf{q}_{i+1}), \ldots\} \tag{1.9}$$

where $(\mathbf{p}_{i+1}, \mathbf{q}_{i+1}) = T(\mathbf{p}_i, \mathbf{q}_i)$ for any integer i.

Any map describing a Hamiltonian system is called *symplectic* or *canonical.* It preserves the so-called *Poincaré integral invariant*, or just *Poincaré invariant*, which is defined below.

In particular, the symplectic maps are *volume-preserving*:

$$\det \begin{pmatrix} \frac{\partial \mathbf{p}_{i+1}}{\partial \mathbf{p}_i} & \frac{\partial \mathbf{p}_{i+1}}{\partial \mathbf{q}_i} \\ \frac{\partial \mathbf{q}_{i+1}}{\partial \mathbf{p}_i} & \frac{\partial \mathbf{q}_{i+1}}{\partial \mathbf{q}_i} \end{pmatrix} = 1. \tag{1.10}$$

The two-dimensional symplectic maps are therefore *area-preserving*.

The symplectic map theory can be developed starting from the *action principle* (Meiss 1992, 2007). The action is defined as a functional on an orbit segment $(\mathbf{q}(t), \mathbf{p}(t))$ at $t_0 < t < t_1$:

$$S = \int_{t_0}^{t_1} (\mathbf{p} \cdot \dot{\mathbf{q}} - \mathcal{H}(\mathbf{p}, \mathbf{q}, t))\, dt\,, \tag{1.11}$$

where the upper dot ($\dot{}$) is the time derivative. The action on a closed loop in the phase space is an invariant of the Hamiltonian flow; it is just the mentioned above *Poincaré integral invariant.* By introducing a parameter λ, the loop can be described as the set $\{\mathbf{q}(\lambda), \mathbf{p}(\lambda), t(\lambda); 0 \leq \lambda \leq 1\}$. The action of the loop $\mathcal{L}$ is the integral

$$S_{\mathcal{L}} = \int_0^1 \left(\mathbf{p} \cdot \frac{\mathrm{d}\mathbf{q}}{\mathrm{d}\lambda} - \mathcal{H}\frac{\mathrm{d}t}{\mathrm{d}\lambda}\right) d\lambda \tag{1.12}$$

(Meiss 1992).

Assume the loop $\mathcal{L}$ to be a set of initial conditions for Hamiltonian equations. The closed loop $\mathcal{L}$ is defined as a loop in the extended phase space ($\mathbf{q}$, $\mathbf{p}$, t). One may find the loop's evolution by integrating the equations of motion. Consider a time-independent function $\mathcal{H}$ and an arbitrary loop $\mathcal{L}$ on the energy surface $\mathcal{H} = E$. Since $\mathcal{H}$ is constant on E, the term $\mathcal{H}dt$ does not contribute to the contour integral; therefore, on the surface of constant energy one has

$$S_{\mathcal{L}} = \oint_{\mathcal{L}} \mathbf{p} \cdot d\mathbf{q} = \sum_{k=1}^{N} \sigma_k \int_{\mathcal{A}_k} dp_k dq_k \, , \tag{1.13}$$

where $\sigma_k = 1$ if the projection of $\mathcal{L}$ on the plane k is described clockwise, and $\sigma_k = -1$ if the projection is described counter-clockwise.

This equality is nothing but the generalization of the Stokes theorem to the case of phase space of arbitrary dimensions. Recall that the classical Stokes theorem states that the contour integral of $\mathbf{F}(\mathbf{r})$, where $\mathbf{F}$ is an arbitrary vector function of $\mathbf{r}$, over a closed contour C is equal to the flux of the curl $\nabla \times \mathbf{F}$ through the surface bounded by C.

The quantity $S_{\mathcal{L}}$ is called the symplectic area. It is equal to the sum of areas of N projections of $\mathcal{L}$ on the planes (q_k, p_k); see Fig. 1.7. The symplectic area along the flow specified by the time-independent Hamiltonian is thus preserved, because the Poincaré invariant exists.

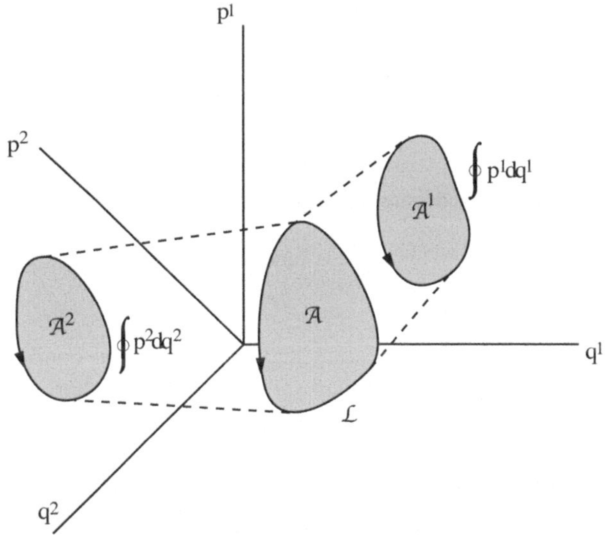

Fig. 1.7 The Poincaré invariant is the sum of oriented areas of the loop projections on the (q_k, p_k) planes (Reprinted Figure 2 with permission from Meiss (1992). © 1992 by the American Physical Society)

Consider a parallelogram contour formed by two vectors $\boldsymbol{\delta}\mathbf{z}^1$ and $\boldsymbol{\delta}\mathbf{z}^2$ (in the phase space $(\mathbf{q},\ \mathbf{p})$). Its symplectic area is given by

$$\omega(\boldsymbol{\delta}\mathbf{z}^1, \boldsymbol{\delta}\mathbf{z}^2) = \boldsymbol{\delta}\mathbf{p}^1 \cdot \boldsymbol{\delta}\mathbf{q}^2 - \boldsymbol{\delta}\mathbf{q}^1 \cdot \boldsymbol{\delta}\mathbf{p}^2 = \sum_{k,l} \delta z_k^1 \omega_{kl} \delta z_l^2 \,. \tag{1.14}$$

In the matrix form, $\boldsymbol{\omega} = \mathbf{J}^{-1}$, where the antisymmetric matrix $\mathbf{J}$ is the *Poisson matrix*, or, the *symplectic unit*:

$$\mathbf{J} = \begin{pmatrix} \mathbf{0} & \mathbf{I} \\ -\mathbf{I} & \mathbf{0} \end{pmatrix} . \tag{1.15}$$

It has dimensions $2N \times 2N$; here $\mathbf{I}$ is the unit matrix of dimensions $N \times N$. The form ω is called the *symplectic form*.

Following Meiss (1992), consider an infinitesimal parallelogram formed by two small arbitrary vectors $\boldsymbol{\delta}\mathbf{z}^1$ and $\boldsymbol{\delta}\mathbf{z}^2$ at a point $\mathbf{z}$ (in the phase space $(\mathbf{q},\ \mathbf{p})$). A symplectic mapping T takes this parallelogram to a point $\mathbf{z}'$:

$$\boldsymbol{\delta}\mathbf{z}' = \frac{\partial T}{\partial \mathbf{z}} \boldsymbol{\delta}\mathbf{z} \,. \tag{1.16}$$

The symplectic area of the image is equal to that of the inverse image:

$$\omega(\boldsymbol{\delta}\mathbf{z}^{1'}, \boldsymbol{\delta}\mathbf{z}^{2'}) = \omega(\boldsymbol{\delta}\mathbf{z}^1, \boldsymbol{\delta}\mathbf{z}^2) \tag{1.17}$$

and

$$\mathbf{M}^{\mathrm{tr}} \boldsymbol{\omega} \mathbf{M} = \boldsymbol{\omega} \,, \tag{1.18}$$

where “tr” means “transposed,” and $\mathbf{M}$ is the Jacobi matrix with elements

$$M_{kl} = \frac{\partial \mathbf{z}_k'}{\partial \mathbf{z}_l} \,. \tag{1.19}$$

Equation (1.18) represents the necessary and sufficient local condition for a map to possess the Poincaré invariant. A map satisfying it everywhere is called *locally symplectic* (Meiss 1992, 2007). This matrix equality is conveniently used to verify the symplectic property of any map.

Taking determinants of the right and left parts of Eq. (1.18), one has

$$\mathrm{Det}(\mathbf{M}^{\mathrm{tr}} \boldsymbol{\omega} \mathbf{M}) = \mathrm{Det}(\boldsymbol{\omega}) \,, \tag{1.20}$$

hence

$$(\mathrm{Det}(\mathbf{M}))^2 = 1\,, \tag{1.21}$$

because $\mathrm{Det}(\boldsymbol{\omega}) \neq 0$. Therefore, $\mathrm{Det}(\mathbf{M}) = \pm 1$. In reality $\mathrm{Det}(\mathbf{M}) = +1$, that is, any symplectic map preserves both volume and orientation. In particular, any two-dimensional symplectic map preserves both area and orientation; and, conversely, any two-dimensional map preserving both area and orientation is locally symplectic.

Any numerical procedure for solving a system of differential equations consists in iterations of a map. It is important to be aware that, if the system of differential equations is Hamiltonian, it is desirable to ensure that the integration scheme is symplectic.

1.5 The Standard Map

Let $T : (x, y) \to (x', y')$ be a two-dimensional symplectic map of a cylinder (with x being the angular coordinate) onto itself, and assume that T is differentiable. Then T is a twist map (with the twist to the right) if there exists K such that

$$\frac{\mathrm{d}x'}{\mathrm{d}y} \geq K > 0 \tag{1.22}$$

(Meiss 1992). In other words, x' is a monotonically increasing function of y.

The twist condition is physically natural (Meiss 1992, 2007), as soon as y represents the momentum, and a larger momentum usually implies a higher speed. Therefore, points with a larger value of y will move farther along x. In fact, the twist property (1.22) is ubiquitous: it naturally emerges when one of the variables is momentum-like, and the other one is angle-like.

It often happens that one and the same map arises in both physical and abstract mathematical models. For example, the *standard map* arises in “microtron” and Frenkel–Kontorova physical models (Chirikov 1979, 2000; Lichtenberg and Lieberman 1992; Meiss 1992); while, on the other hand, it arises purely mathematically as a local (in the action-like variable) approximation of more complicated maps, e.g., separatrix maps.

The standard map (sometimes called the Chirikov map or the Chirikov–Taylor map) is given by

$$\begin{aligned} y_{i+1} &= y_i + K \sin x_i\,, \\ x_{i+1} &= x_i + y_{i+1}\,, \end{aligned} \tag{1.23}$$

where K is the so-called *stochasticity parameter* (Chirikov 1979; Lichtenberg and Lieberman 1992).

The Hamiltonian of the dynamical system described by the standard map is

$$\mathcal{H} = \frac{p^2}{2} + \frac{K}{4\pi^2} \sum_{k=-\infty}^{k=+\infty} \cos(x - kt) \tag{1.24}$$

(Chirikov 1979). The variables $x(t_i)$, $p(t_i) = y_i/(2\pi)$, taken at $t = t_i = 2\pi i$ (where i are integers) correspond to the variables of the standard map (1.23). In fact, the standard map is a *symplectic integrator* for the equations of motion defined by the Hamiltonian (1.24); i.e., the standard map can be regarded as a symplectic scheme for numerical integration of the given equations of motion.

The stochasticity parameter K of the standard map (1.23) relates to the adiabaticity parameter of the original system (1.24) by the equation

$$K = (2\pi/\lambda)^2. \tag{1.25}$$

The standard map (1.23) describes the motion in an infinite multiplet of equal-amplitude equally-separated resonances, explicitly represented by the trigonometric terms in (1.24). It has the twist property (1.22).

At small values of the parameter K, $K \ll 1$, the phase space of the standard map is mostly regular, with the chaotic component measure tending to zero with $K \to 1$. On increasing K, the chaotic separatrix layers of resonances swell, resonances start to overlap, and the chaotic behaviour becomes more and more pronounced. The chaotic component of phase space rises in volume. and the maximum Lyapunov exponent (measuring the rate of divergence of nearby trajectories, see Sect. 2.1) of the chaotic motion increases. An important constant in the standard map theory is the critical value of the stochasticity parameter, at which the chaotic orbits become unbounded in momentum, because the *last invariant curve* is destroyed (Chirikov 1979; Greene 1979; Meiss 1992; Lichtenberg and Lieberman 1992). This happens at $K = K_G = 0.971635406\ldots$, as analytically shown by Greene (1979). At this value of K, the relative measures of the chaotic and regular components of the map's phase space are approximately equal, about one half each: ≈ 0.46 and ≈ 0.54, respectively (Shevchenko 2004a).

1.6 The Separatrix Map

For weakly perturbed systems, the chaotic component of the phase space in the vicinity of perturbed separatrices can be described using the notion of a *chaotic layer*, a near-separatrix domain where a dynamical system moves chaotically. The chaotic layer theory has applications in various areas of physics, mechanics and, in particular, in celestial mechanics (Chirikov 1979; Shevchenko 2007a, 2010). The key role in this theory is played by *separatrix maps*. They represent the motion of a system close to a separatrix in a discrete way ("stroboscopically," as in constructing the phase space sections): the system state, set by the "time" and "energy" variables, is fixed not continuously, but discretely at passages by the model pendulum (describing the resonance) of the positions of equilibrium.

The separatrix map was deduced in an implicit form at the end of 1960s by Zaslavsky and Filonenko (1968), and in the modern form in the seventies by Chirikov (1977, 1978, 1979). Initially it was named "whisker map," as the perturbed separatrices were graphically called "whiskers."

The classical separatrix map is an effective tool for analysis of the motion in *chaotic layers* formed by splitted separatrices of non-linear resonances (Piftankin and Treshchev 2007), and its theory can be used for estimating parameters of chaotic layers, when the standard map theory, valid for the case of an infinite multiplet of equally-spaced equal-size resonances, is not applicable.

A number of problems on non-linear resonances in mechanics and physics is described by the perturbed pendulum-like Hamiltonian (1.6). We assume that $\mathcal{F} > 0$ and $\mathcal{G} > 0$. The case of symmetric perturbation $a = b$ with $k = 1$ is of especial interest. In this case, the Hamiltonian (1.6) describes, in particular, the pendulum with the vertically oscillating point of suspension. Models with arbitrary non-zero values of integer k and zero either a or b concern the problem of a particle motion in the field of two planar waves (Escande and Doveil 1981; Escande 1985; Zaslavsky et al. 1991; Zaslavsky and Abdullaev 1995). Models with $k = 1/2$ and specific a and b values were applied to describe dynamics in vicinities of the 3/1 orbital resonance in planetary satellite systems (Malhotra 1990; Shevchenko 2000b). A model with $k = 1$ and $b = -a/7$ describes rotational dynamics close to synchronous spin-orbit resonance, of non-spherical satellites in elliptic orbits (Wisdom et al. 1984; Celletti 1990).

The near-separatrix motion of system (1.6) in the symmetric case $a = b$ and $k = 1$ was demonstrated in Chirikov (1977, 1978, 1979) to be effectively described by the separatrix map

$$
\begin{aligned}
w_{i+1} &= w_i - W \sin \tau_i, \\
\tau_{i+1} &= \tau_i + \lambda \ln \frac{32}{|w_{i+1}|} \quad (\mathrm{mod}\ 2\pi).
\end{aligned}
\tag{1.26}
$$

The variable $w \equiv \frac{H_0}{\mathcal{F}} - 1$ denotes the relative (with respect to the unperturbed separatrix value) pendulum energy, and τ is the phase angle of perturbation.

The separatrix map has a *domain-dependent* twist property (Ahn et al. 1996).

The adiabaticity parameter λ (see Sect. 1.3) is equal to the perturbation frequency Ω in units of $\omega_0 = (\mathcal{F}\mathcal{G})^{1/2}$, the frequency of small phase oscillations on resonance:

$$
\lambda = \frac{\Omega}{\omega_0} = \frac{\Omega}{(\mathcal{F}\mathcal{G})^{1/2}}. \tag{1.27}
$$

The second parameter, W, when $a = b$ and $k = 1$, is given by

$$W = \frac{a}{\mathcal{F}} \lambda \alpha_2^c(\lambda), \tag{1.28}$$

or, equivalently,

$$W = \varepsilon\lambda \left(A_2(\lambda) + A_2(-\lambda)\right) = \frac{4\pi\varepsilon\lambda^2}{\sinh\frac{\pi\lambda}{2}} \tag{1.29}$$

(Shevchenko 1998b). Special functions $\alpha_k^c(\lambda)$ with $k = 0, 1, 2, \ldots$ are explicitly given in Appendix C.

The *Melnikov–Arnold integral* ("MA-integral") $W \sin\tau$ gives the model pendulum energy increment over one circulation or half-libration period of the pendulum in its near-separatrix motion, as a function of phase τ at the moment of the pendulum passage of its upper point of equilibrium. The coefficient A_2 is given by the formula

$$A_2(\lambda) = 4\pi\lambda \frac{\exp(\pi\lambda/2)}{\sinh(\pi\lambda)}, \tag{1.30}$$

see Chirikov (1979), Shevchenko (1998b, 2000a). Formula (1.29) differs from that given in Chirikov (1979) and Lichtenberg and Lieberman (1992), as the $A_2(-\lambda)$ term is added here. It is negligible when $\lambda \gg 1$. However, it is significant when $\lambda \ll 1$ and $\lambda \sim 1$ (Shevchenko 1998b).

For arbitrary k, formulas for W are presented in Chirikov (1979) and Shevchenko (2000a), and for arbitrary a and b they are given in Shevchenko (2000a). The accuracy of map (1.26) in describing the original system (1.6) can be estimated, by an order of magnitude, as $\sim \varepsilon$ (Treshchev 1998; Abdullaev 2006). The accuracy can be estimated directly, e.g., by measuring the chaotic layer width.

The expression used in map (1.26) for the phase τ increment is approximate: it is valid only asymptotically at low strengths of perturbation, i.e., at $w \ll 1$. To improve the accuracy at larger perturbation amplitudes, the phase increment logarithmic approximation can be replaced by the increment's original expression through elliptic integrals (Shevchenko 1998b, 1999a).

Another important circumstance that should be taken into account when using Eqs. (1.26), is that the motion of the original system (1.6) is iterated by Eqs. (1.26) asynchronously (Shevchenko 1998b): the relative energy w is taken on passing the upper vertical position $\varphi = \pm\pi$ of the model pendulum, while the perturbation phase τ is taken at its lower vertical position $\varphi = 0$. The map can be synchronized by an especial procedure (Shevchenko 1998b, 2000a). The synchronized map provides authentic phase space sections for the near-separatrix motion at both high and low frequencies of perturbation (i.e., at any value, either greater or less than unity, of the adiabaticity parameter λ), as demonstrated in Shevchenko (2000a) by direct comparisons of the map phase portraits and phase space sections obtained by direct numerical integrations of the original system. We consider the synchronization procedure in detail blow in Sect. 1.8.

The asymptotic formula for W that ensues from Eq. (1.29) at $\lambda \sim 0$ is $W \approx 8\varepsilon\lambda$. Its good agreement with the actual amplitude of the separatrix map, as found by means of numerical integrations of the original system, was verified in Vecheslavov (2004).

An equivalent form of Eqs. (1.26), used, in particular, in Chirikov and Shepelyansky (1984) and Shevchenko (1998a,b), is

$$\begin{aligned} y_{i+1} &= y_i + \sin x_i, \\ x_{i+1} &= x_i - \lambda \ln |y_{i+1}| + c \pmod{2\pi}, \end{aligned} \tag{1.31}$$

where $y = w/W,\ x = \tau + \pi$; and

$$c = \lambda \ln \frac{32}{|W|}. \tag{1.32}$$

An illustrative example of the chaotic layer described by the separatrix map (1.31) is shown in Fig. 1.8. In this example, the adiabaticity parameter $\lambda = 1.34$, and $c = 2.32$.

Another example, with $\lambda = 5$ and $c = 0$, is given in Fig. 1.9. In Fig. 1.10, a zoom of *critical structure* at the layer border is presented. The corresponding area in Fig. 1.9 is shown by a rectangle. The layer border is situated at $|y| \approx \lambda$; at these values of y, as can be easily verified, the linearization of the separatrix map in y results in the standard map (1.23) with $K \approx 1$.

In our pendulum model, the so-called *non-adiabatic chaos* takes place if the perturbation frequency is much greater than the frequency of small-amplitude oscillations on resonance. In other words, the condition is that the distance between resonances in the momentum should be much greater than the sum of their half-

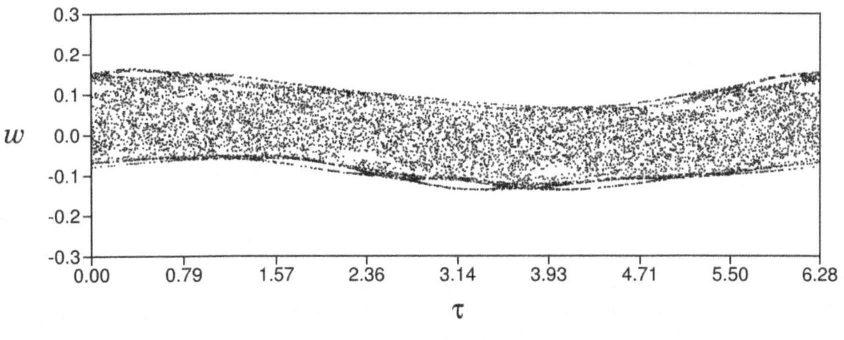

Fig. 1.8 An example of the chaotic layer, as described by the separatrix map (1.31) with $\lambda = 1.34$ and $c = 2.32$ (Figure 1 from Shevchenko (1998b), by permission of IOP Publishing/AAS)

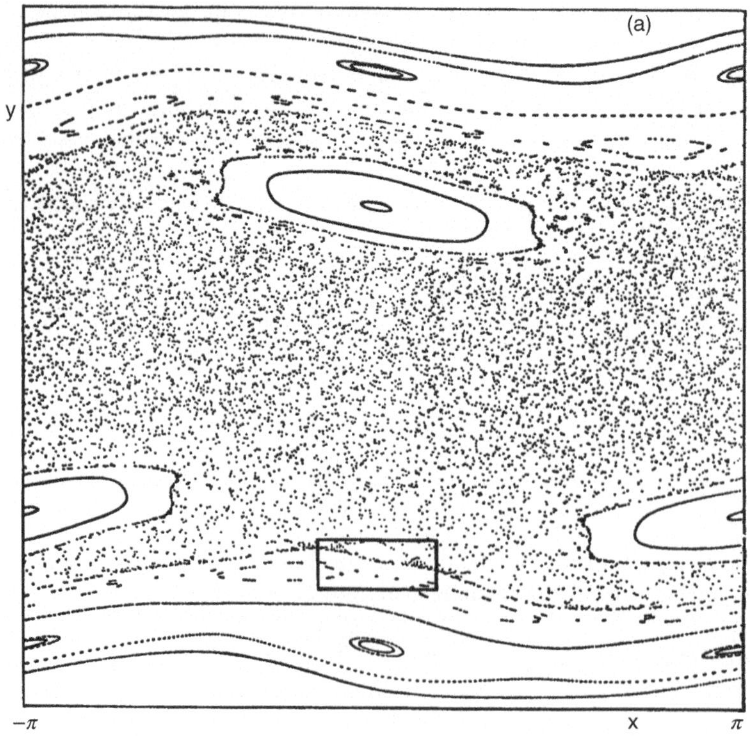

Fig. 1.9 An example of the chaotic layer, as described by the separatrix map (1.31) with $\lambda = 5$ and $c = 0$ (Figure 1a from Chirikov (1991), with permission from Elsevier)

widths. Conversely, *adiabatic chaos*, or *slow chaos*, takes place if the interacting resonances overlap and merge.

As a rule, non-adiabatic chaos is bounded inside a near-separatrix layer. In case of adiabatic chaos, the motion represents a global diffusion across overlapping resonances, merged in a single domain in phase space.

Let us now consider how the ordinary separatrix map (1.26) can be generalized to the case of asymmetric perturbation (Shevchenko 1999a). The introduction of such a map expands the field of application of the separatrix map theory; for example, this map can be directly applied to describe chaotic rotations of a satellite, as discussed further on in Sect. 7.1.

Consider the Hamiltonian (1.6). The separatrix map in case of asymmetric perturbation $a \neq b$ differs from the separatrix map in the symmetric case, because energy increments are different for the prograde and retrograde motions of the pendulum. Therefore, the separatrix map in the asymmetric case represents an algorithm, the *separatrix algorithmic map*. The essence of this algorithm consists in taking into account alternations of values of the parameter W. It alternates when the direction of the motion changes. The separatrix algorithmic map contains

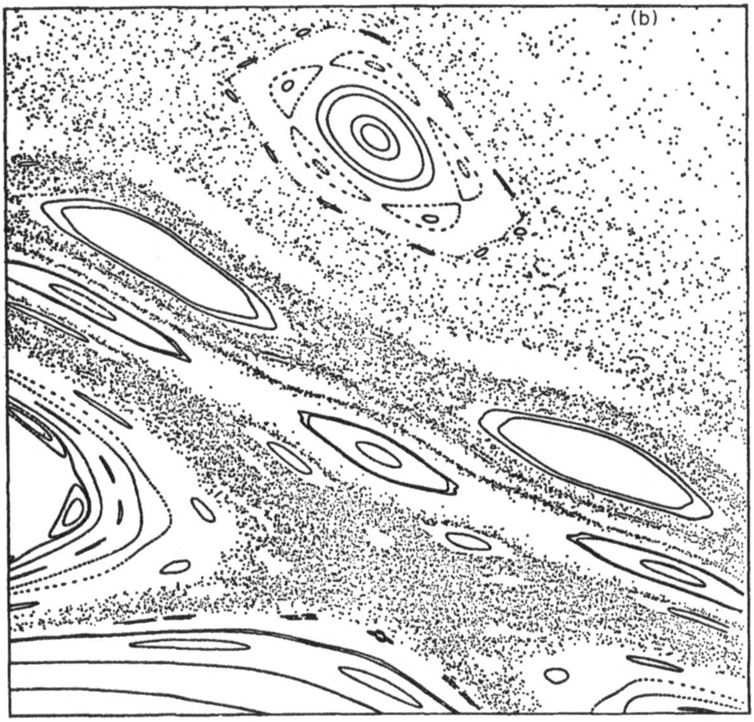

Fig. 1.10 A close-up of the rectangle-delimited area in Fig. 1.9 (Figure 1b from Chirikov (1991), with permission from Elsevier)

conditional transfer statements (Shevchenko 1999a). The map is given by

$$\begin{aligned}
&\text{if } w_i < 0 \text{ and } W = W^- \text{ then } W := W^+, \\
&\text{if } w_i < 0 \text{ and } W = W^+ \text{ then } W := W^-; \\
&w_{i+1} = w_i - W \sin \tau_i, \\
&\tau_{i+1} = \tau_i + \lambda \ln \frac{32}{|w_{i+1}|} \pmod{2\pi};
\end{aligned} \tag{1.33}$$

where λ is the adiabaticity parameter, as given by Eq. (1.27). The former W parameter is generalized to

$$\begin{aligned}
W^+(\lambda, \eta) &= \varepsilon\lambda \left(A_2(\lambda) + \eta A_2(-\lambda)\right), \\
W^-(\lambda, \eta) &= \varepsilon\lambda \left(\eta A_2(\lambda) + A_2(-\lambda)\right),
\end{aligned} \tag{1.34}$$

and $\varepsilon = \frac{a}{\mathcal{F}}$, $\eta = \frac{b}{a}$.

One should be aware that the expression for the increment of the phase τ in the ordinary separatrix map (1.26) is an approximation, valid asymptotically when the near-separatrix rotation/oscillation period on resonance tends to infinity. Therefore, it is applicable at low strengths of perturbation ($W \ll 1$), when the chaotic layer is thin. Conversely, if the perturbation is not weak, one may improve the performance of the separatrix map by means of replacing the logarithmic approximation by the original exact expression through elliptic integrals (Shevchenko 1998b). The expression form depends on whether the phase rotates or oscillates. On performing this substitution, one obtains for the τ increment:

$$\Delta_{n+1}\tau = \begin{cases} 2\lambda \mathbf{K}\left((1 + \frac{w_{n+1}}{2})^{1/2}\right), & \text{if } w_{n+1} < 0, \\ 2\lambda(1 + \frac{w_{n+1}}{2})^{-1/2}\mathbf{K}\left((1 + \frac{w_{n+1}}{2})^{-1/2}\right), & \text{if } w_{n+1} > 0, \end{cases} \tag{1.35}$$

where $\mathbf{K}(k)$ is the elliptic integral of the first kind. The first line in Eq. (1.35) corresponds to libration of the model pendulum, and the second one to its rotation.

The separatrix algorithmic map maps the motion of the system (1.6) (at $k = 1$) on the plane (τ, w) at fixed values of the resonant phase angle φ equal to 0 and $\pm\pi$. When Poincaré sections are constructed numerically in applied problems, it is customary to use another plane, namely the plane (φ, p) taken at a fixed value of the phase angle of perturbation, e.g. $\tau = 0 \pmod{2\pi}$.

Consider the problem how the section of the second kind can be found with the help of the separatrix map. In order to characterize the current state of the system, described by the separatrix algorithmic map, introduce temporary designations $w = w_n$, $\tau = \tau_n$; and $\Delta\tau = \tau_n - \tau_{n-1}$. The value of W, which specifies the current prograde/retrograde state of the system, is taken at the next iteration $n + 1$. The matter is that the values of the variables w_n and τ_n, given by the separatrix algorithmic map (1.33), as well as by the usual separatrix map (1.26), correspond to different instants of time (the property of asynchronism, see Sect. 1.8). The phase τ_n is mapped with a delay in relation to the relative energy w_n. The delay is equal to a half-period of rotation, or a quarter-period of libration of the model pendulum. Due to this delay, the value of W in the formula for w_{n+1} in the separatrix algorithmic map (1.33) specifies the prograde/retrograde state of the system at the preceding instant τ_n; or, inversely, the state of the system at τ_n is determined by the value of W in the formula for w_{n+1}, i.e., at the next iteration.

Let us find the phase point which is connected with the position of the system at the current instant τ (corresponding to $\varphi = 0$) by a trajectory backwards in time, and which is situated at the nearest surface $\tau = 0 \pmod{2\pi}$. The trajectory is assumed to be regular and possessing energy w. For convenience, the current instant τ is taken modulo 2π, while the increment $\Delta\tau$ is not. If $\Delta\tau \leq \tau$, there are no intersections with the plane of interest on the open interval of time backwards in relation to the current state, and no projection is made therefore at such an iteration of the map. Otherwise, the projection to the nearest surface $\tau = 0 \pmod{2\pi}$ is given by the formulae

$$\varphi = \begin{cases} \varphi\left(t = -\frac{\tau}{\Omega}\right), & \text{if } W = W^{+} \text{ (prograde)}, \\ -\varphi\left(t = -\frac{\tau}{\Omega}\right), & \text{if } W = W^{-} \text{ (retrograde)}, \end{cases} \tag{1.36}$$

$$p = \begin{cases} p\left(t = -\frac{\tau}{\Omega}\right), & \text{if } W = W^{+} \text{ (prograde)}, \\ -p\left(t = -\frac{\tau}{\Omega}\right), & \text{if } W = W^{-} \text{ (retrograde)}, \end{cases} \tag{1.37}$$

where $\varphi(t)$ and $p(t)$ represent the explicit solution of equations of the unperturbed non-linear pendulum (see, e.g., Wisdom 1985):

$$\begin{aligned} \cos\varphi(t) &= 1 - 2k^2 \mathrm{sn}^2(\omega_0 t), \\ \sin\varphi(t) &= 2k\ \mathrm{sn}^2(\omega_0 t)(1 - k^2 \mathrm{sn}^2(\omega_0 t))^{1/2}, \\ p(t) &= \frac{2\omega_0 k}{\mathcal{G}}\ \mathrm{cn}(\omega_0 t), \end{aligned} \tag{1.38}$$

for libration, and

$$\begin{aligned} \cos\varphi(t) &= \mathrm{cn}^2(\omega_r t) - \mathrm{sn}^2(\omega_r t), \\ \sin\varphi(t) &= 2\ \mathrm{sn}(\omega_r t)\ \mathrm{cn}(\omega_r t), \\ p(t) &= \frac{2\omega_r}{\mathcal{G}}\ \mathrm{dn}(\omega_r t), \end{aligned} \tag{1.39}$$

for rotation; sn, cn, dn are Jacobi elliptic functions, $\omega_r = \frac{\omega_0}{k}$,

$$k = \begin{cases} \left(1 + \frac{w}{2}\right)^{1/2}, & \text{if } w < 0 \text{ (libration)}, \\ \left(1 + \frac{w}{2}\right)^{-1/2}, & \text{if } w > 0 \text{ (rotation)}, \end{cases} \tag{1.40}$$

is the elliptic modulus (compare with Eq. (1.35)).

One iteration of the separatrix algorithmic map can produce several (or even many) projected points. To find all projected points for a given iteration, it is straightforward to use the following algorithm (Shevchenko 1999a). First, it is verified whether the intersection condition $\Delta\tau > \tau$ is valid, and if yes, a projection is made. Then the interval τ is incremented by 2π and it is verified whether the intersection condition is still valid. If yes, the projection is accomplished once more with the new value of τ, and one more phase point on the plane (φ, p), $\tau = 0 \pmod{2\pi}$, is found. The cycle is repeated until $\Delta\tau \leq \tau$. The complete algorithm can be represented as follows:

$$
\begin{aligned}
&\text{while } \Delta\tau > \tau \text{ do}\\
&\text{evaluate } \varphi\,,\ p \text{ by Equations (1.36, 1.37)}\\
&\tau := \tau + 2\pi\\
&\text{end do}
\end{aligned}
\tag{1.41}
$$

Note that the input value of τ is taken modulo 2π, while the increment $\Delta\tau$ and consequent values of τ are not.

This procedure is applied at each iteration of the separatrix algorithmic map. Since it is based on the regular approximation of the chaotic motion (when time increments are much less than the Lyapunov time; for the definition of the Lyapunov time, see Sect. 2.1), it is called the *regular projection algorithm.*

Now let us consider a non-linear resonance in the perturbed pendulum model with several harmonic perturbations (i.e., in comparison with Hamiltonian (1.6), the number of equally-spaced perturbing harmonics may be arbitrary):

$$
H = \frac{\mathcal{G}p^2}{2} - \mathcal{F}\cos\varphi + \sum_{k=1}^{M} a_k \cos(\varphi - k\tau) + \sum_{k=1}^{M} b_k \cos(\varphi + k\tau). \tag{1.42}
$$

Thus the number of resonances in the multiplet is equal to $2M + 1$.

Let us build a separatrix map for Hamiltonian (1.42) with symmetric perturbations ($a_k = b_k$). If the perturbations are asymmetric, the problem is more complicated, because the separatrix map becomes algorithmic, as in the paradigmatic triplet case (Shevchenko 1999a).

Setting $a_k = b_k$ and calculating the increment of the energy variable (analogously to the triplet case, considered in Chirikov (1979)) gives the result $\sum_{k=1}^{M} W_k \sin(k\tau_i)$, whereas the increment of the time variable remains the same as in the triplet case. Thus the separatrix map (1.26) is generalized to a *multiplet separatrix map*, given by

$$
\begin{aligned}
w_{i+1} &= w_i - \sum_{k=1}^{M} W_k \sin(k\tau_i),\\
\tau_{i+1} &= \tau_i + \lambda \ln \frac{32}{|w_{i+1}|} \pmod{2\pi},
\end{aligned}
\tag{1.43}
$$

where

$$
W_k = 4\pi\varepsilon_k \frac{\lambda_k^2}{\sinh\frac{\pi\lambda_k}{2}},
$$

where $\lambda_k = k\lambda$ and $\varepsilon_k \equiv \frac{a_k}{\mathcal{F}} = \frac{b_k}{\mathcal{F}}$.

The domain of validity of map (1.43) (in describing the near-separatrix motion) is expected to be usually much smaller than that of map (1.26), because the natural condition of validity $|W| \lesssim 1$ generalizes here to the condition $\sum_{k=1}^{M} |W_k| \lesssim 1$. Thus, if there is a lot of perturbing harmonics, the maximum allowed amplitudes ε_k in the multiplet case must be usually much smaller than the maximum allowed amplitude ε in the triplet case, at any given value of λ.

Besides, note that in case of non-adiabatic perturbation ($\lambda \gtrsim 1/2$) the multiplet map (1.43) can be usually replaced by the classical map (1.26) for the "central" triplet (with $W = W_1$), because at high values of λ the coefficients W_k at $k > 1$ are exponentially small with k, with respect to W_1.

As we have seen, the separatrix map techniques allows one to rapidly construct sections of phase space near the separatrix. Experience shows that separatrix maps provide advantages by two or three orders of magnitude in the speed of computation, in comparison with direct integrations. It is even more important that the separatrix map theory provides means for straightforward analytical description of the near-separatrix phase space. It allows one to precalculate the locations of resonances and borders of chaotic layers; it provides conditions for occurrence of marginal resonances and Hamiltonian intermittency, considered further on in Sect. 4.4.

1.7 Generalized Separatrix Maps

Let us write down the classical separatrix map in the form analogous to Eqs. (1.31):

$$\begin{aligned} y_{i+1} &= y_i + \sin x_i, \\ x_{i+1} &= x_i + \lambda \ln |y_{i+1}| + c, \end{aligned} \tag{1.44}$$

where y denotes, as defined above in Sect. 1.6, the normalized relative pendulum's energy, x is normalized time, and λ and c are constant parameters.

Consider a similar map with a power-law increment of the phase x instead of the logarithmic one:

$$\begin{aligned} y_{i+1} &= y_i + \sin x_i, \\ x_{i+1} &= x_i + \lambda |y_{i+1}|^{-\gamma}, \end{aligned} \tag{1.45}$$

or, equivalently,

$$\begin{aligned} w_{i+1} &= w_i + W \sin \tau_i, \\ \tau_{i+1} &= \tau_i + \kappa |w_{i+1}|^{-\gamma}. \end{aligned} \tag{1.46}$$

The map (1.46) is represented in form (1.45) by substituting $w = Wy$ and $\tau = x$; therefore, $\lambda = \kappa W^{-\gamma}$. Form (1.45) is more convenient, because it has only one parameter, λ, apart from the power-law index γ.

A number of dynamical and physical models are described by maps (1.45) and (1.46) with appropriate rational values of γ. The $\gamma = 1/4$ and $1/3$ values give Markeev's maps (Markeev 1994, 1995), describing the near-separatrix motion in two important models of resonances. The $\gamma = 1/2$ value gives the "$\hat{L}$-map" (Zaslavsky et al. 1991) for the motion of a non-relativistic particle in the field of a wave packet. This value also gives a map for the classical Morse oscillator driven by a time-periodic force (Abdullaev 2006). The $\gamma = 1$ value gives the Fermi map (Zaslavsky and Chirikov 1965; Lichtenberg and Lieberman 1992) for the Fermi acceleration mechanism for cosmic rays. The $\gamma = 3/2$ value gives the Kepler map (Petrosky 1986; Chirikov and Vecheslavov 1986; Vecheslavov and Chirikov 1988; Petrosky and Broucke 1988; Abdullaev 2006) for a number of physical and celestial-mechanical applications (Shevchenko 2011b). The $\gamma = 2$ value gives the "ultrarelativistic map" (Zaslavsky et al. 1991) for the motion of a relativistic particle in the field of a wave packet. All these maps describe the near-separatrix (in a broad sense) motion in corresponding models. For example, in case of the Kepler map model, the separatrix separates the bound and unbound states of motion of a particle in the restricted three-body problem.

1.8 Geometry of Chaotic Layers

An important property of the separatrix map in form (1.26), already mentioned above, is that it maps the motion asynchronously (Shevchenko 1998b; Abdullaev and Zaslavsky 1995, 1996): the energy variable is taken at the "pendulum" angle $\varphi = \pm\pi$, whereas the phase angle of perturbation is taken at $\varphi = 0$. This property is a direct outcome of the way how the map is constructed. According to Shevchenko (1998b), it leads to the phase portrait asymmetry with respect to the vertical axes $\tau = 0$ or $\tau = \pi$; see Figs. 1.8 and 1.9, and also Figure 6 in Ahn et al. (1996), or Figure 3b in Jeon et al. (1996)).

Abdullaev and Zaslavsky (1995, 1996) outlined this property for the classical separatrix map, and derived an improved map: a *shifted separatrix map*. In their construction, the phase angle of the perturbation can be taken at any arbitrary cross section, and the energy variable is taken at $\varphi = \pm\pi$, as in the classical setup.

However, the cross section $\varphi = \pm\pi$ (corresponding to the pendulum's unstable equilibrium) does not represent the dynamics completely, because the pendulum angle does not reach $\pm\pi$ point during oscillations. Therefore, it is more suitable to derive a procedure for synchronizing the map to the surface of section $\varphi = 0$ (corresponding to the pendulum's stable equilibrium).

Such a procedure for synchronizing the classical separatrix map was proposed in Shevchenko (1998b) for the case of symmetric perturbation. Here we describe the synchronization procedure as derived in Shevchenko (2000a) for the general case of asymmetric perturbation (1.6).

The synchronization makes meaningful comparisons of phase space sections of the near-separatrix motion obtained by direct numerical integration with the corresponding phase portraits of the separatrix map. Such comparisons allows to directly verify ranges of validity of the separatrix map in its parameter space.

To construct the phase space section at $\varphi = 0$ of the system with the Hamiltonian (1.6), it is necessary to calculate the energy increment not on the $(-\infty, +\infty)$ time interval, but on the $(-\infty, 0)$ or $(0, +\infty)$ time intervals. First of all, let us write down some necessary formulas for the motion on the unperturbed separatrix. As derived, e.g., in Chirikov (1979) and Lichtenberg and Lieberman (1992), the time variation of the pendulum angle is given by

$$\varphi^{\pm}(t) = \pm(4 \arctan \exp(t) - \pi), \tag{1.47}$$

where, as adopted in the following, the "+" upper index corresponds to the prograde motion of the pendulum, and the "−" upper index corresponds to its retrograde motion. The time variation of the momentum is given by

$$p^{+}(t) = \frac{2\omega_0}{\mathcal{G}} \cos \frac{\varphi^{+}(t)}{2} \tag{1.48}$$

for the prograde motion, and

$$p^{-}(t) = -\frac{2\omega_0}{\mathcal{G}} \cos \frac{\varphi^{-}(t)}{2} \tag{1.49}$$

for the retrograde motion.

One finds the required increment of H_0 in the Hamiltonian (1.6) as

$$\Delta H_0 = \int_{-\infty}^{0} \frac{dH_0}{dt} dt, \tag{1.50}$$

where the derivative is given by the Poisson bracket

$$\begin{aligned} \frac{dH_0}{dt} &= \{H_0, H\} = \frac{\partial H_0}{\partial \varphi}\frac{\partial H}{\partial p} - \frac{\partial H}{\partial \varphi}\frac{\partial H_0}{\partial p} = \\ &= ak\mathcal{G}p\left(\sin(k\varphi - \tau) + \eta \sin(k\varphi + \tau)\right), \end{aligned} \tag{1.51}$$

and $\eta = b/a$. The H_0 increment is expressed in terms of the Melnikov–Arnold integrals, presented in detail in Appendix C. In the prograde case the increment is given by

$$\Delta^+ H_0 = -a\frac{\lambda}{2}\left(A_{2k}(\lambda) + \eta A_{2k}(-\lambda)\right)\sin\tau_0 - \\ - a\left(\lambda\left(B_{2k}(\lambda) - \eta B_{2k}(-\lambda)\right) + 1 + \eta\right)\cos\tau_0. \tag{1.52}$$

Deriving the expression in the retrograde case is similar. In general, for the relative energy ($w = (H_0/\mathcal{F}) - 1$) increment one has

$$\Delta^{\pm} w = -\frac{W^{\pm}}{2}\sin\tau_0 - \delta^{\pm} W^{\pm}\cos\tau_0, \tag{1.53}$$

where

$$W^+(k,\lambda,\eta) = \frac{a}{\mathcal{F}}\lambda\left(A_{2k}(\lambda) + \eta A_{2k}(-\lambda)\right), \tag{1.54}$$

$$W^-(k,\lambda,\eta) = \frac{a}{\mathcal{F}}\lambda\left(\eta A_{2k}(\lambda) + A_{2k}(-\lambda)\right), \tag{1.55}$$

$$\delta^+(k,\lambda,\eta) = \frac{\lambda\left(B_{2k}(\lambda) - \eta B_{2k}(-\lambda)\right) + 1 + \eta}{\lambda\left(A_{2k}(\lambda) + \eta A_{2k}(-\lambda)\right)}, \tag{1.56}$$

$$\delta^-(k,\lambda,\eta) = \frac{\lambda\left(\eta B_{2k}(\lambda) - B_{2k}(-\lambda)\right) + 1 + \eta}{\lambda\left(\eta A_{2k}(\lambda) + A_{2k}(-\lambda)\right)}. \tag{1.57}$$

Expressions (1.56, 1.57) for $\delta^{\pm}$ are valid when the relative energy w is defined as $w = (H_0/\mathcal{F}) - 1$. Expressions (1.54) and (1.55) represent a generalization of formulas (1.34) to the case of arbitrary k.

If w is defined for the full energy H, i.e., $w = \frac{H}{\mathcal{F}} - 1$, then the expressions for $\delta^{\pm}$ are somewhat different. As soon as $\frac{dH}{dt} = \frac{\partial H}{\partial t}$, the increment of H in that case is given by

$$\Delta H = \int_{-\infty}^{0} \frac{\partial H}{\partial t} dt. \tag{1.58}$$

Therefore, the expressions for $W^{\pm}$ do not change, but the expressions for $\delta^{\pm}$ become

$$\delta^+(k,\lambda,\eta) = \frac{B_{2k}(\lambda) - \eta B_{2k}(-\lambda)}{A_{2k}(\lambda) + \eta A_{2k}(-\lambda)}, \tag{1.59}$$

$$\delta^-(k,\lambda,\eta) = \frac{\eta B_{2k}(\lambda) - B_{2k}(-\lambda)}{\eta A_{2k}(\lambda) + A_{2k}(-\lambda)}. \tag{1.60}$$

Below the definition of w in terms of H_0 is everywhere adopted; therefore, we do not employ formulas (1.59, 1.60).

Also note that the given formulas are valid for any integer or half-integer value of k, but the numerical illustrations in this section are given solely for $k = 1$.

The value of w at $\varphi = 0$, when its value w_n at $\varphi = \pm\pi$ is known, is found by summing w_n and its increment $\Delta^{\pm} w_n$:

$$\widetilde{w}_n = w_n - \frac{W^{\pm}}{2} \sin \tau_n - \delta^{\pm} W^{\pm} \cos \tau_n = \\ = \frac{w_n + w_{n+1}}{2} - \delta^{\pm} W^{\pm} \cos \tau_n, \tag{1.61}$$

where w_{n+1} is the next energy iteration of the non-synchronized separatrix algorithmic map. The synchronization of the map's phase portrait is achieved by replacing the (w_n, τ_n) pair of values with $(\widetilde{w}_n, \tau_n)$.

The formulas (1.61) synchronize the separatrix algorithmic map (1.33), where the index of δ (plus or minus) is taken to be the same as for the current value of W. In case of symmetric perturbation, corresponding to the classical separatrix map, any one-named quantities with different indices (plus or minus) are equal to each other.

Algorithm (1.33), used together with the synchronization procedure (1.61), allows one to build synchronized phase portraits of the near-separatrix motion of any system with Hamiltonian (1.6). Instead of employing the synchronization procedure at each iteration of the map, the separatrix algorithmic map can be straightforwardly written down in terms of the "synchronized" variables $\widetilde{w}$, τ. The change of variables from w, τ to $\widetilde{w}$, τ is canonical. However, the map constructed in this way is implicit. Therefore, it is inconvenient in practical calculations. We write it down here solely in the symmetric case $a = b$:

$$\begin{aligned} \widetilde{w}_{n+1} &= \widetilde{w}_n - W \sin \frac{\tau_n + \tau_{n+1}}{2} \left(\cos \frac{\tau_{n+1} - \tau_n}{2} - 2\delta \sin \frac{\tau_{n+1} - \tau_n}{2} \right), \\ \tau_{n+1} &= \tau_n + \lambda \ln 32 - \lambda \ln \left| \frac{\widetilde{w}_n + \widetilde{w}_{n+1}}{2} + \right. \\ &\left. + \frac{W}{2} \cos \frac{\tau_n + \tau_{n+1}}{2} \left(\sin \frac{\tau_{n+1} - \tau_n}{2} + 2\delta \cos \frac{\tau_{n+1} - \tau_n}{2} \right) \right| \\ &(\mathrm{mod}\ 2\pi). \end{aligned} \tag{1.62}$$

The synchronization parameter δ can be found by taking the Melnikov–Arnold integrals numerically directly in expressions (1.56) and (1.57), or (1.59) and (1.60). Let $k = 1$. The primitives of the integrand expressions in the formulas for $\alpha_2^c(\lambda)$ and $\beta_2^s(\lambda)$ (see Appendix C), as functions of the independent variable t, oscillate at $t \to +\infty$ with non-zero amplitudes. This implies that the upper limits in time for the numerical integrations should be taken equal to some non-arbitrary discrete values, see Shevchenko (2000a) for details.

Consider the synchronization procedure in case of symmetric perturbation ($a = b$) and $k = 1$ in more detail. This case corresponds to the classical separatrix map (1.26). The synchronization parameter δ in is then given by

$$\delta(\lambda) = \frac{\beta_2^s(\lambda) + \frac{2}{\lambda}}{\alpha_2^c(\lambda)} \tag{1.63}$$

for both prograde and retrograde motions; $\alpha_2^c(\lambda)$ and $\beta_2^s(\lambda)$ are given in Appendix C. Using the recurrent relations presented in Appendix C one finds

$$\delta(\lambda) = \frac{1}{2\pi}\left(\beta_1^c(\lambda) + \frac{2}{\lambda^2}\right)\sinh\frac{\pi\lambda}{2} = \\ = \frac{1}{\pi}\left(\mathrm{Re}\left(\psi\left(i\frac{\lambda}{2}\right) - \psi\left(i\frac{\lambda}{4}\right)\right) + \frac{1}{\lambda^2} - \ln 2\right)\sinh\frac{\pi\lambda}{2}, \tag{1.64}$$

where $\psi(z) = \Gamma'(z)/\Gamma(z)$ is the digamma-function, and, here, i is the imaginary unit.

Let us now see how the synchronized separatrix map performs in comparison with direct numerical integrations of the original system of differential equations.

We take first an intermediate value of the adiabaticity parameter, namely, $\lambda = 0.5$. In Figs. 1.11 and 1.12a, the corresponding phase portraits of the non-synchronized and synchronized separatrix maps are shown. For comparison, the phase space section constructed by a direct numerical integration of the original

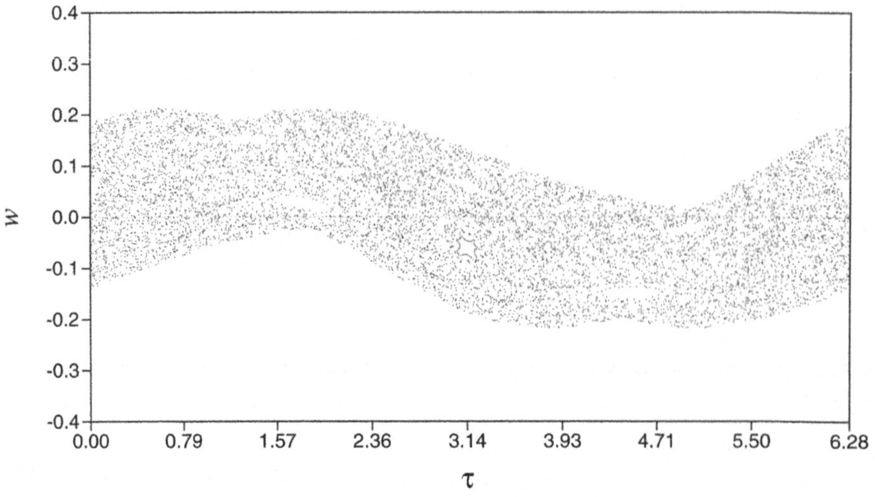

Fig. 1.11 A phase portrait of the separatrix map (1.26); $\lambda = 0.5$ and $W = 0.181$ (Figure 2 from Shevchenko (2000a). With permission from Pleiades Publishing Inc.)

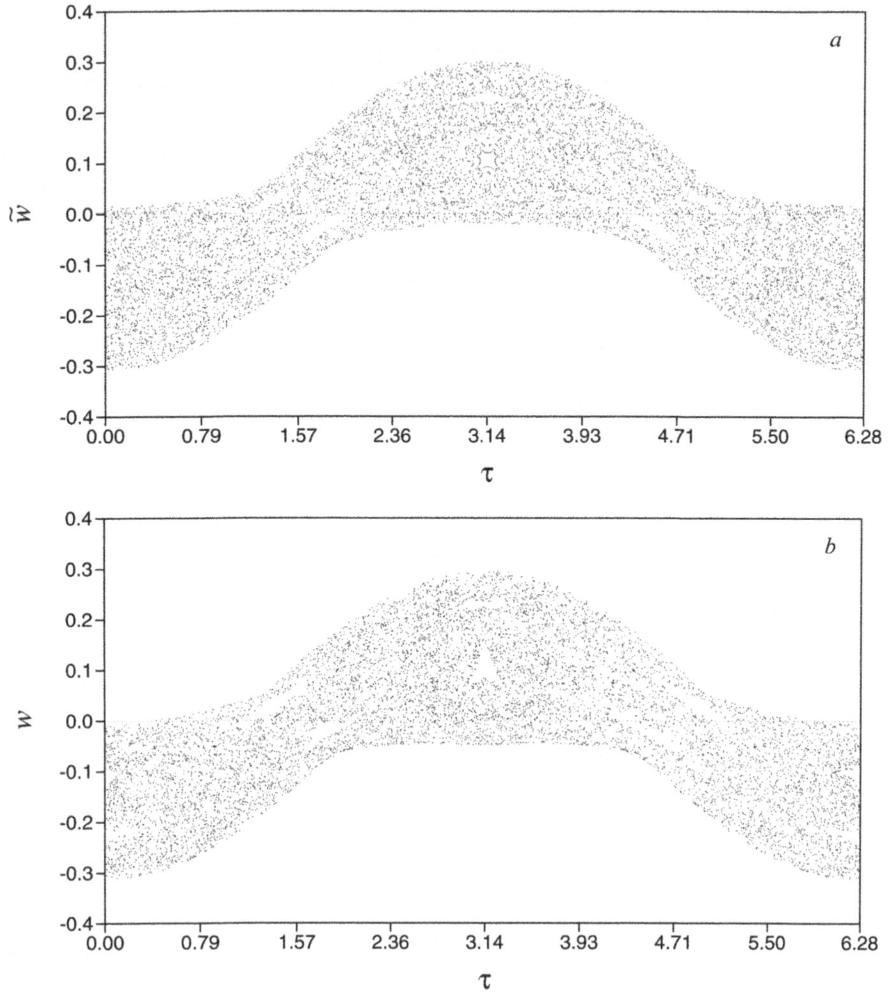

Fig. 1.12 (**a**) The same as in Fig. 1.11, but synchronized to the surface $\varphi = 0$; $\delta = 0.929$. (**b**) The corresponding phase space section obtained by a direct numerical integration. The tilde (synchronization mark) over w is omitted, because the section is inherently synchronized (Figure 3 from Shevchenko (2000a). With permission from Pleiades Publishing Inc.)

system at $\varphi = 0$ is presented in Fig. 1.12b. Only the near-separatrix chaotic layer is shown here and in the figures that follow.

One may see that the asymmetry (with respect to the $\tau = \pi$ line), present in the non-synchronized phase portrait (Fig. 1.11), disappears in the synchronized portrait (Fig. 1.12a), demonstrating that the asymmetry represents an artifact of the non-synchronized map.

Figure 1.12b shows the corresponding phase section obtained by a direct numerical integration of the original system (1.6). The parameters in the Hamiltonian (1.6) correspond to $\lambda = 0.5$ and $W = 0.181$ in the separatrix map (1.26). Namely, the parameters are: $\mathcal{F} = 1$, $\mathcal{G} = 2$, $a = b = 0.05$ (symmetric perturbation) and $\Omega = 2^{-1/2}$. The integration is performed by the eighth order Dormand–Prince (Hairer et al. 1987) method with controlled step-size; for details see Shevchenko (2000a).

During the integration, the w and τ values are taken on the unified surface of section $\varphi = 0$. Therefore, the section in Fig. 1.12b can be compared directly to Fig. 1.12a, where the phase portrait of the synchronized separatrix map is given. One may see that the behaviour of the original system (1.6) is perfectly described by the synchronized separatrix map.

In Figs. 1.13 and 1.14, phase portraits of the synchronized separatrix map are given for low and high frequency perturbation cases ($\lambda = 0.1$ and 10, respectively). Direct numerical integrations (not presented here for brevity) reproduce the behaviour identically.

Figures 1.13 and 1.14 show that the chaotic layer is strongly bent in the both limits of low and high frequencies of perturbation. The bending is present only in the phase portraits of the synchronized map. Conversely, the chaotic layer of the non-synchronized map always engulfs the $w = 0$ axis.

The bending amplitude is given by the product δW, see Eq. (1.61). At $k = 1$ and $\eta = 1$ (symmetric perturbation), from Eqs. (1.54), (1.55), (1.56, and (1.57), if $\lambda = 0$, one has $\delta^+ W^+ = \delta^- W^- = 4\varepsilon$, and if $\lambda = +\infty$ one has $\delta^+ W^+ = \delta^- W^- = 0$. At $k = 1$ and $\eta = 0$ (maximum asymmetry of perturbation), if $\lambda = 0$, then $\delta^+ W^+ = \delta^- W^- = 2\varepsilon$, and if $\lambda = +\infty$, then $\delta^+ W^+ = \delta^- W^- = 0$. Therefore, in the high-frequency perturbation limit, at any η, the bending amplitude always

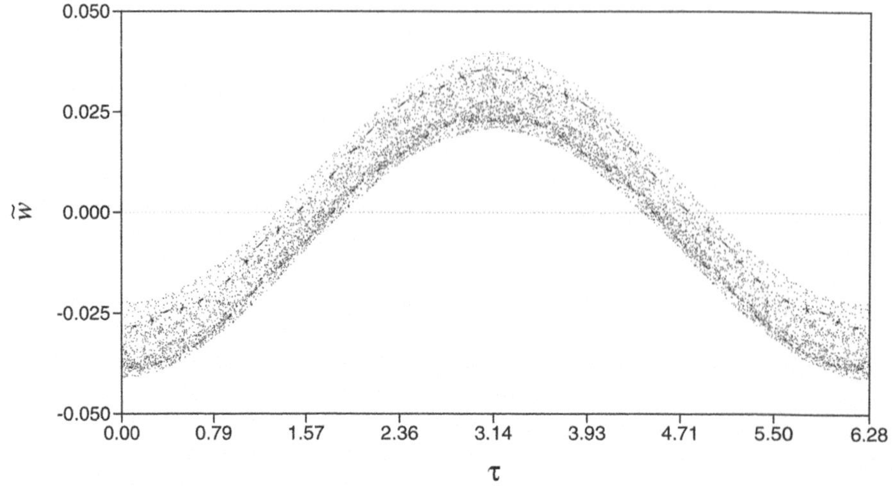

Fig. 1.13 The phase portrait of the synchronized separatrix map (1.26), (1.61) with $\lambda = 0.1$, $W = 0.00797$, and $\delta = 4.99$ (Figure 4 from Shevchenko (2000a). With permission from Pleiades Publishing Inc.)

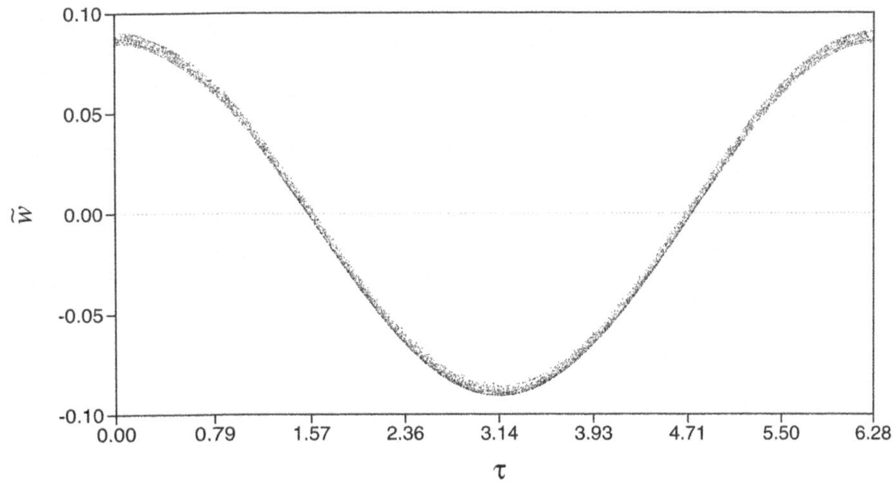

Fig. 1.14 The phase portrait of the synchronized separatrix map (1.26), (1.61) with $\lambda = 10$, $W = 0.000379$, and $\delta = -232.5$ (Figure 5 from Shevchenko (2000a). With permission from Pleiades Publishing Inc.)

tends to zero. However, in comparison with the layer width, the bending cannot be regarded as small. In the high-frequency limit of perturbation, the layer half-width $\approx \lambda W$; see Chap. 5. Therefore, at energies $w > 0$, corresponding to the model pendulum rotation, the ratio of the bending amplitude to the layer half-width is $\approx \frac{\delta^+}{\lambda}$ for the prograde motion, and $\approx \frac{\delta^-}{\lambda}$ for the retrograde motion. At any value of the asymmetry parameter η, the absolute values of these products tend to infinity at $\lambda \to +\infty$. At energies $w < 0$, corresponding to the model pendulum oscillation, the prograde and retrograde motions alternate, and the required ratio can be estimated as $\approx \frac{\delta^+}{\lambda}$ for the prograde motion and $\approx \frac{\delta^- W^-}{\lambda W^+}$ for the retrograde motion.

Therefore, in the high-frequency perturbation limit the bending is strong in the sense that the amplitude of energy deviations with respect to the unperturbed separatrix is much greater than the chaotic layer proper width.

According to Eq. (1.61), the synchronization transforms the unperturbed separatrix $w = 0$ into the curve $\widetilde{w} = \delta^\pm W^\pm \cos\tau$. Let us introduce the polar coordinates τ and $\rho = \widetilde{w} + 2$; here the energy zero level corresponds to the pendulum's lower equilibrium state. When the chaotic layer is thin, $\lambda W^\pm \ll \delta^\pm W^\pm$, its shape in this frame represents a Pascal limaçon: $\rho = 2 + \delta^\pm W^\pm \cos\tau$. The bending effect in case of the maximum asymmetry of perturbation ($\eta = 0$) becomes appreciable (the bending amplitude becomes greater than the layer width) at $\lambda \approx 8$, both for the prograde and retrograde motions.

At perturbations with $\varepsilon \sim \lambda^{-1}$ and $\varepsilon > \lambda^{-1}$ the standard Poincaré–Melnikov method for calculating exponentially small effects associated with the separatrix splitting generally requires corrections (Gelfreich 1997; Treshchev 1998). For a perturbation of arbitrary asymmetry (i.e., for any η) in system (1.6), the correction factor to the separatrix map parameter W, according to Simó's formula (see Gelfreich 1997), is $|f(x)| = \left|\frac{\sinh(x)}{x}\right|$, where $x \equiv (2\varepsilon_1\varepsilon_2)^{1/2} = \frac{(2ab)^{1/2}}{\mathcal{F}}$; here x may be either a real or an imaginary number depending on the signs of a and b. In case of Fig. 1.14, the correction factor $f(\sqrt{2}) = 1.368$.

This correction merely influences the layer width but not its bending amplitude since the bending has no relation to the separatrix splitting effect. In Fig. 1.14, the layer width is small compared to the bending amplitude; therefore, the correction would not influence the general appearance of the phase portrait.

To conclude this topic, one should note that the synchronized separatrix algorithmic map describes motion of the system on the "phase perturbation angle—energy deviation" plane at a fixed angle $\varphi = 0$. When Poincaré sections are constructed numerically in applied problems, a different surface of section is frequently used, namely, the "resonance angle—momentum" plane at a fixed perturbation phase angle, for example 0 (mod 2π). Is it possible to build this second type of section using separatrix map? In a good enough approximation, this is indeed possible by utilizing a procedure which is again an algorithm containing conditional transfer instructions; see algorithms and references in Sect. 1.6.

Chapter 2
Numerical Tools for Studies of Dynamical Chaos

To date, quite a number of useful numerical tools that can be used to explore instabilities and dynamical chaos in planetary systems, including our Solar system and its subsystems, have been developed. The tools are based, in particular, on massive computations of "mean exponential growth factors of nearby orbits" (MEGNO) (Cincotta and Simó 2000; Cincotta et al. 2003; Goździewski 2003), "fast Lyapunov indicators" (FLI) (Pilat-Lohinger and Dvorak 2002), Lyapunov characteristic exponents (LCE) (Melnikov and Shevchenko 1998; Shevchenko and Melnikov 2003; Popova and Shevchenko 2013), fundamental frequencies of motion (frequency analysis) (Correia et al. 2009; Laskar and Correia 2009). A software package for calculating various indicators of chaos (including LCE, FLI, and MEGNO) is described in Carpintero et al. (2014). Dvorak et al. (2004) and Schwarz et al. (2011) proposed and broadly used a "maximum eccentricity method" (MEM), consisting in computation and graphical representation of the largest (during a fixed integration time) eccentricities of a test particle, estimated on grids of values of initial conditions. This method can be regarded as a version of massive numerical assessments of escape/encounter conditions in three-body and many-body systems (Holman and Wiegert 1999; Pilat-Lohinger et al. 2003; Kholshevnikov and Kuznetsov 2011).

In this chapter, we concentrate on numerical tools needed to characterize the chaotic behaviour in problems of celestial mechanics addressed further on in the book. The presentation partially follows (in Sects. 2.1 and 2.7) the lecture by Shevchenko (2011c).

2.1 Lyapunov Exponents

For dynamical chaos to emerge, three basic conditions should be satisfied (Devaney 1986; Meiss 1992, p. 810): (1) Sensitive (essential) dependence on initial conditions, implying positive Lyapunov exponents (i.e., nearby orbits diverge exponentially

I. I. Shevchenko, *Dynamical Chaos in Planetary Systems*, Astrophysics and Space Science Library 463, https://doi.org/10.1007/978-3-030-52144-8_2

in time). (2) The boundedness of the phase space, implying that the exponential divergence does not simply reduce to a smooth expansion to infinity. (3) The set of orbits with such behaviour should have non-zero measure. No strict analytical proofs of the non-zero measure of irregular component in typical Hamiltonian systems are available (Meiss 1992), but computer-generated phase space sections hint on this; the irregular component seems to form so-called "fat fractals" (Umberger and Farmer 1985).

If conditions (2) and (3) are implied to be satisfied, then, if nearby trajectories in the bounded phase space diverge exponentially (in other words, if the distance between two initially close points of these trajectories grows in time exponentially), then the motion is regarded as chaotic.

The rate of divergence of nearby trajectories in the phase space is characterized by the *Lyapunov characteristic exponents* (LCE).[1] Set the initial time moment $t = t_0$ and let $\mathbf{x}(t_0)$ and $\mathbf{x}'(t_0)$ be the starting points of a *nominal* and a *shadow* (i.e., initially nearby) trajectories, respectively, in the phase space of motion. Consider the evolution of the displacement $d(t) \equiv \|\mathbf{x}(t) - \mathbf{x}'(t)\| \ll 1$ with time, starting with $t = t_0$. The Lyapunov exponent is defined as the double limit

$$L = \limsup_{\substack{d(t_0)\to 0 \\ t\to\infty}} \frac{1}{t - t_0} \ln \frac{d(t)}{d(t_0)}. \tag{2.1}$$

Depending on the direction of the initial displacement vector in the phase space of a Hamiltonian system, the quantity L may attain $2N$ generally different values, where N is the system's number of degrees of freedom. The $2N$ values come in N symmetric pairs: for each $L_i > 0$ there exists $L_{i+N} = -L_i < 0$; $i = 1, \ldots, N$ (Lichtenberg and Lieberman 1992). Therefore, it is enough to calculate a half of all exponents, e.g., the set of non-negative ones, $L_i \geq 0$.

The set of all $2N$ exponents is called the spectrum of Lyapunov exponents, or, the *Lyapunov spectrum*. In typical applied problems of Hamiltonian dynamics, a non-zero value of the *maximum Lyapunov exponent* (maximum in the spectrum) indicates the presence of dynamical chaos, in accord with the chaos conditions formulated above, whereas its zero value indicates that the motion is regular, i.e., quasiperiodic or periodic (Chirikov 1979; Lichtenberg and Lieberman 1992).

On an everywhere dense set of starting data for shadow trajectories, the calculated Lyapunov exponent attains its maximum value, which is just the maximum Lyapunov exponent (Lichtenberg and Lieberman 1992; Meiss 1992). We denote it henceforth by L. The inverse of this quantity, $T_{\mathrm{L}} \equiv L^{-1}$, is the so-called *Lyapunov time*. It represents the characteristic time of predictable dynamics. The Lyapunov timescales of celestial-mechanical systems are important, because any exact theory of motion cannot be constructed on times much greater than the Lyapunov time corresponding to given parameters and initial conditions of a considered system.

[1] A general definition of Lyapunov exponents of a function is given in Adrianova (1995). Here we consider LCEs of dynamical systems.

The art of computation of the Lyapunov exponents (and, in the first place, the maximum Lyapunov exponent) has a long history, starting in seventies of the twentieth century (Benettin et al. 1976; Benettin and Galgani 1979; Benettin et al. 1980), and since that time it has become an extensive part of applied mathematics, reviewed, in particular, in Froeschlé (1984) and Lichtenberg and Lieberman (1992). Modern numerical methods for computation of the Lyapunov exponents are effective and precise.

On the other hand, analytical estimating the Lyapunov exponents in practical applications usually has a lower precision, but analytical approaches provide a deeper insight in the causes of chaotic behaviour whenever it arises. The analytical methods were started to be developed relatively recently (Morbidelli and Froeschlé 1996; Holman and Murray 1996; Murray and Holman 1997); they are considered in detail in Chap. 3.

Numerically, the maximum Lyapunov exponent is calculated on a finite time interval (say, of m time units) by means of the formula

$$L(m) = \frac{1}{m \Delta t} \sum_{i=1}^{m} \ln r_i, \tag{2.2}$$

where r_i is the ratio of the current nominal–shadow displacement, $d_i = d(t = i)$, to the preceding one, $d_{i-1} = d(t = i - 1)$: $r_i = d(t = i)/d_{i-1}$, and Δt is the time unit corresponding to one iteration (Benettin et al. 1976; Lichtenberg and Lieberman 1992). The displacements should be periodically renormalized to a small value (preserving the direction of the displacement vector), so that the shadow trajectory is kept in a vicinity of the nominal one; see Fig. 2.1 (in this scheme, T denotes the linearized mapping transformation).

In this way the problem with the first limit $d(t_0) \to 0$ in the defining formula (2.1) is solved numerically. The problem with the second limit $t \to \infty$ is usually solved as follows: the values of $\log L(m)$, calculated consequently, as given by Eq. (2.2), are plotted versus $\log m$, and the value of $\log L$ at which the dependence is *saturated*, i.e., attains a horizontal plateau at some constant non-zero L, is fixed; see Fig. 2.3 for an illustration. If the trajectory is regular, then $\log L$ goes down on average linearly with $\log m$ (because $d(t)$ in Eq. (2.1) oscillates around a constant value), and the curve never attains any plateau. Therefore, chaotic and regular orbits can be distinguished from each other.

Let us summarize the properties of LCEs in a greater detail, following Benettin et al. (1976), Benettin and Galgani (1979), Benettin et al. (1980), and Lichtenberg and Lieberman (1992). Consider the dynamical system

$$\dot{x}_i = V_i(x_1, \ldots, x_M), \quad i = 1, \ldots, M. \tag{2.3}$$

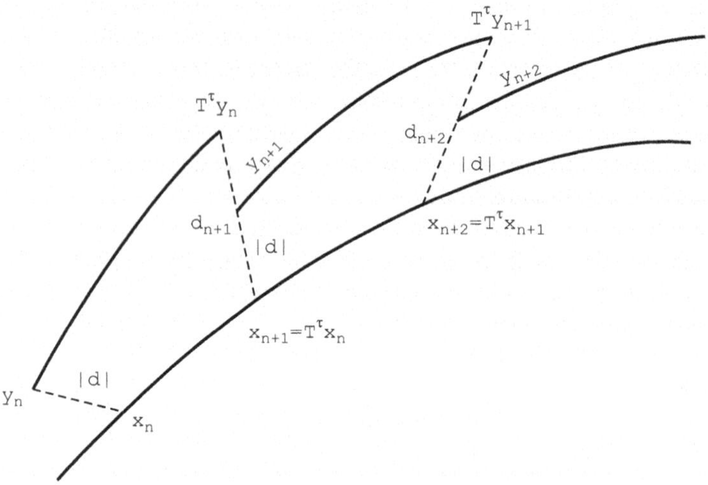

Fig. 2.1 A straightforward numerical procedure for calculation of the maximum LCE. The displacement d_n is periodically renormalized (Reprinted Figure 1 with permission from Benettin et al. (1976). © 1976 by the American Physical Society)

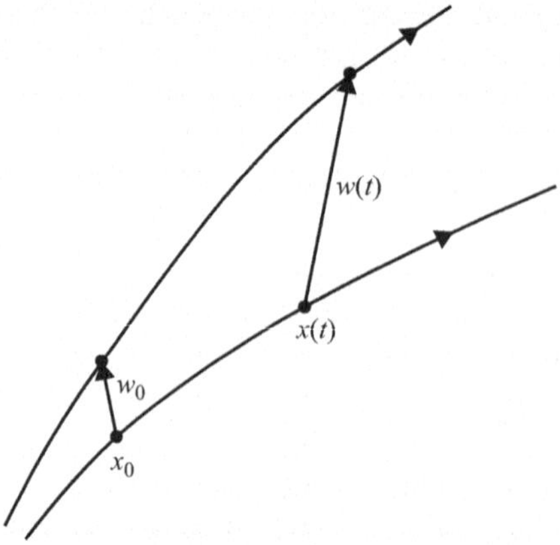

Fig. 2.2 The nominal and shadow trajectories, and the evolution of the tangent (nominal–shadow displacement) vector **w** (Figure 1 from Benettin and Galgani (1979))

Let two nearby trajectories start at points $\mathbf{x}_0$ and $\mathbf{x}_0 + \Delta\mathbf{x}_0$, as shown in Fig. 2.2. The so-called *tangent vector* is $\mathbf{w} \equiv \Delta\mathbf{x}$, and its length is given by

$$d(\mathbf{x}_0, t) = \|\mathbf{w}(\mathbf{x}_0, t)\|. \tag{2.4}$$

Linearizing Eqs. (2.3), we obtain the equations describing the evolution of **w**:

$$\dot{\mathbf{w}} = \mathbf{M}(\mathbf{x}(t)) \cdot \mathbf{w}, \tag{2.5}$$

where the *Jacobi matrix* $\mathbf{M} \equiv \partial \mathbf{V}/\partial \mathbf{x}$. These are the so-called *variational equations*.

Following Eq. (2.1), we define the quantity

$$\sigma(\mathbf{x}_0, \mathbf{w}_0) = \lim_{\substack{t\to\infty \\ d(t_0)\to 0}} \frac{1}{t - t_0} \ln \frac{d(\mathbf{x}_0, t)}{d(\mathbf{x}_0, t_0)}. \tag{2.6}$$

It has the meaning of the mean rate of divergence of nearby trajectories in the scale of the exponential functions $\exp \alpha t$; it is evident that for the latter functions one has $\sigma = \alpha$.

There exists a full system of M fundamental solutions $\|\mathbf{e}_i\|$ of Eqs. (2.5). For each of these solutions the rate σ has a definite (in general, individual) value:

$$\sigma_i(\mathbf{x}_0) = \sigma(\mathbf{x}_0, \mathbf{e}_i). \tag{2.7}$$

These are just the LCEs. Traditionally they are enumerated in the order of the value decreasing: $\sigma_1 \geq \sigma_2 \cdots \geq \sigma_M$. These quantities do not depend on the choice of the phase space metrics (Oseledets 1968).

Consider a periodic trajectory of system (2.3), and, therefore, let the initial conditions $\mathbf{x}_0$ belong to a periodic trajectory. Then Eqs. (2.5) define a linear map

$$\mathbf{w}_{n+1} = \mathbf{A} \cdot \mathbf{w}_n \tag{2.8}$$

with a period τ. The matrix $\mathbf{A}$ has M generally complex eigenvalues λ_i, enumerated as

$$|\lambda_1| \geq |\lambda_2| \cdots \geq |\lambda_M|. \tag{2.9}$$

Setting $\mathbf{w}_0 = \mathbf{e}_i$, where $\mathbf{e}_i$ is the eigenvector corresponding to the eigenvalue λ_i, from Eqs. (2.8) one has

$$\mathbf{w}_n = \lambda_i^n \mathbf{e}_i, \tag{2.10}$$

and, in accord with formula (2.6),

$$\sigma(\mathbf{e}_i) = \frac{n}{n\tau} \ln |\lambda_i| = \frac{1}{\tau} \ln |\lambda_i| = \sigma_i. \tag{2.11}$$

The initial tangent vector

$$\mathbf{w}_0 = c_1\mathbf{e}_1 + c_2\mathbf{e}_2 + \cdots + c_M\mathbf{e}_M \tag{2.12}$$

evolves as determined by the first non-zero coefficient c_i: if $c_1 \neq 0$, then $\sigma(\mathbf{w}_0) = \sigma_1$; if $c_1 = 0$ and $c_2 \neq 0$, then $\sigma(\mathbf{w}_0) = \sigma_2$; and so on. Therefore, each element in the Lyapunov spectrum determines the rate σ in some subspace with the number of dimensions smaller by unity, in comparison with the preceding element. Therefore, for almost all $\mathbf{w}_0$ one has $\sigma(\mathbf{w}_0) = \sigma_1$.

Oseledets (1968) generalized the notions of the eigenvalues and eigenvectors of the tangent map to the case of non-periodic trajectories. The possibility of such generalization is determined by the possibility of approximation of the non-periodic trajectories by periodic ones with large enough periods.

For any trajectory given by Eqs. (2.3), at least one of the LCEs, namely that corresponding to the eigenvector along the trajectory, is equal to zero. In Hamiltonian systems, the LCEs have the symmetry

$$\sigma_i = -\sigma_{2N-i+1}, \tag{2.13}$$

where N is the number of degrees of freedom, and $2N$ is equal to M in Eqs. (2.3). For any autonomous Hamiltonian system two Lyapunov exponents are therefore equal to zero.

Mappings For the maps, the rate (2.6) of divergence is straightforwardly reformulated by substituting time t with the current iteration number n:

$$\sigma^{\mathrm{map}}(\mathbf{x}_0, \mathbf{w}_0) = \lim_{\substack{n\to\infty \\ d(0)\to 0}} \frac{1}{n} \ln \frac{d(\mathbf{x}_0, n)}{d(\mathbf{x}_0, 0)}. \tag{2.14}$$

Consider an M-dimensional map

$$\mathbf{x}_{n+1} = \mathbf{F}(\mathbf{x}_n), \tag{2.15}$$

and let $\lambda_i(n)$ $(i = 1, \ldots, M)$ be the eigenvalues of the matrix

$$\mathbf{A}_n = [\mathbf{M}(\mathbf{x}_n) \cdot \mathbf{M}(\mathbf{x}_{n-1}) \cdot \cdots \cdot \mathbf{M}(\mathbf{x}_1)]^{1/n}, \tag{2.16}$$

where $\mathbf{M} = \partial\mathbf{F}/\partial\mathbf{x}$ is the Jacobi matrix for the map $\mathbf{F}$. Then the LCEs are given by

$$\sigma_i^{\mathrm{map}} = \lim_{n\to\infty} \ln |\lambda_i(n)|. \tag{2.17}$$

The LCEs of a map on a Poincaré section of the phase space of an original dynamical system are directly proportional to the LCEs of the original system:

$$\sigma_i^{\mathrm{map}}(\mathbf{x}_0) = \overline{\tau}\sigma_i(\mathbf{x}_0), \tag{2.18}$$

where $i = 1, 2, \ldots, M$. The proportionality coefficient $\overline{\tau}$ is equal to the mean time interval between consecutive crossings of the surface of section.

The map constructed for an autonomous Hamiltonian system with N degrees of freedom, therefore, has $M = 2N - 2$. The two zero LCEs of the original system are eliminated by reducing the continuous system to the discrete map.

Reviews of modern methods of the computation of the LCE spectra can be found in Geist et al. (1990) and Skokos (2010); here we describe a modern algorithm proposed and developed by Von Bremen et al. (1997). It is based on the QR-decomposition of the tangent map matrix. The algorithm utilizes the Householder transformation, and is therefore known as the HQR method. Von Bremen et al. (1997) proved it to be numerically more stable than the method based on the Gram–Schmidt orthogonalization originally proposed to compute LCEs in Benettin et al. (1976). Let

$$\mathbf{x}_{i+1} = \mathbf{F}_i(\mathbf{x}_i) \tag{2.19}$$

be the original map, then

$$\delta\mathbf{x}_{i+1} = \mathbf{J}_i(\delta\mathbf{x}_i) \tag{2.20}$$

is its tangent map. Here $\mathbf{x}_i$ is the system state at the ith iteration step, $\delta\mathbf{x}_i$ is the corresponding tangent vector, and $\mathbf{J}_i$ is the Jacobi matrix. The approximation for each LCE is given by

$$L^k(m) = \frac{1}{m\Delta t}\sum_{j=1}^{m} \ln |r_j^k|, \tag{2.21}$$

where $r_j^k = R_j^{kk}$ with $k = 1, \ldots, 2N$ are the diagonal elements of the right-triangular matrix $\mathbf{R}_j$ obtained iteratively by the QR-factorization of the product $\mathbf{J}_j\mathbf{Q}^{(j-1)}$, i.e., $\mathbf{Q}_j\mathbf{R}_j = \mathbf{J}_j\mathbf{Q}^{(j-1)}$, where the matrix $\mathbf{Q}$ is orthogonal, and $\mathbf{Q}^{(0)}$ is the identity matrix.

The QR-decomposition of the matrix product $\mathbf{J}_m\mathbf{J}_{m-1}\cdots\mathbf{J}_1$ is obtained in the following sequence of factorizations:

$$\begin{aligned}
\mathrm{qr}[\mathbf{J}_m\mathbf{J}_{m-1}\cdots\mathbf{J}_1] &= \mathrm{qr}[\mathbf{J}_m\mathbf{J}_{m-1}\cdots\mathbf{J}_2(\mathbf{J}_1\mathbf{Q}_0)] = \\
&= \mathrm{qr}[\mathbf{J}_m\mathbf{J}_{m-1}\cdots\mathbf{J}_3(\mathbf{J}_2\mathbf{Q}_1)][\mathbf{R}_1] = \\
&= \mathrm{qr}[\mathbf{J}_m\mathbf{J}_{m-1}\cdots\mathbf{J}_4(\mathbf{J}_3\mathbf{Q}_2)][\mathbf{R}_2\mathbf{R}_1] = \\
&= \cdots = \\
&= \mathrm{qr}[\mathbf{J}_m\mathbf{J}_{m-1}\cdots\mathbf{J}_{j+1}(\mathbf{J}_j\mathbf{Q}_{j-1})][\mathbf{R}_{j-1}\mathbf{R}_{j-2}\cdots\mathbf{R}_2\mathbf{R}_1] = \\
&= \cdots = \\
&= \mathbf{Q}_m[\mathbf{R}_m\cdots\mathbf{R}_2\mathbf{R}_1] = \mathbf{Q}_m\mathbf{R},
\end{aligned} \tag{2.22}$$

where $\mathbf{Q}_0 = \mathbf{I}$ (the identity matrix), and qr[·] denotes the QR-factorization process.

At each step j of the algorithm (2.22) ($j = 1, 2, \ldots, m$), a premultiplication $\mathbf{B}_j = \mathbf{J}_j\mathbf{Q}_{j-1}$ is made followed by QR-factorization of $\mathbf{B}_j = \mathbf{J}_j\mathbf{Q}_{j-1} = \mathbf{Q}_j\mathbf{R}_j$. Therefore, $\mathbf{R} = \mathbf{R}_m\cdots\mathbf{R}_2\mathbf{R}_1$, and the diagonal elements of $\mathbf{R}$ are products of

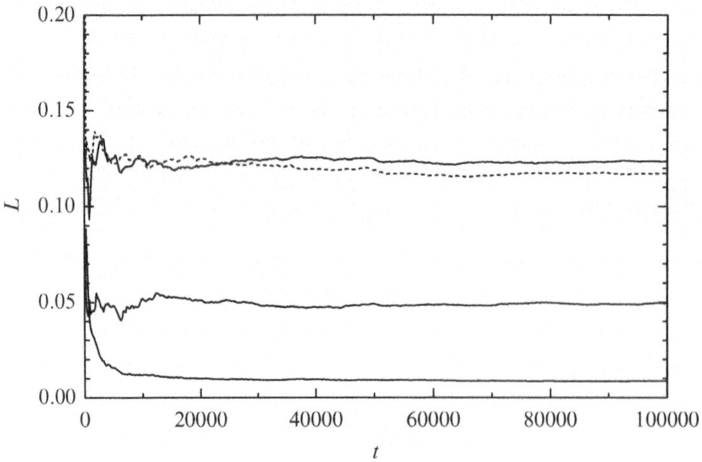

Fig. 2.3 Time dependences of the elements of the current Lyapunov spectrum (obtained by the HQR method; solid curve) in a model for the chaotic rotation of Hyperion, the seventh satellite of Saturn. The current maximum LCE, obtained by the shadow trajectory method, is also presented (dashed curve); note its accord with the first element of the Lyapunov spectrum obtained by the HQR method (Figure 3 from Shevchenko and Kouprianov (2002), reproduced with permission © ESO)

the corresponding diagonal elements of the matrices $\mathbf{R}_j$. The actual Lyapunov exponents are the limiting values of $L^k(m)$, when $m \to \infty$. In practice, dependences of $\log L^k_m$, on $\log m$ are constructed, and one finds the value of $\log L_m$ at which the dependences "saturate," i.e., reach a horizontal plateau. An example of computation of the Lyapunov spectrum of a chaotically rotating satellite orbiting a planet is presented in Fig. 2.3.

If the tangent map matrix $\mathbf{J}_i$ is not available analytically, there are two ways to compute it (Shevchenko and Kouprianov 2002). First, one may replace the tangent vectors $\delta\mathbf{x}_i$ in Eq. (2.20) with small displacement vectors: $\Delta\mathbf{x}_i = \mathbf{x}'_i - \mathbf{x}_i$, where $\mathbf{x}_i$ and $\mathbf{x}'_i$ are the nominal and shadow trajectories, respectively, at the integration step i. Then both points are iterated independently using Eq. (2.19). This gives $\Delta\mathbf{x}_{i+1} = \mathbf{x}'_{i+1} - \mathbf{x}_{i+1}$. After repeating this procedure for $2N$ linearly independent vectors $\Delta\mathbf{x}_i$, the equation

$$\Delta\mathbf{x}_{i+1} = \mathbf{J}_i \Delta\mathbf{x}_i \tag{2.23}$$

can be solved with respect to the tangent map matrix $\mathbf{J}_i$.

The tangent map matrix can be found with a higher precision, but at a much greater computation cost, by computing it directly, by means of simultaneous integration of the original and variational systems. This approach does not need the introduction of an auxiliary small shift parameter $\Delta\mathbf{x}_0$. The computational efficiency of this method is decreased by the necessity to perform $2N$ additional iterations of the flow map at each step. However, an extra opportunity exists to compute the tangent map matrix. This opportunity is appropriate when map (2.19) is defined by an original continuous dynamical system

$$\mathbf{x}' = \mathbf{F}(\mathbf{x}, t). \tag{2.24}$$

Then the tangent map matrix is approximated by

$$\mathbf{J}_i = \mathbf{I} + \nabla_{\mathbf{x}}\mathbf{F}(t_i) \cdot \Delta t, \tag{2.25}$$

where $\nabla_{\mathbf{x}} F(t_i)$ is the Jacobi matrix of the system at $t = t_i = i\Delta t$, and $\mathbf{I}$ is the unity matrix. This approximation, however, is valid only when the iteration step Δt is sufficiently small.

An appropriate test of accuracy for calculating full Lyapunov spectra is to compute the sum of all exponents $\sum_{k=1}^{2N} L^k$; this sum must be close to zero with sufficient precision. Possible deviations can be conditioned by the choice of the initial shadow particle shift, the iteration step size, or the quality of the integrator employed.

2.2 Fast Lyapunov Indicators

When it is needed to massively compute LCEs over grids of values of parameters or initial conditions, the LCE method turns out to be rather time consuming; see, e.g., a discussion in Shevchenko and Kouprianov (2002). To reduce the computational costs, simplified analogues of the Lyapunov exponents were proposed and developed. Nowadays, the most popular among them are the "fast Lyapunov indicators" (FLI) (Froeschlé et al. 2012) and the "mean exponential growth factors of nearby orbits" (MEGNO) (Cincotta and Simó 2000; Cincotta et al. 2003).

The main idea in the FLI concept, as proposed in Froeschlé et al. (2012), is to track, in the course of a numerical integration, the distance between two trajectories that are initially close to each other, and fix the time when the distance exceeds an initially prescribed constant (a threshold, which is chosen in preliminary test runs). When the dynamics is explored globally, i.e., on large sets of values of parameters and/or initial data, these time values can serve as a relative measure of the stability in different domains of phase space. Indeed, when the time is large (or even never fixed on the time interval of integration), it signals the orbit's regularity; when it is short, it usually indicates chaoticity. In contrast to genuine LCE, FLI depend on the choice of the variables of motion, the length of the initial displacement of the shadow orbit, and on the choice of the threshold displacement; but this does not matter much when we are interested in comparative properties of large sets of orbits. Examples of massive computations of FLI are given in Pilat-Lohinger and Dvorak (2002).

Generally, it should be noted that simplifications and assumptions in defining various LCE analogues may, of course, lead to erroneous assessments in type of individual trajectories, but, in turn, the global picture of dynamics for large samples of orbits can be qualitatively described in relatively simple and rapid ways.

2.3 MEGNO Charts

The MEGNO techniques has gained a great popularity in numerical studies of global qualitative dynamics of planetary systems, due to its large potential for providing clear graphical representations of the stability properties and, what is more, for providing major economies in computing resources and computation time. It is introduced in the following way (Cincotta and Simó 2000; Cincotta et al. 2003).

Let $\mathcal{H}(\mathbf{p}, \mathbf{q})$ be the Hamiltonian of a system, and $\mathbf{p}^{(1)}(t), \mathbf{q}^{(1)}(t)$ and $\mathbf{p}^{(2)}(t), \mathbf{q}^{(2)}(t)$ are two initially close trajectories with initial conditions $\mathbf{p}^{(1)}(0)$, $\mathbf{q}^{(1)}(0)$ and $\mathbf{p}^{(2)}(0)$, $\mathbf{q}^{(2)}(0)$; t is time.

Let $\delta\mathbf{p}(t) = \mathbf{p}^{(2)}(t) - \mathbf{p}^{(1)}(t)$ and $\delta\mathbf{q}(t) = \mathbf{q}^{(2)}(t) - \mathbf{q}^{(1)}(t)$. The linearized equations of relative motion are then given by

$$\dot{\delta p_i} = -\sum_j \frac{\partial^2\mathcal{H}}{\partial p_j \partial q_i}(\mathbf{p}^{(1)}(t), \mathbf{q}^{(1)}(t))\delta p_j - \sum_j \frac{\partial^2\mathcal{H}}{\partial q_j \partial q_i}(\mathbf{p}^{(1)}(t), \mathbf{q}^{(1)}(t))\delta q_j\,,$$

$$\dot{\delta q_i} = \sum_j \frac{\partial^2\mathcal{H}}{\partial p_j \partial p_i}(\mathbf{p}^{(1)}(t), \mathbf{q}^{(1)}(t))\delta p_j + \sum_j \frac{\partial^2\mathcal{H}}{\partial p_i \partial q_j}(\mathbf{p}^{(1)}(t), \mathbf{q}^{(1)}(t))\delta q_j \quad (2.26)$$

(Morbidelli 2002). Here δp_i, δq_i, p_i, q_i are the components of $\delta\mathbf{p}$, $\delta\mathbf{q}$, $\mathbf{p}$, $\mathbf{q}$; and the upper dot is for time derivative.

The basic formula (2.1) for the maximum Lyapunov exponent can be formally rendered in the integral form

$$L = \lim_{t\to+\infty} \frac{1}{t} \int_0^t \frac{\dot{\delta}(t')}{\delta(t')} \mathrm{d}t'\,, \quad (2.27)$$

where $\delta = \|\delta\mathbf{p}, \delta\mathbf{q}\|$, and $\delta\mathbf{p}(t)$, $\delta\mathbf{q}(t)$ are solutions of (2.26).

The MEGNO factor is defined as

$$Y(t) = \frac{2}{t} \int_0^t \frac{\dot{\delta}(t')}{\delta(t')} t' \mathrm{d}t' \quad (2.28)$$

(Cincotta and Simó 2000; Cincotta et al. 2003). The quantity defined in this way clearly retains an important property of the Lyapunov exponent: it is independent of the adopted metrics and coordinate system.

It is also straightforward to see that if a trajectory with given initial conditions is quasiperiodic, then δ is linear with time, and therefore $Y(t)$ oscillates around 2. If a trajectory is chaotic, then δ is exponential with time, and therefore $Y(t)$ oscillates around the linear solution $y = Lt$, where L is the maximum Lyapunov exponent.

Accordingly, at $t \to +\infty$, for the current time mean MEGNO one has

$$\overline{Y}(t) = \frac{1}{t}\int_0^t Y(t')\mathrm{d}t' = 2 \tag{2.29}$$

in the first case, and

$$\overline{Y}(t) = \frac{L}{2}t \tag{2.30}$$

in the second case.

As noted in Morbidelli (2002), a computational advantage of MEGNO over LCE is that the computed $L(t)$ keeps a long memory on the initial transient behaviour of the tangent vector $\delta\mathbf{p}$, $\delta\mathbf{q}$, whereas in the integral (2.28) the later-on evolution of the tangent vector is taken with a greater relative weight. This technical shift allows one to suppress the prominence of the initial "noisy" behaviour.

However, one should take into account that when computations are performed on relatively short time scales, the obtained numerical MEGNO values cannot be put, generally speaking, in correspondence with the genuine maximum LCE, because any exact numerical value of MEGNO (as well as LCE) cannot be obtained on any time scales less than the characteristic time of diffusion over the whole connected chaotic region to which the initial conditions belong.

2.4 Frequency Analysis

A major spectral method is the method of frequency analysis (FA). Its description and theoretical justification are given in Laskar (1990), Laskar et al. (1992), and Laskar (1993). For regular orbits, the fundamental frequencies are constant, whereas for chaotic orbits they vary in time. The FA method exploits this difference to distinguish between different types of orbits.

Performing the FA at separate time intervals, one can numerically determine the current fundamental frequencies and find out whether they vary with time, and, in this way, to determine the character of the dynamics. Examples of implementation of the FA technique, as proposed in Laskar (1990), Laskar et al. (1992), and Laskar (1993) in the form of a numerical analysis of fundamental frequencies, can be found in Laskar et al. (1992), Laskar (1993), Correia et al. (2009), Laskar and Correia (2009), and Valluri et al. (2012). In particular, in Laskar et al. (1992), FA is developed and used to demonstrate its accuracy in determining numerically the critical value (already mentioned above in Sect. 1.5) of the stochasticity parameter K of the standard map.

In addition to the general opportunity of identification of regular and chaotic domains in phase space, FA allows one to identify locations of resonances. However, FA is laborious; it may require much more computing time than that required by the LE method in one and the same problem (Maffione et al. 2013).

Massive frequency analysis was performed in Correia et al. (2009) and Laskar and Correia (2009) in application to selected exoplanet systems. In these two works, two-planet systems of Solar-like stars HD 45364 and HD 60532 were explored, respectively. They were both shown to reside in 3/2 mean-motion resonance. These were the first confirmed cases of 3/2 resonance (analogous to the Neptune–Pluto resonance in the Solar system) ever detected in exoplanet systems.

2.5 Sections of Phase Space

As we have seen above in Sect. 1.4, a paradigmatic type of a map is provided by the phase space transformation defined by integrating points in the phase space a time step forward. Another type is given by a *return map* (Poincaré 1899, 1905; Lichtenberg and Lieberman 1992; Meiss 1992; Abdullaev 2006) defined on sections of phase space. In case of two degrees of freedom, the sections provide a graphical and clear picture of the motion.

The return maps can be considered in the following setup (Meiss 1992). Let the Hamiltonian $\mathcal{H}$ be time-independent. Since the energy is conserved, the Hamiltonian flow covers the $(2N-1)$-dimensional energy surface $\mathcal{E}$ corresponding to the value $E = \mathcal{H}$.

Let us introduce another, $(2N-1)$-dimensional surface $\mathcal{Q}$, transversal (nowhere parallel) to the flow locally. The *Poincaré section* of $\mathcal{P}$ is defined as the $(2N-2)$-dimensional intersection of $\mathcal{E}$ with $\mathcal{Q}$. Therefore, the first return map $\mathbf{z}' = T(\mathbf{z})$ (where $\mathbf{z}$ are the points in the phase space of canonical coordinates $\mathbf{q}$ and momenta $\mathbf{p}$) maps the initial $\mathbf{z}$ taken on $\mathcal{P}$ to $\mathbf{z}'$, to which the starting point returns on $\mathcal{P}$ for the first time, following the flow; see Fig. 2.4.

The first return map is symplectic and has an action. For example, let $\mathcal{Q}$ be the surface of $q_N = \text{const}$. It is transversal to the flow if

$$\frac{\mathrm{d}q_N}{\mathrm{d}t} = \frac{\partial \mathcal{H}}{\partial p_N} \neq 0 \tag{2.31}$$

on $\mathcal{Q}$.

The Poincaré section $\mathcal{P}$ can be specified in $2(N-1)$ variables, $(q_1, p_1, \ldots, q_{N-1}, p_{N-1})$. As soon as the energy level is chosen, the transversality condition (and the implicit function theorem) implies that the equality $\mathcal{H}(q_1, p_1, \ldots, q_N, p_N) = E$ can be rendered as

$$p_N = p_N(q_1, p_1, \ldots, q_{N-1}, p_{N-1}; q_N, E)\,. \tag{2.32}$$

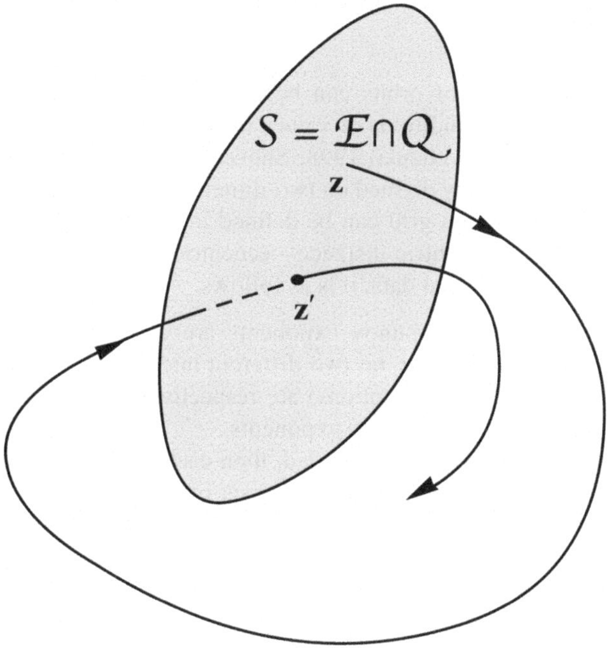

Fig. 2.4 A scheme for the first return map. (Reprinted Figure 5 with permission from Meiss (1992). © 1992 by the American Physical Society)

The first return map T is parameterized by choosing the values of E and q_N. In these frames, the action is reduced to

$$S = \sum_{i=1}^{N-1} \oint p_i dq_i \,. \tag{2.33}$$

In a particular case of a Hamiltonian system with two degrees of freedom ($N = 2$), the map T acts on a two-dimensional phase space, namely, (q_1, p_1). Therefore, the method of Poincaré sections is particularly valuable for systems of two degrees of freedom.

The effectiveness of the method of sections is guaranteed by the *Poincaré return theorem*. It implies that, if the energy surface is bounded (compact), almost all trajectories, except those defined on a set of measure zero, starting from the surface $\mathcal{P}$, sooner or later intersect again with this surface.

By the method of sections it was first ever graphically demonstrated, in 1964, the emergence of chaos in a Hamiltonian system, namely, in the *Hénon–Heiles system* (Hénon and Heiles 1964). The system models the motion of a star in the Galactic potential. Relevant diagrams of stability, resembling the original sections of phase space in this model, are presented below in Sect. 2.7.

2.6 LCE Charts

Regular and chaotic kinds of orbits can be distinguished numerically one from another, on sets of initial conditions or values of parameters, by using a statistical method (Melnikov and Shevchenko 1998; Shevchenko and Melnikov 2003). The representative sets are usually defined on two-dimensional grids. In many problems of celestial mechanics such a grid can be defined in the plane "semimajor axis—eccentricity," or, say, "pericentric distance—eccentricity." The procedure consists of three steps. For sets of initial data, it is as follows.

(1) Finite-time maximum Lyapunov exponents are computed for one and the same set of initial data twice, on two different integration time intervals. Two histograms (differential distributions) are respectively built. Thus one obtains two distributions of the finite-time exponents.
(2) If the phase space of motion is divided, then each of these histograms has at least two peaks, corresponding to two types of motion, regular and chaotic. The peak that shifts (moves in the direction of the Lyapunov exponent smaller values) with increasing the integration time corresponds to the regular motion; that stays fixed, to the chaotic one.
(3) The location of the minimum of the distribution between the peaks provides the critical finite-time Lyapunov exponent value that can be used to separate the regular and chaotic domains of the initial data.

In any further integrations on finer data grids, integrations can be performed on smaller time intervals, using the same critical maximum Lyapunov exponent to separate orbits of different types.

An example of employing the method is given in Fig. 2.5. For this Figure, the histograms are constructed on a grid of the eccentricity and pericentric distance starting values for the *Kepler*-16b exoplanet, discussed further on in Sect. 14.2.1. The time intervals of integration are set to 10^3 yr (red histogram) and 10^4 yr (blue histogram). In the plot's vertical axis, f designates the normalized (by the full sample volume) number of orbits in the $(\log_{10} T_{\rm L}, \log_{10} T_{\rm L} + \Delta \log_{10} T_{\rm L})$ bin, where $\Delta \log_{10} T_{\rm L} = 0.02$; the Lyapunov time $T_{\rm L}$ is the inverse of the maximum Lyapunov exponent. In Fig. 2.5, the shift of the histogram peak with increasing the integration time is evident and pronounced. It is straightforward to see that the threshold value of $\log_{10} T_{\rm L}$, distinguishing chaotic orbits from regular ones, can be set to ≈ 2.5, in the given example. The resulting stability diagrams are shown in Fig. 14.3 and discussed in Sect. 14.2.1.

Generally, the described method for separating chaotic and regular trajectories can be as well used on grids of values of parameters, at fixed initial conditions (Melnikov and Shevchenko 2008).

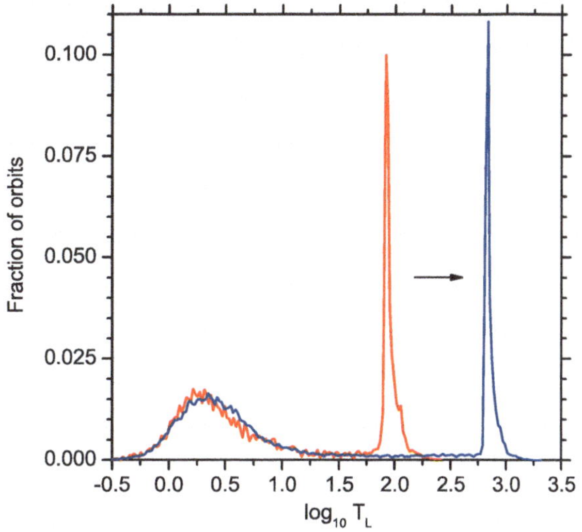

Fig. 2.5 Histograms of the numerical-experimental Lyapunov times computed on two different integration time intervals. The histograms are constructed on a grid of the eccentricity and pericentric distance starting values for the *Kepler*-16b exoplanet. The time intervals of integration are 10^3 yr (for the red histogram) and 10^4 yr (for the blue histogram). The shift of the distribution peak is evident (Figure 1 from Popova and Shevchenko (2013), by permission of IOP Publishing/AAS)

2.7 Poincaré Recurrences

The Poincaré recurrence notion has a major methodological value, due to the applicability potential of the *Poincaré recurrence theorem* (Poincaré 1890). The theorem is valid in a broad class of dynamical systems, including Hamiltonian ones. Assume one has a volume-conserving continuous one-to-one mapping g, transforming a bounded domain D of Euclidian space in itself ($gD = D$). Then, the theorem states (Arnold 1989) that in any neighbourhood U of any point of D there exists a point x that returns to U: $g^n x \in U$ at some n.

In other words, any dynamical system of certain kind (in particular, with the phase space bounded) recurs eventually to any neighbourhood of its initial state. The return time can be large, but it is finite. Although the theorem is valid for systems with the bounded phase space, the notion of Poincaré recurrence is defined for any dynamical system.

The *Poincaré recurrence method* (PRM) is based on an analysis of the Poincaré recurrence statistics on massive grids of initial data or values of parameters (Shevchenko et al. 2020). The Poincaré recurrences are calculated as follows. First, the grid for a system under study is chosen. Then, at each node of the grid, a neighbourhood of the initial point of motion of fixed size ε is defined. The equations of motion of the system are integrated numerically at each node. The integration is stopped when either the first Poincaré recurrence takes place or the specified integration time interval is over. In the first case, the time instant T_r is fixed when the trajectory first returns to the given neighbourhood of the initial point. Then, the recurrence times are represented in a colour or tone grade graphically on the defined grid.

To illustrate the effectiveness of PRM, let us consider the Hénon–Heiles problem. The Hamiltonian of the problem is given by

$$\mathcal{H} = \frac{1}{2}(p_1^2 + p_2^2 + q_1^2 + q_2^2) + q_1^2 q_2 - \frac{1}{3}q_2^3, \tag{2.34}$$

where q_1, q_2 are the canonical coordinates, and p_1, p_2 are the conjugate canonical momenta (Hénon and Heiles 1964). Poincaré sections of the system's phase space were first ever constructed in Hénon and Heiles (1964).

Using notations defined in Sect. 2.5, we set $q_1 = 0$ to define the transversal surface $\mathcal{Q}$ and choose the inequality $p_1 > 0$ to unambiguously define the function $p_1(q_2, p_2; E, q_1)$, where $E \equiv \mathcal{H}$. The domain of allowed motion is bounded by a closed curve, given by the equation $q_2^2 + p_2^2 - \frac{2}{3}q_2^3 = 2E$, if $E < 1/6$. With increasing the energy, E, the chaotic domain grows in volume, and at $E = 1/6$ one finds almost all the phase space of the possible motion be chaotic (Hénon and Heiles 1964; Shevchenko and Melnikov 2003).

In Shevchenko et al. (2020), Poincaré recurrences were computed for a set of initial data defined on a uniform grid in the plane (p_2, q_2); the section was defined at $q_1 = 0$, and p_1 were calculated using Eq. (2.34) at $E = 0.1$. As shown in Shevchenko and Melnikov (2003), at $E = 0.1$ the chaotic domain occupies $\approx$20% of the whole phase space.

Figure 2.6 shows the (p_2, q_2) diagrams with the Poincaré recurrence and Lyapunov times indicated in colour grades. One may see that the two diagrams represent the global dynamics of the Hénon–Heiles almost identically. However, a major advantage of the PRM is that it allows one to characterize the local diffusion timescales (Shevchenko et al. 2020). What is more, PRM is algorithmically simple, in comparison with computations of Lyapunov exponents, and it is straightforward to apply.

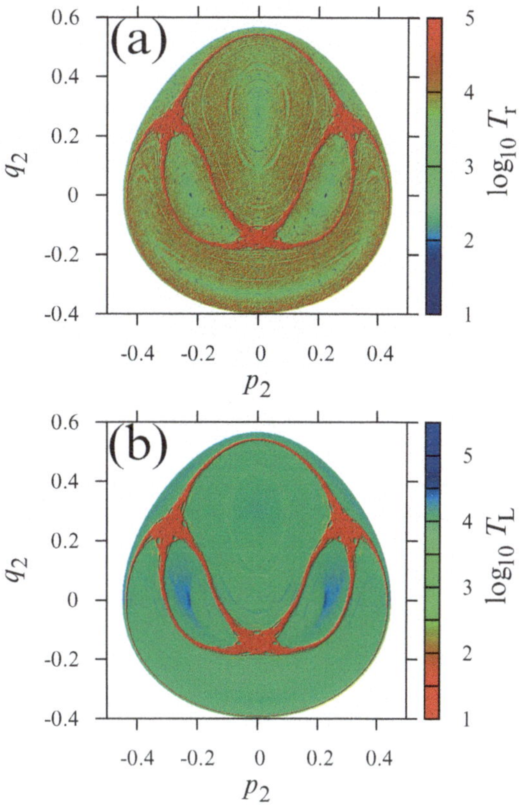

Fig. 2.6 (**a**) The Poincaré recurrence chart for the Hénon–Heiles system, in the (p_2, q_2) plane, at the energy $E = 0.1$. Red colour corresponds to $T_r > 10^5$. (**b**) The Lyapunov time chart for the same system. Red colour corresponds to $T_L < 10$ (Figure 1 from Shevchenko et al. (2020), with permission from Elsevier)

Chapter 3
Lyapunov Timescales

Generally, estimating the Lyapunov exponents, considered above in Sect. 2.1, is one of the most important tools in studies of chaotic motion (Lichtenberg and Lieberman 1992), in particular in celestial mechanics. A non-zero value of the maximum Lyapunov exponent typically indicates the chaotic type of motion, and the exponent equal to zero signals its regular type. The *Lyapunov time* (the quantity reciprocal to the maximum Lyapunov exponent) characterizes the timescale of predictable dynamics. In this Chapter, we consider basic analytical methods how Lyapunov timescales can be estimated. The discussed concepts include: Chirikov's constant, adiabatic chaos, non-adiabatic chaos, Lyapunov exponents in resonance doublets, triplets, and, generally, in resonance multiplets. The presentation is partially based (in Sections 3.2, 3.3 and 3.4) on the papers by Shevchenko (2008b) (by permission from © Oxford University Press), Shevchenko (2007a) (by permission from Cambridge University Press), and Shevchenko (2014) (by permission from Elsevier).

Morbidelli and Froeschlé (1996) and Nesvorný and Morbidelli (1999, p. 256) suggested to estimate the Lyapunov time by taking it equal, by the order of magnitude, to the libration/circulation period of the resonant angle, or, in practice, to the period of small-amplitude oscillations on resonance, i.e., $T_\mathrm{L} \sim \omega_0^{-1}$, where ω_0 is the frequency of small-amplitude oscillations on resonance. This formula is approximately valid at the adiabaticity parameter values $\lambda \simeq 1$. Note that $L/\omega_0 \to 0$ in the both limits $\lambda \to 0$ and $\lambda \to \infty$. Moreover, L/ω_0 may strongly depend on other parameters, such as the perturbation amplitude ε.

An approach, based on a standard-type map theory, was proposed by Holman and Murray (1996), Murray and Holman (1997) for the case of a triplet of overlapping resonances. They introduced an effective overlap parameter K_eff instead of ω_0 or the perturbation frequency Ω. In some way K_eff is analogous to the stochasticity parameter K of the standard map (see, e.g., Chirikov 1979; Lichtenberg and Lieberman 1992). Holman and Murray (1996) considered cases of moderate overlap, when $K_\mathrm{eff} \sim 1$, and strong overlap (adiabatic case), when $K_\mathrm{eff} \gg 1$. In the first case, the maximum Lyapunov exponent was estimated as $L \approx \omega_0$ (the frequency of

I. I. Shevchenko, *Dynamical Chaos in Planetary Systems*, Astrophysics and Space Science Library 463, https://doi.org/10.1007/978-3-030-52144-8_3

small oscillations on resonance), and in the second case as $L \approx \Omega$ (the external perturbation frequency).

In Shevchenko (2002a), Shevchenko (2008b, 2014), a method based on the separatrix map theory was proposed and developed; it allows one to obtain analytical estimates of the maximum Lyapunov exponent in a number of problems on dynamics of the Solar system bodies (Shevchenko 2007a).

The notion of Lyapunov exponents is closely related to that of *dynamical entropy* (Pesin 1977; Benettin et al. 1976; Chirikov 1978, 1979; Meiss 1992). For Hamiltonian systems with 3/2 and 2 degrees of freedom with bounded phase space, Benettin et al. (1976) gave a proof of the relationship

$$h \approx L\mu \tag{3.1}$$

(Benettin et al. 1976, Equation (6)), where h is the dynamical entropy, L is the maximum Lyapunov exponent, and μ is the relative measure of the connected chaotic domain where the motion takes place. Benettin et al. (1976) used Eq. (3.1) in their study of dynamical chaos in the Hénon–Heiles system.

3.1 Chirikov's Constant

Let us estimate the least upper bound for the maximum Lyapunov exponent of the separatrix map (1.31). We call it *Chirikov's constant,* because an analogue of this quantity was introduced in Chirikov (1979). Based on results of massive numerical experiments, Chirikov (1979) found out that the maximum Lyapunov exponent, referred to the mean half-period of phase libration (or the mean period of its circulation), of the motion in the chaotic layer of a non-linear resonance subject to symmetric periodic perturbation is approximately constant in a wide range of the relative frequency of the perturbation.

We will see that, in the perturbed pendulum model of non-linear resonance, Chirikov's constant coincides with the value of the maximum Lyapunov exponent in the limit of infinitely high frequency of perturbation, and it does not depend on the perturbation amplitude, i.e., it is defined robustly. The knowledge of Chirikov's constant is important for accurate analytical estimating the maximum Lyapunov exponent in applications in mechanics and physics (Shevchenko 2002a).

It is instructive to see how measure μ of the main connected chaotic domain of the standard map (1.23), its maximum Lyapunov exponent L, and the product of μ and L, evolve with increasing the stochasticity parameter K (Shevchenko 2004a).

Figure 3.1 illustrates discontinuities in the $\mu(K)$ function. The function was found numerically in Shevchenko (2004a) by means of computing the number of cells explored by a single trajectory on a grid of cells on the phase plane $(x, y) \in [0, 2\pi] \times [0, 2\pi]$ of the standard map (1.23).

A prominent bump in the dependence, shown in detail in Fig. 3.1b, is conditioned by the process of disintegration of the half-integer resonance, while K is increasing from ≈ 2 to ≈ 2.5. The disintegration is due to a sequence of period-doubling

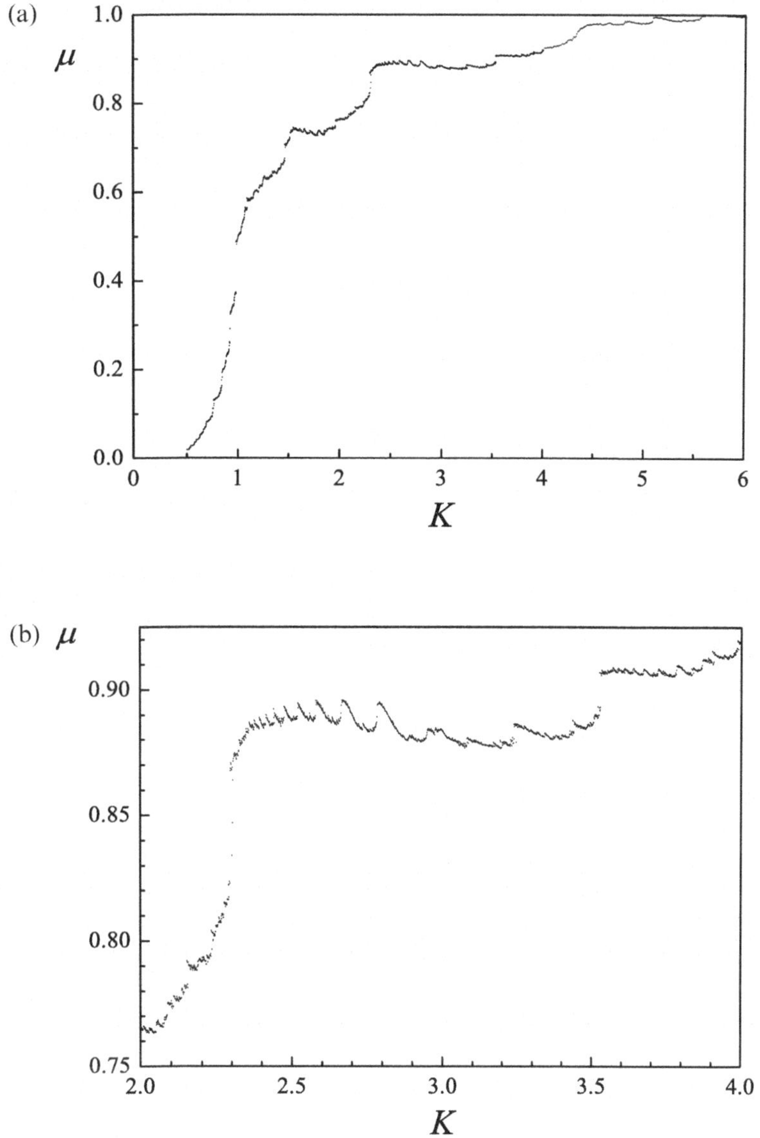

Fig. 3.1 The chaotic component measure $\mu(K)$ (**a**), and its detail zoomed (**b**) (Figure 1 from Shevchenko 2004a. With permission from Pleiades Publishing Inc.)

bifurcations. A similar but less pronounced bump is seen in Fig. 3.1a at $4 < K < 4.5$; this one is due a sequence of period-doubling bifurcations of the integer resonance. At $K \lesssim 6$, the discontinuities are mostly conditioned by the process of absorption of minor chaotic domains by the main chaotic domain around the integer resonance, as K increases.

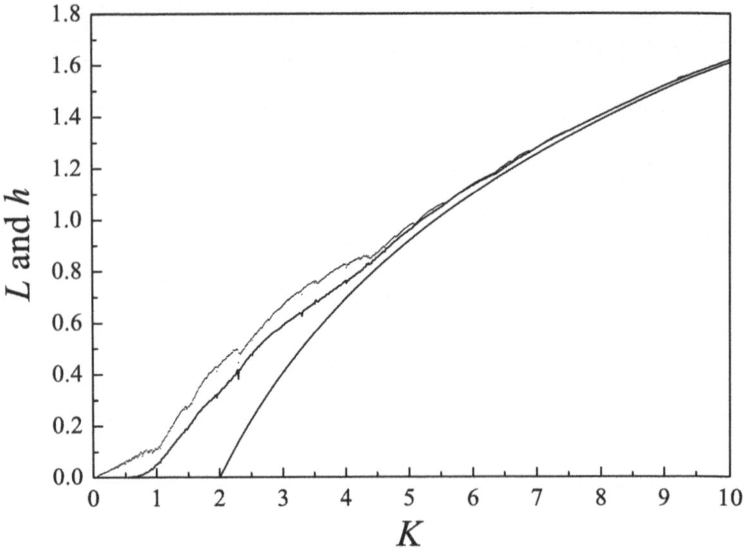

Fig. 3.2 Standard map: maximum Lyapunov exponent and dynamical entropy. Upper curve: $L(K)$; middle curve: $h(K) = \mu(K)L(K)$; lower curve: $\ln \frac{K}{2}$ (Figure 2 from Shevchenko 2004a. With permission from Pleiades Publishing Inc.)

In Fig. 3.2, K dependences of the maximum Lyapunov exponent L and dynamical entropy h are presented in a broader range of K. The Lyapunov exponents were computed in Shevchenko (2004a) by the tangent map method. (Its description is given in Chirikov (1979).)

The K dependence of the product $L\mu$, which is just the dynamical entropy h, is given in Fig. 3.2. The dynamical entropy h looks continuous and monotonic in K, in contrast to the discontinuous behaviour of the maximum Lyapunov exponent. This is no wonder, because the dynamical entropy is a more fundamental quantity (Shevchenko 2004a).

Chirikov (1979) derived analytically that asymptotically, at $K \to \infty$,

$$L(K) = \ln \frac{K}{2}. \tag{3.2}$$

This formula is obtained by averaging the largest eigenvalue of the tangent map, taking into account that the chaotic component dominates. From Fig. 3.2, it is clear that this logarithmic function describes well the asymptotic behaviour of both $L(K)$ and $h(K)$.

Downward spikes, prominent in Fig. 3.2, are due to the *stickiness effect* (Shevchenko 1998a): a chaotic trajectory may stick for a long time to the borders of the chaotic domain, where the motion is close to regular and, therefore, local rates of divergence of nearby orbits are small. The computation times are always finite, that is why the stickiness effect may provide a computed finite-time Lyapunov exponent much smaller than the actual one. This effect conditions a particular power-law relationship between Lyapunov and recurrence timescales (Shevchenko 1998a).

In Figs. 3.1 and 3.2 we see that the $\mu(K)$ and $L(K)$ functions are discontinuous and obviously elude any simple analytical representation. However, at $K \gg 1$, $h(k)$ can be approximated by the function

$$h(K) = \ln \frac{K}{2} + \frac{1}{K^2}, \quad \text{if } K > 4.5, \tag{3.3}$$

with absolute accuracy better than 0.01. Therefore, the K asymptotics of the $h(K)$ function contains a power-law component, in addition to the logarithmic law given by Eq. (3.2). The same can be said for $L(K)$, if one ignores the small (and local in K) distortions emerging due to accelerator modes and periodic solutions of higher orders.

Chirikov's constant can be found by averaging the local maximum Lyapunov exponent over the whole chaotic layer of the separatrix map (1.31) at $\lambda \to \infty$. The local Lyapunov exponent should be taken with a weight directly proportional to the time that the trajectory is spending in the given locality; this time is directly proportional to the local relative measure of the chaotic component. Therefore,

$$C_\mathrm{h} = \lim_{\lambda \to \infty} \frac{\int_0^{y_\mathrm{b}} \tilde{L}_\mathrm{sx}(y) \tilde{\mu}_\mathrm{sx}(y)\, dy}{\int_0^{y_\mathrm{b}} \tilde{\mu}_\mathrm{sx}(y)\, dy}, \tag{3.4}$$

where

$$y_\mathrm{b} = \frac{\lambda}{K_\mathrm{G}} \tag{3.5}$$

is the value of y at the layer's border (Chirikov 1990; Shevchenko 1998b), $\tilde{L}_\mathrm{sx}(y)$ is the local (with respect to y) value of the maximum Lyapunov exponent, and $\tilde{\mu}_\mathrm{sx}(y)$ is the local chaos measure. The tilde cap marks that the quantities are local. This formula is valid in the limit $\lambda \to \infty$, because only in this limit the sum over all integer resonances inside the layer can be rendered as an integral. Besides, Eq. (3.5) is accurate also in this limit.

Setting $y = \lambda/K$, we introduce the new independent variable K, which is nothing but the stochasticity parameter of the standard map that locally approximates our separatrix map. The accuracy of the approximation improves with increasing λ (Chirikov 1979), therefore, at $\lambda \gg 1$, one has $\tilde{L}_\mathrm{sx}(y = \lambda/K) \to L(K)$ and $\tilde{\mu}_\mathrm{sx}(y = \lambda/K) \to \mu(K)$. The λ dependence in the limit $\lambda \to \infty$ is eliminated, and Eq. (3.4) is reduced to

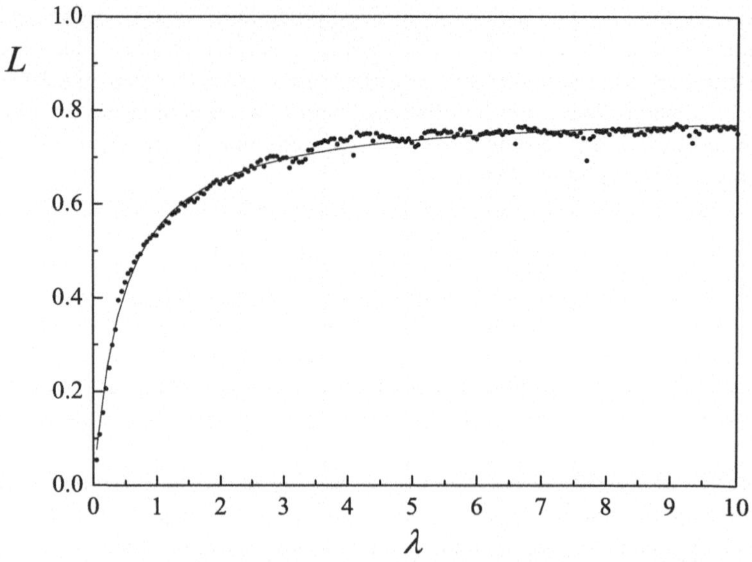

Fig. 3.3 The λ dependence for the maximum Lyapunov exponent of the separatrix map, and its rational approximation (Figure 3 from Shevchenko 2004a. With permission from Pleiades Publishing Inc.)

$$C_{\mathrm{h}} = \frac{K_{\mathrm{G}}}{\sigma} \int_{K_{\mathrm{G}}}^{\infty} L(K)\mu(K) \frac{dK}{K^2}, \tag{3.6}$$

where

$$\sigma = \lim_{\lambda \to \infty} y_{\mathrm{b}}^{-1} \int_0^{y_{\mathrm{b}}} \tilde{\mu}_{\mathrm{sx}}(y)\, dy = K_{\mathrm{G}} \int_{K_{\mathrm{G}}}^{\infty} \mu(K) \frac{dK}{K^2} \tag{3.7}$$

is the ratio of the chaotic component area to the layer's total area. Therefore, one minus σ is the relative total area of all regular islands inside the layer.

In Shevchenko (2004a), Eqs. (3.6) and (3.7) were integrated numerically, with $L(K)$ and $\mu(K)$ taken in tabulated form at $K \in [K_{\mathrm{G}}, 10]$. The remainders at $K > 10$ were estimated analytically, with $h(K) = L(K)\mu(K)$ described by Eq. (3.3), and $\mu(K) = 1$. It was found that

$$C_{\mathrm{h}} = 0.80, \quad \sigma = 0.78, \tag{3.8}$$

with accuracy of two significant digits.

The given estimate of Chirikov's constant can be verified by computing straightforwardly the maximum Lyapunov exponent L of the separatrix map (1.31) in the limit $\lambda \gg 1$. (Note that we use the same designation L in both cases of standard and separatrix maps; this should not cause a confusion). The λ dependence for L is given in Fig. 3.3. It was obtained in Shevchenko (2004a) for the case of the least perturbed border of the layer, minimizing the contribution of marginal resonances, whose emergences are local in varying the parameter c of the map. This case is generic in applications. The dependence is well described by the fitting rational function

$$L(\lambda) = \frac{b + c\lambda}{1 + a\lambda} \tag{3.9}$$

where $a = 2.097 \pm 0.033$, $b = 0$, and $c = 1.691 \pm 0.024$ (Shevchenko 2004a). Here b is set to zero so that $L(0) = 0$.

Chirikov's constant is obtained in the limit $L(\lambda \to \infty)$: $C_{\mathrm{h}} \approx 0.806$, in accord with the estimate (3.8).

3.2 Adiabatic Chaos

Adiabatic chaos in Hamiltonian systems takes place in conditions of strong resonance overlap, see Sect. 1.3. In the perturbed pendulum model of interacting resonances, the adiabaticity parameter λ, Eq. (1.27), is therefore small, $\lambda \ll 1$.

Adiabatic chaos can be analysed by various analytical means, e.g., using methods of theory of adiabatic invariants; see Timofeev (1978), Neishtadt et al. (1997), Chirikov and Vecheslavov (2000a), Chirikov and Vecheslavov (2000b), Arnold et al. (2006), Elskens and Eskande (1991), Elskens and Eskande (1993) and references therein. Here we use a different approach, based on the separatrix map theory (Shevchenko 2002a; Shevchenko 2008b). The applicability of the separatrix map theory for describing the near-separatrix motion in the perturbed-pendulum model of non-linear resonance in the full range of the relative frequency of perturbation, including its low values, was discussed and shown to be legitimate in Shevchenko (2000a).

Following Shevchenko (2002a), we represent the maximum Lyapunov exponent L of the motion in the main chaotic layer of system (1.6) as the maximum Lyapunov exponent L_{sx} of the system's separatrix map divided by the time-averaged rotation period (or, equivalently, libration half-period) T of the resonance phase φ. For convenience, we introduce $T_{\mathrm{sx}} = \Omega T$, which is non-dimensional. One has

$$L = \Omega \frac{L_{\mathrm{sx}}}{T_{\mathrm{sx}}} \tag{3.10}$$

and, as usual, the Lyapunov time $T_L = L^{-1}$. For convenience, let us define two special functions:

$$\Lambda(\lambda, c) = L_{sx}(\lambda, c)/\lambda, \tag{3.11}$$

$$\Theta(\lambda, c) = (\langle \Delta x \rangle - c)/\lambda \equiv (\langle x_{i+1} - x_i \rangle - c)/\lambda = (T_{sx} - c)/\lambda, \tag{3.12}$$

where $\langle \cdots \rangle$ means averaging over the chaotic domain.

The $\Lambda(\lambda, c)$ and $\Theta(\lambda, c)$ functions are 2π-periodic in c. They can be easily numerically tabulated in any intervals of variation of λ and c. In terms of these two functions, formula (3.10) takes the form

$$L = \Omega \frac{\Lambda(\lambda, c)}{\Theta(\lambda, c) + \frac{c}{\lambda}}, \tag{3.13}$$

where c is given by Eq. (1.32). Note that here it is taken modulo 2π only in the arguments of Λ and Θ.

Equation (3.13) is valid for any frequency of perturbation in system (1.6) with $a = b$. In the case of adiabatic chaos, when $\lambda \ll 1$, the diffusion across the chaotic layer is slow, and on relatively short time intervals (during which the transport across the layer can be neglected) the map's trajectory can be approximated by a current *guiding curve*. The guiding curve with an irrational winding number (irrational ratio of frequencies) that is far enough from major rational numbers can be described analytically, in a straightforward way (Shevchenko 2008a). Approximating the winding number by rational numbers m/n, we set $c \approx 2\pi m/n$. At the map's iteration n, the trajectory hits in a small neighbourhood of the starting point; therefore, one has

$$\begin{aligned}\frac{dy}{dx} &= \frac{1}{nc - 2\pi m} \sum_{k=0}^{n-1} \sin(x + kc) = \\ &= \frac{1}{nc - 2\pi m} \sin \frac{nc}{2} \operatorname{cosec} \frac{c}{2} \sin \left(x + \frac{n-1}{2} c \right).\end{aligned} \tag{3.14}$$

Here formula (1.341.1) from Gradshtein and Ryzhik (1962) was used. Integrating, in the limit $n \to \infty$ one has

$$y = -\frac{1}{2} \operatorname{cosec} \frac{c}{2} \cos \left(x - \frac{c}{2} \right) + C, \tag{3.15}$$

where C is an arbitrary constant of integration.

Any trajectory is chaotic if its guiding curve (3.15) crosses the curve $y = \sin x$, because the latter curve is singular; see Eqs. (1.31). Therefore, the tangency of the guiding and singular curves corresponds to the chaotic layer's border, and the layer's half-width is then given by

$$y_{\mathrm{b}} = \left|\mathrm{cosec}\frac{c}{2}\right| \tag{3.16}$$

(Shevchenko 2008a). In terms of the relative energy variable w in Eqs. (1.26), it is

$$w_{\mathrm{b}} = \left|W\ \mathrm{cosec}\left(\frac{\lambda}{2}\ln\frac{32}{|W|}\right)\right|, \tag{3.17}$$

where W, in the considered case of $k = 1$ and $a = b$, is given by formula (1.28).

Conditions for validity of the approximation (3.15) are simple. In deriving the increment in x, the term $\lambda \ln|y_{i+1}|$ in Eqs. (1.31) is neglected; therefore, one has to set $c \gg \lambda \ln|y_{\mathrm{b}}|$ and $c \gg \lambda \ln\left|\mathrm{cosec}\frac{c}{2}\right|$, at $\lambda \ll 1$. We find that the c value should be far enough from the main resonance, i.e., c should not be close to $0 \mod 2\pi$. Clearly, it should also not correspond to other prominent resonances.

The quantity $-\ln|y_{i+1}|$ is equal to $(\langle\Delta x\rangle - c)/\lambda$ (see Eqs. (1.31)), where y_{i+1} is replaced by y of Eq. (3.15). Averaging this quantity over the chaotic layer in the boundaries given by Eq. (3.17), by taking the integral, one finds analytically

$$\Theta(c) \approx \ln\left|4\sin\frac{c}{2}\right|. \tag{3.18}$$

Also the $\Lambda(\lambda, c)$ function is needed. For an illustration, a $\Lambda(c)$ function, computed numerically at a fixed λ (namely, at $\lambda = 0.01$), is shown in Fig. 3.4a. The $\Lambda(c)$ function in Fig. 3.4a has a lot of peaks and, at a first glance, may seem not to fit any analytical description. However, in a higher resolution in c (see Fig. 3.4b and c), the resonant peaks turn out to be very thin; therefore, in the generic non-resonant case they can be neglected, and one has

$$\Lambda \approx 1. \tag{3.19}$$

Using Eq. (3.18), we may estimate the winding numbers of the resonances corresponding to the peaks of the $\Lambda(c)$ function:

$$Q = \frac{1}{2\pi}\langle\Delta x\rangle \approx \frac{1}{2\pi}\left(c + \lambda\ln\left|4\sin\frac{c}{2}\right|\right). \tag{3.20}$$

The peaks appear when regular islands inside the chaotic layer emerge.

In Fig. 3.4b and c, an anticorrelation between Λ and μ (the measure of the chaotic component inside the chaotic layer's outer borders) is clearly seen. In case of the standard map, the dependence of the maximum Lyapunov exponent L on the stochasticity parameter K has similar narrow local depressions arising due to emergence of regular islands in the map's phase space at specific values of K; see Shevchenko (2004b).

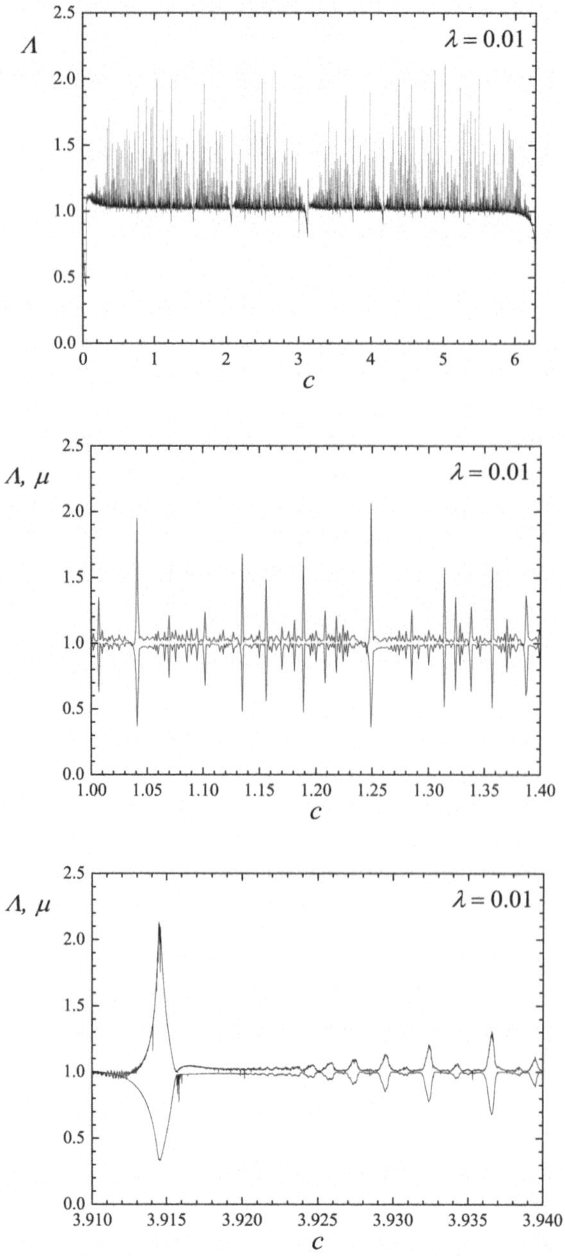

Fig. 3.4 Upper panel: the c dependence of Λ, at $\lambda = 0.01$. Middle panel: a close-up of the upper graph near the 1/6 and 1/5 resonances, which correspond to the two highest peaks in the graph. Bottom panel: a close-up near the 5/8 resonance, which corresponds to the highest peak in the graph. In the middle and bottom panels, the total measure μ of the chaotic component is shown by the lower curve. The measure μ was estimated by computing the number of cells explored by a single trajectory on a grid exposed on the phase plane (Figure 1 from Shevchenko 2008b, © Oxford University Press)

Based on Fig. 3.4a, b, and c, one may conjecture that when chaos is adiabatic and resonances are absent, $\Lambda \approx 1$; equivalently, $L_{sx} \approx \lambda$ (Shevchenko 2008b). This can be verified by constructing the λ dependence of L_{sx} in case of the *least perturbed layer*, or, *complete ergodicity*, at which the average winding number of the motion inside the chaotic layer is far from any resonances leading to significant perturbations in the $\Lambda(c)$ dependence. Therefore, the average winding number should be chosen approximately equal to the *golden number* $(3 - \sqrt{5})/2 \approx 0.382$. The λ dependence of the separatrix map maximum Lyapunov exponent L_{sx} in this generic non-resonant case is presented in Shevchenko (2008b). Its fitting by the linear law $L_{sx}(\lambda) = a\lambda$ gives $a = 1.011 \pm 0.001$. The linear fit is valid at $\lambda \lesssim 0.3$. At greater values of λ, including medium ($\lambda \sim 1$) and high ($\lambda \gg 1$) ranges, the $L_{sx}(\lambda)$ dependence can be described by a rational function, as illustrated in Fig. 3.3.

The linear law $L_{sx} = \lambda$ has a straightforward physical interpretation (Shevchenko 2008b). Indeed, the motion on the guiding curve can be regarded as regular only until the trajectory hits in a neighbourhood of the singular curve, and the trajectory is momentarily shifted to another guiding curve. The relative area μ_{sn} of the "scattering" neighbourhood of the singular curve scales as λ, while the average maximum modulus of the tangent map eigenvalues for the motion in the neighbourhood remains constant. The average return time to the scattering neighbourhood in case of the non-resonant motion is inversely proportional to μ_{sn}; therefore, the linear law, $L_{sx} \propto \lambda$, naturally arises.

Setting $\Lambda = 1$, and Θ as given by Eq. (3.18), from Eqs. (3.13) and (3.18) one has

$$L \approx \frac{\Omega}{\ln\left|4 \sin\frac{c}{2}\right| + \frac{c}{\lambda}}, \tag{3.21}$$

where $c = \lambda \ln \frac{32}{|W|}$, see Eq. (1.32).

At $k = 1$ and $\lambda \ll 1$, Eq. (1.28) gives $W \approx 8\lambda\varepsilon$, hence

$$L \approx \frac{\Omega}{\ln\left|\frac{16}{\lambda\varepsilon} \sin\left(\frac{\lambda}{2}\ln\frac{4}{\lambda|\varepsilon|}\right)\right|}. \tag{3.22}$$

For estimating L analytically, by means of Eq. (3.13), the special functions $\Lambda(c)$ and $\Theta(c)$ can be used as tabulated in advance at any given values of λ.

To confront the theory with any numerical data, the maximum Lyapunov exponent of system (1.6) can be directly computed as a function of the perturbation amplitude ε; see Fig. 3.5. In this Figure, a perfect agreement is observed in the ε range covering six orders of magnitude. However, small peaks perturbing the smooth theoretical curves are present. Their emergence is conditioned by the influence of resonances, as ε is varied.

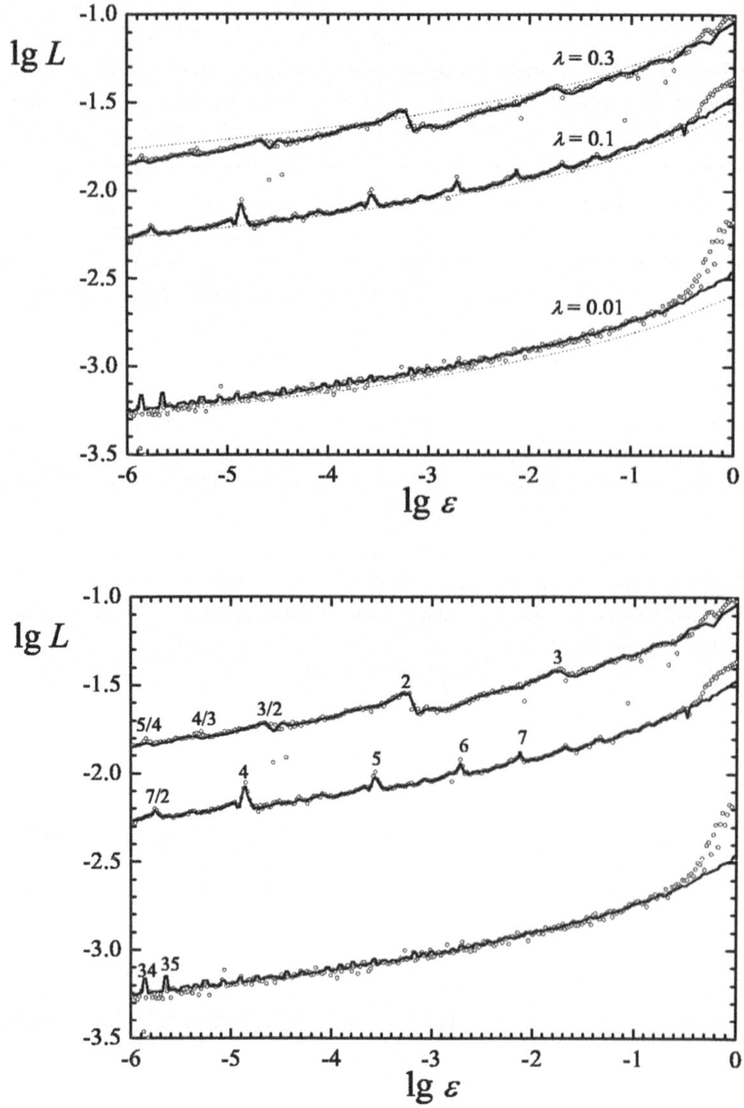

Fig. 3.5 Upper panel: the maximum Lyapunov exponent of system (1.6), in dependence on the perturbation amplitude; numerical data (circles) and analytical functions given by Eqs. (3.13) (solid curves) and (3.22) (dotted curves). Lower panel: the same, but with inverse winding numbers, Q^{-1}, indicated at resonant peaks (Figure 5 from Shevchenko 2008b, © Oxford University Press)

3.3 Lyapunov Exponents in Resonance Doublets and Triplets

Lyapunov exponents can be analytically estimated in a number of generic models of resonance interaction and overlap, though often rather approximately, first of all because one has to deal with relations smoothed over original fractal-like dependences, as we will see in many examples below.

In this Section, we consider four generic resonance interaction types, which can be used as models approximating many actual dynamical situations. These four types include: non-adiabatic chaotic resonance triplet, non-adiabatic chaotic resonance doublet, adiabatic chaotic resonance triplet, and adiabatic chaotic resonance doublet.

A fundamental model of perturbed non-linear resonance is given by the perturbed pendulum Hamiltonian (1.6) with $k = 1$:

$$H = \frac{\mathcal{G}p^2}{2} - \mathcal{F}\cos\varphi + a\cos(\varphi - \tau) + b\cos(\varphi + \tau). \tag{3.23}$$

A section of the phase space of system (3.23), visualizing a non-adiabatic chaotic resonance triplet, is given in Fig. 1.5. The section is taken at $\tau = 0 \bmod 2\pi$, and the system parameters are: $\Omega = 5$ and $\omega_0 = 1$, i.e., the adiabaticity parameter $\lambda = 5$, $a = b$, the perturbation amplitude $\varepsilon = \frac{a}{\mathcal{F}} = 0.5$.

Non-adiabatic Chaotic Resonance Triplet Choosing $a = b$ and $\lambda > 1/2$, one has a symmetric triad of interacting resonances, and chaos is non-adiabatic. Following Shevchenko (2002a), we take the dependence of the maximum Lyapunov exponent of the separatrix map (1.31) upon λ in the form

$$L_{\mathrm{sx}}(\lambda) \approx C_{\mathrm{h}} \frac{2\lambda}{1 + 2\lambda}, \tag{3.24}$$

where $C_{\mathrm{h}} \approx 0.80$ is Chirikov's constant, see Sect. 3.1.

For the average increment of τ per the separatrix map iteration inside the chaotic layer one has

$$T_{\mathrm{sx}}(\lambda, W) \approx \lambda \ln \frac{32e}{\lambda|W|} \tag{3.25}$$

(Chirikov 1979; Shevchenko 2002a); here e is the base of natural logarithms, and W is given by formula (1.28) or formula (1.29). The basic Eq. (3.10) gives the Lyapunov time of the original system:

$$T_{\mathrm{L}} = \frac{T_{\mathrm{pert}}}{2\pi} \frac{T_{\mathrm{sx}}}{L_{\mathrm{sx}}} \approx T_{\mathrm{pert}} \frac{(1 + 2\lambda)}{4\pi C_{\mathrm{h}}} \ln \frac{32e}{\lambda|W|}, \tag{3.26}$$

where $T_{\mathrm{pert}} = 2\pi/\Omega$ is the period of perturbation in the original system time units.

Non-adiabatic Chaotic Resonance Doublet The just given analysis of the symmetric case $a = b$ allows one to proceed to the more general asymmetric case, $a \neq b$. Indeed, we shall see that the maximum Lyapunov exponent in the asymmetric case can be found by averaging contributions of separate components of the chaotic layer. To perform the averaging, one should assess ratios of the mean times of the system's residence in three different components of the layer; these three components correspond to the model pendulum's direct rotation, reverse rotation, and libration. The ratios of the mean times of residence depend on the perturbation asymmetry. Let us consider the limiting case, that with of a or b equal to zero. In this case, one of the two perturbing resonances simply does not exist; thus, instead of the resonance triad we have a resonance duad. If $\lambda > 1/2$ and $b = 0$, then $|W^-| \ll |W^+|$, and, vice versa, if $\lambda > 1/2$ and $a = 0$, then $|W^-| \gg |W^+|$. The negligible W^- or W^+ is further on set to zero, and the dominating W^+ or W^- is further on designated as W.

Consider first the chaotic layer's libration component. Then, W^- and W^+ alternate (replace each other) at each iteration of the map (1.33). If one of $W^\pm$ is equal to zero, the separatrix algorithmic map (1.33) on any doubled iteration step reduces to the classical separatrix map (1.26) with the doubled value of λ and the same non-zero value of the dominating W (here the λ and W parameters are regarded as mutually independent). One iteration of the new map corresponds to two iterations of the old one. The half-width of the chaotic layer of map (1.26) is known to be $\approx \lambda W$; see Chap. 5. Therefore, the size of the libration component of the chaotic layer in w in the asymmetric perturbation case effectively doubles and becomes $\approx 2\lambda W$.

In the circulation case, the layer's component corresponding to the reverse rotations does not exist, if $W^- = 0$; conversely, the component corresponding to the direct rotations does not exist, if $W^+ = 0$. The remaining component is described by the ordinary separatrix map (1.26) with the same λ and W equal to non-zero W^+ or W^-; the component's size in w is $\approx \lambda W$.

The averaged (over the whole layer) maximum Lyapunov exponent can be calculated as the sum of weighted contributions of the chaotic layer components corresponding to the model pendulum's librations, direct rotations, and reverse rotations. The weights are directly proportional to the relative times of the system's residence in these three components. Assuming ergodicity, they are therefore proportional to measures (squares) of the components in the phase plane of the map. Based on the just given estimates of the layer's components widths, we find that, in the duad case, the weight ratio of the libration and circulation components is 4:1.

Hence, one obtains the maximum Lyapunov exponent and the Lyapunov time of the original system:

$$L = \frac{\Omega}{\mu_{libr} + 1}\left[\mu_{libr}\frac{L_{\mathrm{sx}}(2\lambda)}{T_{\mathrm{sx}}(2\lambda, W)} + \frac{L_{\mathrm{sx}}(\lambda)}{T_{\mathrm{sx}}(\lambda, W)}\right], \tag{3.27}$$

$$T_{\mathrm{L}} = \frac{T_{\mathrm{pert}}}{2\pi} \cdot \frac{\mu_{libr} + 1}{\mu_{libr}\frac{L_{\mathrm{sx}}(2\lambda)}{T_{\mathrm{sx}}(2\lambda, W)} + \frac{L_{\mathrm{sx}}(\lambda)}{T_{\mathrm{sx}}(\lambda, W)}}, \tag{3.28}$$

where $\mu_{libr} \approx 4$, and W, L_{sx}, T_{sx} are given by Eqs. (1.28), (3.24), and (3.25).

Adiabatic Chaotic Resonance Triplet If $\lambda < 1/2$, the diffusion across the layer can be regarded as slow, and any trajectory of the ordinary separatrix map (1.31) can be regarded as following close to some current curve called the *guiding curve*, until it hits in vicinities of the map's *singular curve*; see Sect. 3.2. Using an analytical expression for the guiding curve, the following approximate formula for the maximum Lyapunov exponent was derived above in Sect. 3.2:

$$L \approx \frac{\Omega}{\ln\left|4\sin\frac{c}{2}\right| + \frac{c}{\lambda}}, \tag{3.29}$$

where $c = \lambda \ln \frac{32}{|W|}$. For $\lambda \ll 1$ one has $W \approx 8\varepsilon\lambda$, hence the Lyapunov time

$$T_{\mathrm{L}} \approx \frac{T_{\mathrm{pert}}}{2\pi} \ln\left|\frac{16}{\varepsilon\lambda}\sin\left(\frac{\lambda}{2}\ln\frac{4}{|\varepsilon|\lambda}\right)\right|. \tag{3.30}$$

Adiabatic Chaotic Resonance Doublet Using the approximation of the Melnikov–Arnold integral $A_2(\lambda) \approx 2\pi\lambda + 4$ at $\lambda \ll 1$, $\eta = 0$, one has: $W^{\pm} \approx \varepsilon\lambda(4 \pm 2\pi\lambda) \approx 4\varepsilon\lambda$. Therefore, in this case, the separatrix algorithmic map (1.33) degenerates to the ordinary separatrix map (1.26) with $W \approx 4\varepsilon\lambda$, i.e., mathematically the case is equivalent to the "slow triad" case, but with a different (halved) value of W. It is then straightforward to write down the formula for the Lyapunov time:

$$T_{\mathrm{L}} \approx \frac{T_{\mathrm{pert}}}{2\pi} \ln\left|\frac{32}{\varepsilon\lambda}\sin\left(\frac{\lambda}{2}\ln\frac{8}{|\varepsilon|\lambda}\right)\right|. \tag{3.31}$$

Summarizing the analytical results, we may state that for the four generic resonance interaction types, namely, non-adiabatic chaotic resonance triplet, non-adiabatic chaotic resonance doublet, adiabatic chaotic resonance triplet, and adiabatic chaotic resonance doublet, the Lyapunov timescales of motion are given, respectively, by Eqs. (3.26), (3.28), (3.30), and (3.31).

These analytical results, to assess their accuracy, can be confronted with results of direct numerical integrations of the original Hamiltonian system (3.23), as performed in Shevchenko (2007a). The program package (Shevchenko and Kouprianov 2002; Kouprianov and Shevchenko 2003), realizing the HQRB method by Von Bremen et al. (1997), was used in Shevchenko (2007a) to compute the maximum Lyapunov exponents. Its capabilities are by far greater than that necessary in the given task. It allows one to compute the full Lyapunov spectrum of a multidimensional dynamical system. The Lyapunov spectrum of the given perturbed pendulum system consists of a pair of Lyapunov exponents, namely, the maximum one and its negative counterpart.

The computation results, obtained at $\lambda = 0.1$ (adiabatic chaos) and 2 (non-adiabatic chaos), are shown in Fig. 3.6. In the Figure, they are accompanied by theoretical curves, specified by Eqs. (3.26), (3.28), (3.30), and (3.31), and the relation $L = 1/T_{\mathrm{L}}$. An agreement of the numerical results with the analytical formulas is clearly present.

The computed dependences possess, however, minor local wave-like disturbances. These patterns are natural and inherent, as they are conditioned by emergence of system's resonances, as ε is incremented.

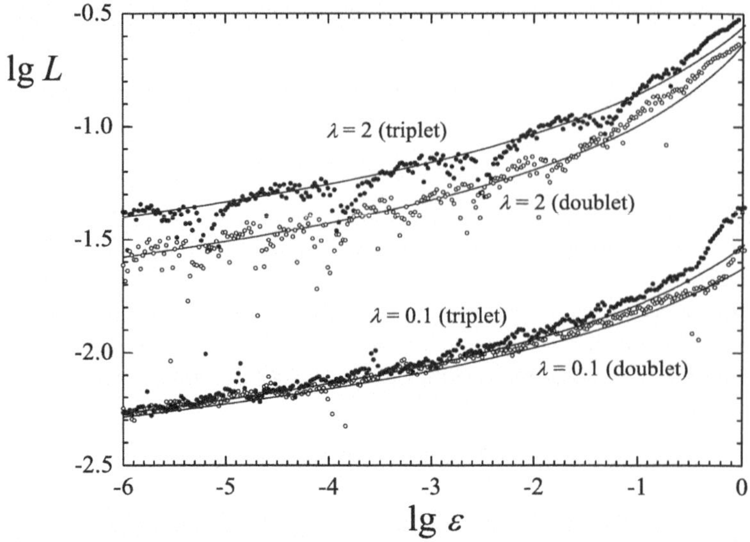

Fig. 3.6 The maximum Lyapunov exponent L in the chaotic layer of system (3.23), in dependence on the perturbation amplitude ε: results of direct integrations (circles) and theoretical (solid) curves, specified by Eqs. (3.26), (3.28), (3.30), and (3.31) (Figure 2 from Shevchenko 2007a, by permission from © Cambridge University Press)

3.4 Lyapunov Exponents in Resonance Multiplets

In this Section, we consider a dynamical situation, when there is a lot of interacting resonances. An infinite multiplet of equally-sized equally-spaced interacting resonances is given by the standard map Hamiltonian (1.24) and is described by the standard map (1.23) (Chirikov 1979; Lichtenberg and Lieberman 1992).

An analytical approximation for the standard map's maximum Lyapunov exponent was derived in Chirikov (1979). By means of linearizing the standard map in the momentum variable y in a vicinity of its fixed value and solving the characteristic equation for the linearized map, one gets for the maximum eigenvalue at $K \gg 1$:

$$l^+ \approx K|\cos x|. \tag{3.32}$$

Assuming complete ergodicity of the motion at $K \gg 1$, the maximum Lyapunov exponent is obtained by averaging l^+ over the map's phase plane, or, equivalently over the phase variable x:

$$L_{\mathrm{st}} \approx \frac{1}{2\pi}\int_0^{2\pi} \ln K|\cos x|dx \tag{3.33}$$

and, therefore,

$$L_{\mathrm{st}} \approx \ln\frac{K}{2}. \tag{3.34}$$

Equation (3.34) provides the needed approximation. Chirikov (1979) observed that the difference between the theoretical and actual numerical-experimental values of L_{st}, on increasing K, becomes less than $\approx 2\%$ already at $K = 6$. For a graphical comparison of the theory and numerics in a different numerical experiment, see Fig. 3.2.

Therefore, for any multiplet of equally-sized equally-spaced resonances in model (1.24), one has

$$L \approx \ln\frac{2\pi^2}{\lambda^2}, \tag{3.35}$$

where the system's adiabaticity parameter $\lambda = 2\pi K^{-1/2}$, and K is the stochasticity parameter of the standard map.

In the standard map dynamics, the onset of global chaos takes place at $K \approx 1$; at this K value, integer resonances start to overlap and the unbounded transport in the momentum variable becomes possible (Chirikov 1979; Lichtenberg and Lieberman 1992). Judging by the data presented in Fig. 3.2, Eqs. (3.34) and (3.35) can be used for satisfactory estimations of Lyapunov exponents already at $\lambda \lesssim 6$ and, with much higher precision, at $\lambda \lesssim 3$. Therefore, the domain of validity of these approximations covers not only the realm of adiabatic chaos, but also an actual portion of the realm of non-adiabatic one.

However, for the standard map model to be realistic in any application, the number of resonances in the multiplet of a system under study should be great, because in the standard map model it is infinite. Let us assume that the number of resonances in a resonance multiplet is greater than three. In celestial-mechanical applications, the number of subresonances of a resonance can be very large, see, e.g., Sects. 8.2.3 and 8.2.4. In such cases, the multiplet is called a *supermultiplet*. If chaos is non-adiabatic ($\lambda \gtrsim 1/2$), then, to estimate Lyapunov timescales, one may take into account solely the resonances that are closest to the guiding one, because the interaction with all other resonances in the multiplet is exponentially small with λ. Therefore, Eqs. (3.26) and (3.28), derived for resonance triplets and doublets, can be used.

However, if chaos is adiabatic ($\lambda \lesssim 1/2$), the triplet or doublet models are not adequate anymore and one has to develop a different approach. In the limiting case of infinitely many interacting equally-sized equally-spaced resonances, formulas based on the standard map theory can be straightforwardly used.

Precise fitting formulas for the maximum Lyapunov exponent L of the standard map, as a function of the stochasticity parameter K, are given in Shevchenko (2004b); Shevchenko (2004a) for $K < 1$ and $K > 4.5$:

$$L = \frac{1}{T_{\text{pert}}} \cdot \begin{cases} 0.1333K, & \text{if } K < 1, \\ \ln \frac{K}{2} + \frac{1}{K^2}, & \text{if } K > 4.5. \end{cases} \tag{3.36}$$

Formulas (3.36) are graphically presented in Fig. 3.7. The asymptotic curves, if extrapolated to intermediate K values, intersect at $K = 2$. Using the asymptotic formulas in the $1.0 \lesssim K \lesssim 4.5$ interval (corresponding to $3.0 \lesssim \lambda \lesssim 6.3$) would obviously underestimate actual Lyapunov exponents. To describe $L(K)$ at any K from zero to infinity, the following fitting formulas were proposed in Shevchenko (2014):

$$L = \frac{1}{T_{\text{pert}}} \cdot \begin{cases} 0.1333K, & \text{if } K < 1.1, \\ 0.469(K - 1.037)^{1/2}, & \text{if } 1.1 \le K < 4.4, \\ \ln \frac{K}{2} + \frac{1}{K^2}, & \text{if } K \ge 4.4. \end{cases} \tag{3.37}$$

Therefore, the Lyapunov time for the *infinitet* (infinite multiplet) of interacting resonances is given by

$$T_{\text{L}} \approx T_{\text{pert}} \cdot \begin{cases} \dfrac{7.50}{K} (\approx 0.190\lambda^2), & \text{if } K < 1.1 \text{ (or, if } \lambda > 6.0), \\ 2.133(K - 1.037)^{-1/2}, & \text{if } 1.1 \le K < 4.4 \text{ (or, if } 3.0 < \lambda \le 6.0), \\ \left(\ln \dfrac{K}{2} + \dfrac{1}{K^2}\right)^{-1}, & \text{if } K \ge 4.4 \text{ (or, if } \lambda \le 3.0), \end{cases} \tag{3.38}$$

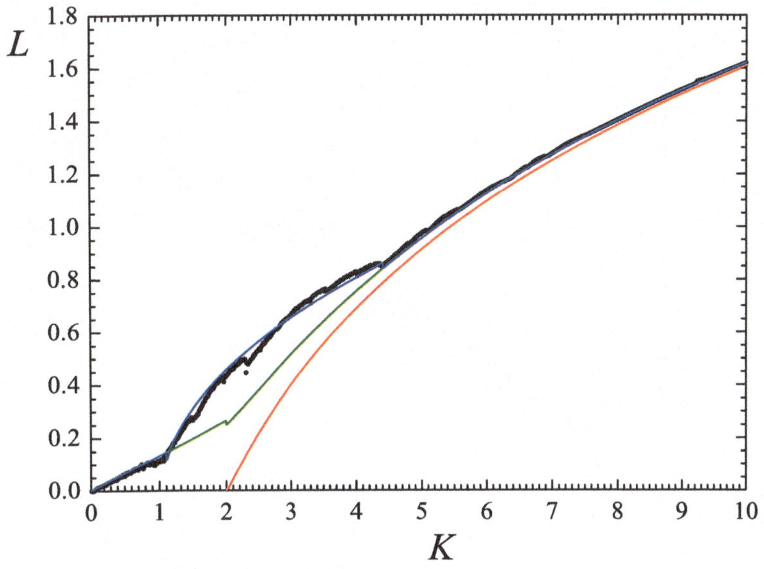

Fig. 3.7 Dots: the numerical-experimental dependence $L(K)$ for the maximum Lyapunov exponent of the standard map, based on data of Shevchenko (2004b); Shevchenko (2004a). Lower (red) curve: $\ln(K/2)$. Middle (green) curve: functions (3.36) in junction. Upper (blue) curve: formula (3.37). In the both formulas, T_{pert} is set to unity (Figure 4 from Shevchenko 2014, with permission from Elsevier)

where

$$K = (2\pi/\lambda)^2. \tag{3.39}$$

It is obvious from Figs. 3.7 and 3.8 that at $K \lesssim 1$, i.e., at K less than its critical value $K_{\text{G}} \approx 1$, the upper bound of $L(K)$ is close to linear with K.

This is explainable in the separatrix map theory framework Shevchenko (2004b). Indeed, $L(K)$ is expected here to be described by Eq. (3.26), corresponding to the non-adiabatic chaotic resonance triplet, because at $K \lesssim 1$ one has $\lambda \gtrsim 6$ and, therefore, the resonances close to the guiding one can be solely taken into account, as a first approximation.

Equation (3.10) gives $L = \Omega L_{\text{sx}}/T_{\text{sx}}$, where L_{sx} and T_{sx} are described by Eqs. (3.24) and (3.25), respectively. From Eq. (3.24) it follows that $L_{\text{sx}} \approx$const, if $\lambda \gtrsim 6$. On the other hand, $T_{\text{sx}} \propto K^{-1}$, if $K \ll 1$, see Eq. (3.25) (or Equation (6.18) in Chirikov 1979). Combining the relations, one finds $L \propto K$, if K is small enough. The slope of this theoretical linear asymptotic law is, however, somewhat less than that of the numerical linear fit $L \approx 0.1333K$, presented by Eq. (3.37). An inspection of Fig. 3.8 reveals that the slope of the actual numerical dependence decreases with K; this explains the difference between the theoretical and adopted numerical slopes.

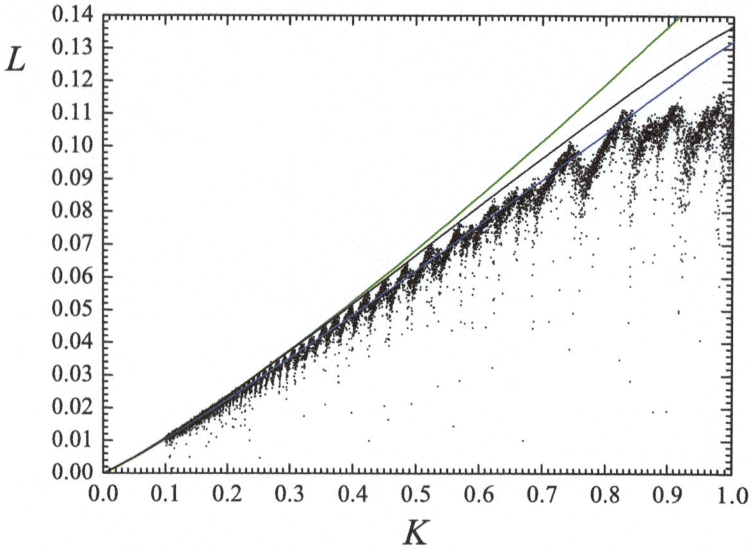

Fig. 3.8 Dots: the numerical-experimental dependence $L(K)$ for the maximum Lyapunov exponent of the standard map, based on data of Shevchenko (2004b); Shevchenko (2004a), in the interval $0 < K < 1$. Lower (blue) curve: the separatrix map theory without any correction to the MA-integral. Upper (green) curve: the separatrix map theory with Chirikov's zero-order correction to the MA-integral. Middle (black) curve: the separatrix map theory with the Chirikov–Lazutkin–Gelfreich correction to the MA-integral (Figure 5 from Shevchenko 2014, with permission from Elsevier)

The actual numerical $L(K)$ dependence at $K < 1$ has a lot of sharp local minima, accumulating to $K = 0$. Their emergence is due to *marginal resonances* at the borders of the chaotic layer (Shevchenko 2014). Marginal resonances also play a major role in defining the chaotic layer width (Shevchenko 1998b, 2012; Soskin et al. 2012).

Let us derive a formula for the $L(K)$ upper bound. The basic relation (3.10) gives $L = \Omega L_{sx}/T_{sx}$, where $\Omega = 2\pi$; L_{sx} and T_{sx} are described by Eqs. (3.24) and (3.25), respectively.

We modify the expression for W, which enters in Eq. (3.25), substituting W with $W_{st} = R_{st}W$, where R_{st} is a correction factor, specific for the standard map (Chirikov 1979). T_{sx} is therefore given by

$$T_{sx} = \lambda \ln \frac{32e}{\lambda R_{st}|W|}. \tag{3.40}$$

Using Eq. (3.39), an equivalent form can be obtained:

$$T_{\rm sx} = \Omega \left(\frac{\pi^2}{K} - K^{-1/2} \ln \frac{2R_{\rm st}\pi^4}{eK^{3/2}} \right) \quad (3.41)$$

(Chirikov 1979).

The correction factor $R_{\rm st}$ was warranted to improve the separatrix map performance in describing the chaotic layers of integer resonances of the standard map, and at first it was numerically estimated (in numerical experiments with the standard and separatrix maps) as $R_{\rm st} \approx 2.15$ (Chirikov 1979). Later on, this factor was estimated with an unbounded numerical accuracy:

$$R_{\rm st} = \frac{f_0}{16\pi^3} = 2.255244393\ldots, \quad (3.42)$$

as it had been expressed through the so-called *Lazutkin splitting constant* $f_0 = 1118.8277059409008\ldots$ (Vecheslavov and Chirikov 1998; Vecheslavov 1999).

If $K \neq 0$, the stable and unstable separatrices of any integer resonance of the standard map intersect transversally. To describe this phenomenon, Lazutkin (2005) derived an asymptotic formula for the separatrix splitting angle, valid at $K \ll 1$. The splitting angle at the first intersection of the separatrices with the $x = \pi$ axis is given by

$$\alpha = \frac{\pi}{h^2} \exp\left(-\frac{\pi^2}{h}\right) \sum_{m=0}^{\infty} c_m h^{2m}, \quad (3.43)$$

where

$$h = \ln\left(1 + \frac{K}{2} + \left(K + \frac{K^2}{4}\right)^{1/2}\right), \quad (3.44)$$

and for the first three coefficients c_m one has the formulas

$$c_0 = f_0, \quad c_1 = f_1 - \frac{c_0}{4}, \quad c_2 = f_2 - \frac{c_1}{4} - \frac{25c_0}{72}, \quad (3.45)$$

where

$$f_0 = 1118.8277059\ldots, \quad f_1 = 18.59891\ldots, \quad f_2 = -2.17205\ldots \quad (3.46)$$

(Lazutkin 2005; Gelfreich 1999). Using Eq. (3.43), we arrive at

$$R_{\rm st} \approx \frac{1}{16\pi^3}(c_0 + c_1h^2 + c_2h^4), \quad (3.47)$$

where $h \approx K^{1/2}$.

Combining Eqs. (3.10), (3.24), (3.40), and (3.47), it is straightforward to construct a theoretical $L(K)$ dependence, which, therefore, embodies the separatrix map theory with the Chirikov–Lazutkin–Gelfreich correction to the Melnikov–Arnold integral (MA-integral). The corresponding curve is presented in Fig. 3.8 in black.

For comparison, the lower (blue) curve in this Figure embodies the separatrix map theory without any correction to the MA-integral, i.e., the theory with R_{st} set to unity. The upper (green) curve embodies the separatrix map theory with Chirikov's zero-order (in h) correction to the MA-integral, i.e., the theory with R_{st} set to 2.2552.

As one would have expected, the middle (black) curve, which is based on the most advanced theory, is the best one in approximating the upper bound for the numerically found dependence.

Now we have all analytical tools that are necessary to estimate Lyapunov exponents in multiplets of resonances, in the first fundamental (pendulum) model of resonance. Let us see how these analytical tools perform versus numerical results. The original dynamical system is specified by the Hamiltonian (1.42). For computing the Lyapunov exponents in direct integrations of the original system, the HQRB numerical method by Von Bremen et al. (1997) is used, as described in Shevchenko (2014).

Let us consider the ultimately large perturbation amplitude, namely, $\varepsilon_k = \varepsilon = 1$; in other words, let us consider equally-sized equally-spaced multiplets. We call the amplitude $\varepsilon = 1$ ultimately large, because the case of $\varepsilon > 1$ can be reduced to the case of $\varepsilon < 1$ by changing the choice of the guiding resonance.

In addition to the above considered theory valid in the case of small-amplitude perturbations, the standard map theory, given by Eqs. (3.37) and (3.38), can be applied for estimating the maximum Lyapunov exponents in multiplets of equally-sized equally-spaced resonances, if the number of resonances in a multiplet is large. Thus, it is assumed that the dynamics in the infinitet ($M = \infty$) is approximately the same as in any multiplet with $M \gg 1$.

λ dependences, both theoretical and numerical-experimental, of the maximum Lyapunov exponent L (normalized by ω_0) for four kinds of multiplets (doublet, triplet, septet, and infinitet) of equally-sized equally-spaced resonances are shown in Fig. 3.9. The dependences for the triplet and the septet obviously occupy intermediate positions between the dependences for the doublet and the infinitet. The numerical data for the doublet (black dots) agrees well with the separatrix map theory (black curve), although the perturbation amplitude is maximal, $\varepsilon = 1$.

Comparing the maxima of the curves for ε equal to 0.01 and 1 (see Shevchenko 2014), one finds that L/ω_0 depends on the perturbation amplitude ε rather strongly: the difference is about three times. This difference highlights the fact that, when estimating L, it is usually insufficient to take into account solely the frequencies (Ω and ω_0), because the perturbation amplitudes may also play an important role.

Designating the Lyapunov exponents for the doublet, triplet and infinitet cases as $L^{(2)}$, $L^{(3)}$, and $L^{(\infty)}$, respectively, one may see that at $\lambda \sim 1$–3 (i.e., in the interval where L/ω_0 are maximal) the ratios $L^{(\infty)}/L^{(2)}$ and $L^{(\infty)}/L^{(3)}$ both are ≈ 2. (At the curves maxima, they are equal to 2.2 and 1.7, respectively.) How do the ratios

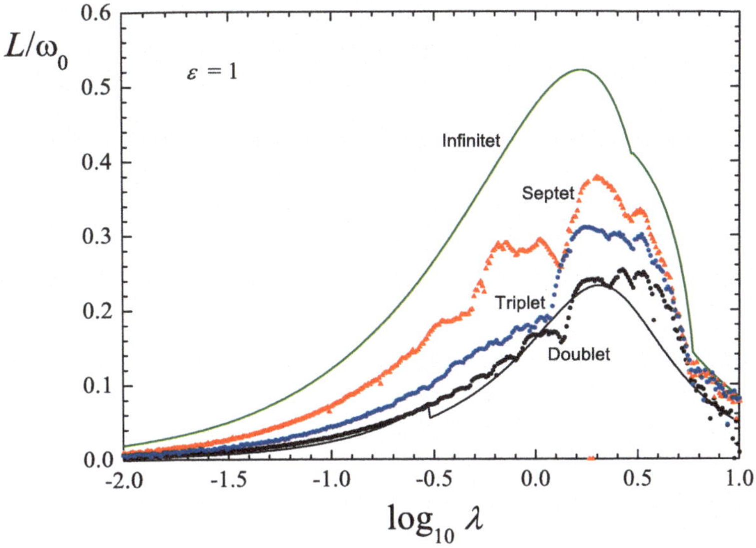

Fig. 3.9 λ dependences of the maximum Lyapunov exponent L, normalized by ω_0, for multiplets of equally-sized equally-spaced resonances. Dots: numerical-experimental data. Upper (green) solid curve: the standard map theory, given by Eqs. (3.37) and (3.39), for the infinitet. Lower (black) solid curve: the separatrix map theory, given by Eqs. (3.28) and (3.31), for the doublet (Figure 7 from Shevchenko 2014, with permission from Elsevier)

$L^{(\infty)}/L^{(2)}$ and $L^{(\infty)}/L^{(3)}$ behave in the limits $\lambda \to 0$ and $\lambda \to \infty$? Note that L/ω_0 tends to zero in the both limits.

Consider first the limit $\lambda \to \infty$. From Eqs. (3.28) and (3.26) for the non-adiabatic chaotic doublets and triplets, respectively, one may easily find that

$$\frac{T_{\mathrm{L}}^{(2)}}{T_{\mathrm{pert}}} = \frac{\mu_{\mathrm{libr}} + 1}{2C_{\mathrm{h}}(\mu_{\mathrm{libr}} + 2)}\lambda^2 \approx \frac{5}{12C_{\mathrm{h}}}\lambda^2 \approx 0.521\lambda^2 \tag{3.48}$$

and

$$\frac{T_{\mathrm{L}}^{(3)}}{T_{\mathrm{pert}}} = \frac{1}{4C_{\mathrm{h}}}\lambda^2 \approx 0.313\lambda^2. \tag{3.49}$$

Therefore,

$$\frac{L^{(3)}}{L^{(2)}} = 2\frac{\mu_{\mathrm{libr}} + 1}{\mu_{\mathrm{libr}} + 2} \approx 5/3 \approx 1.67 \tag{3.50}$$

asymptotically.

If $\lambda \to \infty$, then $L^{(\infty)} = L^{(3)}$ and $L^{(\infty)}/L^{(2)} = L^{(3)}/L^{(2)} \approx 1.67$.

Therefore, if $\lambda \gg 1$, the Lyapunov exponents in multiplets of equally-sized equally-spaced resonances do not depend much on the number of resonances in a multiplet: they differ by no more than a factor of two. Indeed, the Lyapunov exponent in the infinitet is only about 70% greater than that in the doublet.

The differences in the opposite adiabatic limit ($\lambda \to 0$) are not so small. Indeed, Eqs. (3.31), (3.30), and (3.38) in this limit give $L^{(\infty)}/L^{(2)} \to \infty$, and $L^{(3)}/L^{(2)} \to 1$.

The differences are unimportant in applications, because $L/\omega_0 \to 0$ at $\lambda \to 0$. For the effect of the perturbation amplitude, the following remarks are in order (Shevchenko 2014).

In the range of amplitudes $\varepsilon \sim \lambda^{-1}$ and above, the standard Poincaré–Melnikov method for calculating the separatrix splitting effects requires corrections (Gelfreich 1997; Treshchev 1998). Assume that the perturbation is maximal, i.e., $\varepsilon = 1$. Then, in the doublet case, the perturbation is completely asymmetric ($\eta = 0$), therefore, according to Gelfreich (1997), the correction is zero. For the triplet of arbitrary asymmetry, Simó's hypothetical formula (Gelfreich 1997) gives the correction factor R to the W parameter of the separatrix map for system (1.6):

$$|R(x)| = \left|\frac{\sinh(x)}{x}\right|, \tag{3.51}$$

where $x = (2\varepsilon_1\varepsilon_2)^{1/2} = \frac{(2ab)^{1/2}}{\mathcal{F}}$. The value of x may be either real or imaginary, depending on the signs of a and b. The value of W is corrected by means of multiplying it by R.

In the symmetric triplet case, $\eta = 1$ and the correction factor $R(\sqrt{2}) \approx 1.3683$. Therefore, the correction factor for the triplet of equally-sized equally-spaced resonances is significantly less than that for the infinitet of equally-sized equally-spaced resonances, because, in the latter case, $R \approx 2.2552$.

Concluding this Section, it is important to underline the numerical result that, independent from the adiabaticity parameter λ value (controlling the degree of interaction/overlap of resonances in multiplets), the maximum Lyapunov exponent value in a multiplet of equally-spaced equally-sized resonances is minimal in the doublet case and maximal in the infinitet case. This is consistent with the separatrix map and standard map theories: it is clear from Fig. 3.9 that the theoretical curves for the doublet and the infinitet serve as the lower and upper bounds for all Lyapunov exponent values obtained numerically for all kinds of multiplets.

Another important fact worth mentioning is that, for almost all λ at the horizontal axis of Fig. 3.9, the resonance overlap condition in the infinitet is satisfied (except at $\lambda \gtrsim 6.4$, i.e., at $\log_{10}\lambda \gtrsim 0.8$). Indeed, the resonances in the infinitet start to overlap, on decreasing λ, at $\lambda = 2\pi/\sqrt{K_G} \approx 6.37$, as follows from the standard map theory (Chirikov 1979; Meiss 1992).

Chapter 4
Diffusion Timescales

Timescales of chaotic transport are no less important than Lyapunov timescales. They are even more important, because they are directly related to observable properties of planetary systems and their subsystems. In view of further applications, in this Chapter we discuss properties of chaotic transport, diffusion rates first of all, in multiplets of interacting and overlapping resonances. As a model for the multiplet, the standard map Hamiltonian is taken. Relationships between Lyapunov and diffusion timescales are analysed; in this connection, two kinds of Hamiltonian intermittency are discussed. The presentation is partially based (in Sections 4.2, 4.3, and 4.4) on the papers by Shevchenko (2011a), Shevchenko (1998a) (by permission from Elsevier), and Shevchenko (2010).

4.1 Diffusion Rates in Resonance Multiplets

In this Section we discuss properties of chaotic transport in multiplets of interacting and overlapping resonances. As a model for the multiplet, we take the standard map Hamiltonian (1.24).

From the standard map's (1.23) first equation, we see that the variation in y per step is $K \sin x_i$; therefore, at each iteration of the map, an $O(\epsilon)$ uncertainty in x propagates to an uncertainty $O(K\epsilon)$ in y, and, consequently, in the next value of x. Therefore, the escalation of error is exponential (Meiss 1992). The error rises if the stochasticity parameter $K > 1$, and, therefore, one may suppose that the change in the map's behaviour at this value of K is critical. And this is indeed so: at $K \approx 1$ the last *rotational invariant curve* in the map's phase plane is destroyed and the motion becomes unbounded in y (Chirikov 1979; Greene 1979; Lichtenberg and Lieberman 1992; Meiss 1992). As Meiss (1992) notes, the exact threshold value $K = K_{\mathrm{G}} \approx 0.97$, is close to the given simple estimate.

I. I. Shevchenko, *Dynamical Chaos in Planetary Systems*, Astrophysics and Space Science Library 463, https://doi.org/10.1007/978-3-030-52144-8_4

In view of the rapid loss of phase coherence at $K \gg 1$ one may expect that a statistical description of the motion becomes plausible at such values of K (Chirikov 1979). If the motion of the system within the phase plane square $2\pi \times 2\pi$ were completely ergodic (no regular islands present), then the distribution of x in the $[0, 2\pi]$ segment would be flat. The ergodicity was observed in many numerical experiments, accomplished at $K \gg 1$ (Chirikov 1979).

Since the Lyapunov time $\approx \ln \frac{K}{2}$ (see Eq. (3.34), one may assume that even initially adjacent values of phase variables become statistically independent in a few iterations. This is the so-called assumption of *random-phase approximation*.

The *diffusion coefficient* D is defined, formally for any kind of discrete motion, as the mean-square spread in a selected variable per unit of time. In case of the standard map this variable is y, which is then intentionally not taken modulo 2π and, therefore, its variation is unbounded, if $K > K_G$. Therefore,

$$D \equiv \lim_{t\to\infty} \frac{\langle (y_t - y_0)^2 \rangle}{t}, \tag{4.1}$$

where the angular brackets denote averaging over some set of starting values, and the discrete time $t \equiv k$ in Eqs. (1.23) (Meiss 1992; Lichtenberg and Lieberman 1992).

The cumulative variation in y, achieved at time $t \equiv k$, is

$$\Delta y = K \sum_0^t \sin x_t. \tag{4.2}$$

The time averaging and the phase-space averaging (denoted in the following by the bar and by the angle brackets, respectively) are equivalent, due to the assumed ergodicity. Therefore, one has

$$\overline{\Delta y} = \langle \Delta y \rangle = 0, \qquad \overline{(\Delta y)^2} = \langle (\Delta y)^2 \rangle = \frac{1}{2} K^2 t \tag{4.3}$$

(Chirikov 1979). As directly follows from Eqs. (4.3), the motion in y represents normal diffusion with the rate

$$D_{\rm QL} = \frac{\langle (\Delta y)^2 \rangle}{t} = \frac{1}{2} K^2. \tag{4.4}$$

This formula embodies the so-called *quasilinear diffusion* approximation (Meiss 1992). The validity of approximation (4.4) was confirmed in many numerical experiments: it was numerically validated that, when K is large, the diffusion coefficient $D \approx D_{\rm QL}$ (Chirikov 1979; Lichtenberg and Lieberman 1992). Corrections to

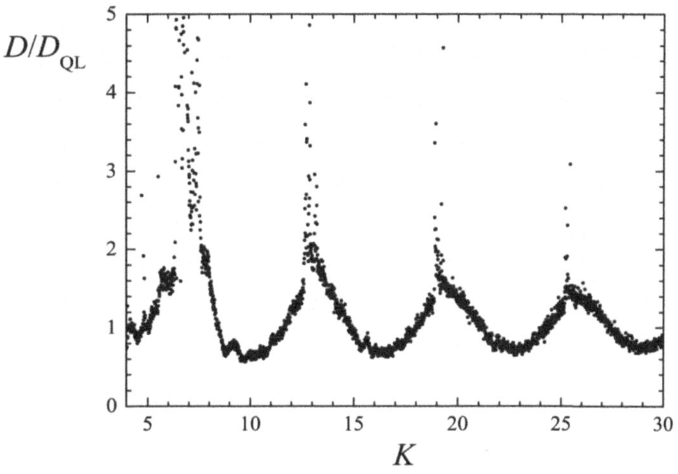

Fig. 4.1 A numerical-experimental dependence $D(K)$, normalized by the quasilinear law

D_{QL}, especially prominent at $K \sim 1$, were analytically calculated in Rechester et al. (1981); Cary et al. (1981), where they were expressed in the form of series of products of Bessel functions. They perfectly reproduce wave-like variations of $D(K)$ observed in numerical experiments; see Fig. 4.1.

Another prominent feature in the $D(K)$ dependence (Fig. 4.1) is a sequence of periodic bursts of the diffusion rate to infinity on small limited intervals of K. They are due to sticking of trajectories to small islands corresponding to the so-called *accelerator modes*, which represent the periodic orbits that satisfy $x_n = x_0 + 2\pi k$ and $y_n = y_0 + 2\pi l$ for integer k, l, n. Sticking of chaotic trajectories to these islands leads to ballistic flights: on long time intervals the y variable varies linearly with time. In the presence of the accelerator modes the diffusion coefficient D appears to be infinite (Karney et al. 1982; Meiss et al. 1983). This regime of motion is a kind of *anomalous diffusion* (Chirikov 1996).

The emergence of small regular islands, corresponding to periodic orbits at particular values of K, results in local (in K) minima in the $\mu(K)$ dependence, where μ is the relative measure of the chaotic component in the phase space; see Fig. 4.2. The accelerator modes, emerging at $K \approx 2\pi m$ ($m = 1, 2, \ldots$) (Chirikov 1979, 1996), are responsible for the most prominent double minima in the graph. Comparing with Figs. 4.1 and 4.2, one may verify that the bursts of D are due to the accelerator modes, as their K locations coincide with the K locations of the most prominent minima in μ.

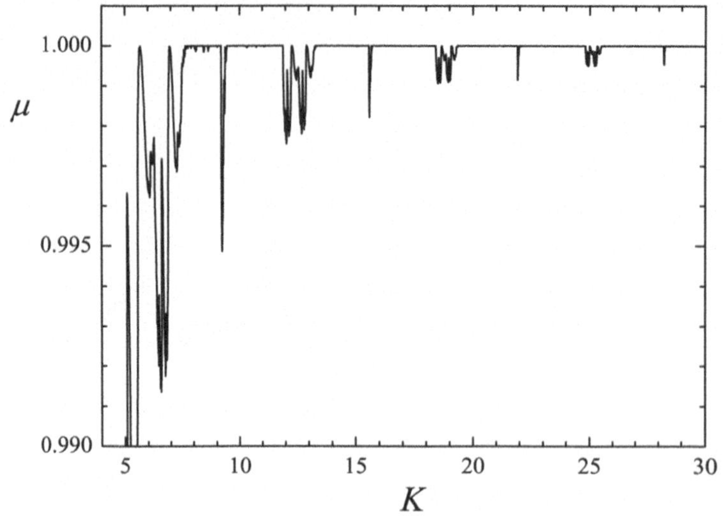

Fig. 4.2 Relative measure of the chaotic component in the standard map's phase space, as a function of K (Figure 5 from Shevchenko 2007b, with permission from Elsevier)

4.2 Diffusion Rates in Resonance Doublets

Diffusion rates and timescales in infinitets[1] of overlapping resonances, as well as in supermultiplets[2] of overlapping resonances can be analytically estimated using formulas derived for the standard map, as shown above in Sect. 4.1. In the opposite limit of multiplets comprising only two or three resonances, how the diffusion timescales can be found? Here we describe how this problem can be solved, following Shevchenko (2011a) and based on the *separatrix algorithmic map* theory. The separatrix algorithmic map (1.33) was introduced above in Sect. 1.6.

A general approach for such kind of problems was originally proposed in Chirikov and Vecheslavov (1986, 1989), where it was applied to describe long-term cometary dynamics. Chirikov (2000) employed a similar method in a study of anomalous diffusion in a general separatrix map model; see p. 11 in Chirikov 2000 and, in particular, formulas (20) and (21) therein.

First of all, a reservation should be made that it is often only approximate that one may consider chaotic transport in any celestial-mechanical problem, in which solely two or three resonances overlap, as diffusive. The matter is that the value of the adiabaticity parameter λ in most of applications is equal, by order of magnitude, to unity: $\lambda \sim 1$. This is due to selection biases: indeed, if $\lambda \gg 1$, then chaos is simply insignificant, because the chaotic layers are thin, see Sect. 5.3; on the other hand,

[1]Multiplets comprising an infinite number of resonances.

[2]Multiplets comprising a finite but large number of resonances.

if $\lambda \ll 1$, chaos can be too slow to arise significantly within prescribed finite time intervals. However, in the second case, estimating the diffusion times is nevertheless important, in order to be able to compare them with other important physical or dynamical timescales, in any problem under study.

As explained in (Chirikov 2000, p. 12–13), when $\lambda \sim 1$, "...the layer width is reduced down to the size of a single kick. ... Hence, the diffusion approximation becomes inapplicable. Instead, the so-called ballistic relaxation comes into play which is much quicker. In other words, a slow diffusive motion ... is replaced now by rapid jumps of a trajectory over the whole layer ...."[3] Therefore, the diffusion rate estimates presented below in this Section should be regarded as an extrapolation of diffusive description. Actually they may mostly serve as upper bounds for chaotic transport timescales.

Chirikov and Vecheslavov (1986, 1989) considered chaotic dynamics in the *Kepler map* model. The Kepler map is a kind of a generalized separatrix map; it is discussed in Sects. 1.7 and 11.1.5. On general grounds, Chirikov and Vecheslavov (1986, 1989) took the diffusion rate (in the map's action-like variable), in the chaotic domain of the map's phase space, to be approximately equal to the mean (over time) squared energy increment per one iteration of the map. Analogously, in case of the ordinary separatrix map (1.26), we set the diffusion rate (in the energy variable) to be equal to the mean squared energy increment, i.e., $\langle W^2 \sin^2 \tau_i \rangle$. Averaging over the phase interval $0 \leq \tau_i < 2\pi$, we find the diffusion rate to be

$$D_{\mathrm{map}} \approx \frac{W^2}{2}, \tag{4.5}$$

where W is given by Eq. (1.28).

In case of the separatrix algorithmic map (1.33), the chaotic layer consists of three components, corresponding to prograde rotations, retrograde rotations, and librations of the phase variable. They should be considered separately. In the prograde rotation case, the diffusion rate D_{map} in the energy variable obviously equals $\approx (W^+)^2/2$; and in the retrograde rotation case, it equals $\approx (W^-)^2/2$.

Using formulas (1.34), it is straightforward to calculate the parameters W^+ and W^- of the separatrix algorithmic map (1.33). If $\lambda > 1/2$, the equality $b = 0$ implies $|W^-| \ll |W^+|$, and the equality $a = 0$ implies $|W^-| \gg |W^+|$. If $W^- = 0$, the chaotic layer component corresponding to retrograde rotations does not exist, its measure is zero. Conversely, if $W^+ = 0$, the layer component corresponding to direct rotations does not exist.

The third component, that corresponding to circulations, is described by the ordinary separatrix map (1.26) with two parameters, λ and $W = W^{\pm}$ (the non-zero value among W^+ and W^-). The component extent in w is equal to half-width of the map's chaotic layer in case of non-adiabatic chaos, i.e., $\approx \lambda|W|$, see Sect. 5.3.

[3]The "single kick" is a particle's energy increment per one iteration of the separatrix map.

Consider the layer's libration side. On this side, the W^- and W^+ values of the $W^\pm$ parameter alternate (replace each other) at each iteration of the separatrix algorithmic map (1.33). As shown above in Sect. 3.3, if W^- or W^+ is equal to zero, then on any doubled iteration step the separatrix algorithmic map (1.33) reduces to the ordinary separatrix map (1.26) with λ doubled and W equal to the largest (non-zero) $W^\pm$. Since one iteration of the new map corresponds to two iterations of the original one, one finds that the diffusion rate referred to the original map time units is given by

$$D_{\mathrm{map}} \simeq \frac{1}{4}(W^\pm)^2, \tag{4.6}$$

where $W^\pm$ is the largest (non-zero) $W^\pm$.

For the circulation component, one has

$$D_{\mathrm{map}} \simeq \frac{(W^+)^2}{2}, \tag{4.7}$$

if $b = 0$, and

$$D_{\mathrm{map}} \simeq \frac{(W^-)^2}{2}, \tag{4.8}$$

if $a = 0$.

The libration component extent in w in the given asymmetric perturbation case effectively doubles and becomes $\approx 2\lambda W$, see Sect. 3.3. The parameters λ and W are regarded here as independent from each other. Therefore, the chaotic domain corresponding to libration dominates in the layer extent in w, and, to get a rough estimate of the diffusion rate across the entire layer, it is sufficient to estimate the rate across the libration component alone. Let us consider the doublet; then $b = 0$, $\eta = 0$, and $W^\pm = W^+$, where, according to Eq. (1.54) with $k = 1$,

$$W^+ = \varepsilon\lambda A_2(\lambda) = 4\pi\varepsilon\lambda^2 \frac{\exp(\pi\lambda/2)}{\sinh(\pi\lambda)}. \tag{4.9}$$

Therefore,

$$D_{\mathrm{map}} \simeq \left(2\pi\varepsilon\lambda^2 \frac{\exp(\pi\lambda/2)}{\sinh(\pi\lambda)}\right)^2. \tag{4.10}$$

To get the diffusion rate referred to real time units, the map time units should be expressed through the original time units. This is achieved by dividing the diffusion rate referred to the map time units by the mean period of phase rotations (equivalently, the mean half-period of librations) inside the chaotic layer, since the latter quantity is nothing but the average time interval corresponding to one map iteration. Therefore, the diffusion rate referred to real time units is

$$D = \frac{|\Omega| D_{\mathrm{map}}}{T_{\mathrm{sx}}}, \tag{4.11}$$

where T_{sx} is given by formula (3.25).

The characteristic diffusion time across the chaotic layer is defined here as the inverse of the diffusion rate. It is just the time needed for a diffusing particle to cover the relative energy interval equal to one.

However note that the maximum possible deviation in the relative energy w from zero on the layer's libration side is equal to -2 (Chirikov 1979); therefore, the defined diffusion time gives an appropriate time estimate for the global mixing inside the chaotic layer.

In the doublet case, $b = 0$ and $W^{\pm} = W^{+}$. For the diffusion time one gets

$$T_{\mathrm{d}} = \frac{1}{D} = \frac{T_{\mathrm{sx}}(\lambda, W^{+})}{|\Omega| D_{\mathrm{map}}} \simeq \frac{4 T_{\mathrm{sx}}(\lambda, W^{+})}{|\Omega| (W^{+})^2}, \tag{4.12}$$

where

$$T_{\mathrm{sx}}(\lambda, W^{+}) \approx \lambda \ln \frac{32e}{\lambda |W^{+}|}, \tag{4.13}$$

see Eq. (3.25); here e is the base of natural logarithms. The W^{+} parameter is given by formula (4.9). Finally,

$$T_{\mathrm{d}} \simeq \frac{4\lambda}{|\Omega| (W^{+})^2} \ln \frac{32e}{\lambda |W^{+}|}. \tag{4.14}$$

Conversely, on the layer's rotation side, in case of prograde rotation, the diffusion rate $D_{\mathrm{map}} \simeq (W^{+})^2/2$; therefore, the diffusion time is two times less than that given by formula (4.14).

The diffusion rates in resonance doublets can be estimated by means of formula (4.11), and the diffusion rates in resonance infinitets by using formula (4.4). In any intermediate case (triplets, etc.), by means of the same formulas, two bounding limits for the diffusion rate can be estimated. When estimates by an order of magnitude are needed, assessment of the limits is usually sufficient.

Due to the possibility of ballistic flights mentioned in the beginning of this Section, the chaotic mixing at $\lambda \sim 1$ (and this is the usual case in actual applications) might be far more effective and rapid than in the considered case of normal diffusion. In studies of transport in resonance multiplets, the effect of ballistic relaxation has not been yet explored.

4.3 Lyapunov and Diffusion Timescales: Relationships

Considering Lyapunov and diffusion timescales in any chaotic system, one is naturally led to the question whether these two timescales are related anyway. In this Section, we explore statistical properties of the motion inside the near-separatrix chaotic layer of a non-linear resonance and discuss how statistical relations may arise between the two timescales, and what are the conditions for their emergence. Generally, we follow approaches proposed and developed in Shevchenko (1998a, 2010).

Specifically, we consider two timescales: the chaotic transport time T_r needed for a sudden orbital change and the Lyapunov time T_L, taking into account that the latter's values are computed (as the inverse of the numerically found maximum Lyapunov exponent) in any application on a finite time interval, usually not longer than T_r. The time of a sudden orbital change is dynamically equivalent to the time of a recurrence to some domain of the motion phase space, that is why it is designated here as T_r.

In Soper et al. (1990), Lecar et al. (1992), Murison et al. (1994), Levison and Duncan (1993), a prominent relationship in statistics of sudden orbital changes in a number of celestial-mechanical models (in particular, in asteroidal dynamics and dynamics of TNOs) was revealed and confirmed. This relationship consists in a close-to-quadratic character of the dependence of the time T_r, needed for a sudden orbital change, on the measured Lyapunov time T_L:

$$T_\mathrm{r} \propto T_\mathrm{L}^{\beta}, \tag{4.15}$$

where $\beta \approx 1.7$–1.9 typically. However, a considerable dispersion of the statistical data was usually present.

Similar statistical relationships were revealed by Ferraz-Mello (1997), Shevchenko (1999b), Shevchenko (1998a); they considered behaviour of asteroids in vicinities of the 2/1 and 3/1 mean motion resonances with Jupiter. Tsiganis et al. (2005) studied chaotic diffusion and actual stability of the Jovian Trojan asteroids; for the motion inside the chaotic band around the Trojan stability zone, they obtained the T_L–T_r statistical relationship similar to the typical close-to-quadratic one.

Numerical simulations by Mikkola and Tanikawa (2007), Urminsky and Heggie (2008) of the disruption process in three-body systems revealed a new kind of the T_L–T_r statistical relationship, which turned out to be quasilinear.

Mikkola and Tanikawa (2007) investigated the way of disruption of an equal-mass three-body system. They performed extensive numerical integrations and computed the disruption times T_d and the Lyapunov times T_L for orbits with randomized initial conditions. It was found that the system lifetimes, as a bound triple, and the finite-time Lyapunov exponents (computed until the system decays) were correlated: on long timescales, the T_d–T_L relationship was close to linear.

Urminsky and Heggie (2008) explored correlations between the disruption times T_{d} and the Lyapunov times T_{L} in a hierarchical three-body problem. The three-body problem setting was different from that used in Mikkola and Tanikawa (2007): the Sitnikov problem (Sitnikov 1960) was considered. As an outcome of extensive numerical integrations, a two-component power-law relationship was obtained, whose second component (at longer timescales) turned out to be close to linear. This component is thus analogous to that revealed in Mikkola and Tanikawa (2007) in the equal-mass three-body problem.

Later on, the quasilinear behaviour was also observed in numerical simulations of the disruption process in the restricted three-body problem (Shevchenko 2010).

On shorter timescales, generally no T_{d}–T_{L} correlation at all can be observed; see examples in Shevchenko et al. (2020). For the existing correlations (on longer timescales), two kinds of the T_{L}–T_{r} relationships, as listed above, are known to have been revealed to date in numerical simulations: the close-to-quadratic one (Soper et al. 1990; Lecar et al. 1992; Murison et al. 1994; Levison and Duncan 1993; Shevchenko 1999b, 2008a; Ferraz-Mello 1997; Tsiganis et al. 2005) and the quasilinear one (Mikkola and Tanikawa 2007; Urminsky and Heggie 2008; Shevchenko 2010).

Why these two different kinds of relationships arise? We will consider this problem in the following.

Morbidelli and Froeschlé (1996) discussed conditions for emergence of statistical relationships between the Lyapunov times and the *macroscopic diffusion* times in general nearly integrable Hamiltonian systems. The perturbation strength was considered to be the main controlling parameter, through which the both times were expressed. It was qualitatively shown that in the multiple-resonance overlap regime the relationship should be polynomial.

In Shevchenko (1998a), it was shown that a specific close-to-quadratic T_{L}–T_{r} relationship should arise if (1) the phase space of any Hamiltonian system under study is *divided* (in terminology of Chirikov 1979), i.e., there are *chaos borders* present in the phase space; (2) the maximum Lyapunov exponent for any trajectory is computed on a recurrence. The matter is that when the Lyapunov exponent is computed on a finite time interval, a chaotic trajectory may explore only a finite local neighbourhood inside the whole connected chaotic domain, but not necessarily the whole domain.

Let us associate the time of a sudden orbital change with the recurrence time T_{r}. By the recurrence we imply the first recurrence, i.e., the first entering by a trajectory in some specified domain in the phase space. If the phase space is divided, i.e., if it comprises both chaotic and regular components, the longest recurrences of a chaotic trajectory can be due to its sticking to any border between the chaotic and regular components; on the sticking phenomenon, see Chirikov and Shepelyansky (1981), Chirikov and Shepelyansky (1984), Chirikov and Shepelyansky (1986), Chirikov (1990), Chirikov (1996) and Meiss (1992). Sporadic sticking events result in an intermittent behaviour (Shevchenko 1998b, 2010).

As illustrated above in Sect. 2.1, the maximum Lyapunov exponent is traditionally computed by building the $\log L(m)$ dependence, given by Eq. (2.2), versus

$\log m$, and by finding a $\log L$ value at which the dependence is *saturated*, i.e., attains a horizontal plateau.

Therefore, on one hand, the value of the maximum Lyapunov exponent is computed on the time intervals not less than the time of saturation; on the other hand, the computation time cannot be infinite. These limitations from below and above impose certain selection effects, which are necessary to take into account when calculating Lyapunov exponents.

The near-separatrix dynamics of the non-linear pendulum (serving here as the model of non-linear resonance) is described by the separatrix map (1.26). If the adiabaticity parameter $\lambda \gg 1$, then the separatrix map can be locally linearized in the energy w variable and, thus, be reduced to the standard map (Chirikov 1979). In this way, the chaotic layer of non-linear resonance can be locally described by the standard map (1.23).

Consider chaotic dynamics in a vicinity of the critical curve that separates regular and chaotic domains in the phase space of the map. This kind of dynamics is called *critical* (Chirikov 1990), because it is takes place close to the critical curve. Following designations by Chirikov and Shepelyansky (1986); Chirikov (1990), let $r_n = p_n/q_n$ be the continued fraction convergents to the critical curve's winding number. These convergents are nothing but the *winding numbers* of the principal resonances that are situated close to the critical curve. The stability of the periodic trajectory with winding number r_n can be characterized by the *Greene residue* (Greene 1968, 1979). Its value R_n for a principal resonance close to the critical curve is given by the formula

$$R_n = R^{(1)} \exp\left(1.20 q_n^{1+\alpha}(K - K_{\mathrm{G}})\right) \tag{4.16}$$

(Chirikov and Shepelyansky 1986), where K is the stochasticity parameter of the approximating standard map, $K_{\mathrm{G}} = 0.971635406\ldots \approx 1$ is its critical value, $R^{(1)} \approx 1/4$ is the critical value of the Greene residue; the numerical parameter $\alpha = 0.013$. Formula (4.16) follows from Greene's relation $R \propto K^q$ (Greene 1979) applied at the chaos border; but in Chirikov and Shepelyansky (1986) it was numerically improved: based on extensive simulation data, the numerical coefficient 1.20 was introduced instead of 1.

As noted in Chirikov and Shepelyansky (1986), formula (4.16) can be interpreted in a rather simple way: the locally defined Lyapunov exponents

$$l_n \approx \frac{\ln(4R_n)}{q_n} \tag{4.17}$$

practically do not depend on q_n, i.e., on a particular trajectory, and are equal to the locally defined *Kolmogorov–Sinai entropy* (also called *dynamical entropy*, see Chap. 3) $h \propto \Delta K \equiv K - K_{\mathrm{G}}$. Namely, Eqs. (4.16) and (4.17) give

$$l_n \approx 1.20 \Delta K. \tag{4.18}$$

Formula (4.18) is valid for the convergents to the critical curve. However, as soon as it is based on Greene's formula $R \propto K^q$ (Greene 1979), which is valid for all periodic orbits, it is natural to assume that it is as well valid in the latter general case. Periodic orbits densely fill the map's phase space; therefore, one may use Eq. (4.18) to characterize the maximum Lyapunov exponent local values, l, for all orbits residing in the chaotic layer not far from the chaos border. Hence

$$l \propto \Delta K. \tag{4.19}$$

In this way, the local (finite-time) maximum Lyapunov exponent is specified by the stochasticity parameter K of the approximating standard map.

The ΔK dependence of the transport time (equivalently, *recurrence time* or *sticking time*) T_r near the chaos border can be derived based on Chirikov's *resonance theory of critical phenomena* (Chirikov 1990, Section 4.3). This dependence is derived in the following way Shevchenko (1998a). The recurrence time $T_\mathrm{r} \sim \tau_n$, where τ_n is the transition time from scale q_n, with n maximal for a given recurrence, to the neighbouring scale, because τ_n rapidly diminishes with decreasing n (Chirikov and Shepelyansky 1986). The averaged relationship $\Delta K \propto \rho$ (where $\rho = |r - r_c|$ is the detuning of the winding number r with respect to the winding number r_c of the critical curve) leads to the dependence $\Delta K \propto q_n^{-2}$. Such a dependence is not sufficient to destroy the principal critical scales q_n; instead, narrow (with size $\sim q_n^{-4}$) chaotic layers emerge between them. The flux balance condition, in the statistical equilibrium, gives $\tau_n \propto q_n^4$ (Chirikov and Shepelyansky 1986; Chirikov 1990). The recurrence time $T_\mathrm{r} \sim \tau_n$, and the dependence of ΔK on the winding number detuning is set to be linear; therefore

$$T_\mathrm{r} \propto \Delta K^{-2}. \tag{4.20}$$

Using Eqs. (4.19) and (4.20), the recurrence time T_r is readily expressed through the Lyapunov time T_L, which is the inverse of the locally defined maximum Lyapunov exponent. Finally,

$$T_\mathrm{r} \propto T_\mathrm{L}{}^2. \tag{4.21}$$

Our analysis has been performed here for the perturbed non-linear pendulum, which represents a system with one and a half degrees of freedom, i.e., one degree of freedom plus time dependence. Could the resulting formula be applied to describe chaotic dynamics in systems with many degrees of freedom?

In such systems, the *Arnold diffusion* takes place (Chirikov 1979, 1990). This universal instability is not possible in systems with the number of degrees of freedom less than or equal to two. In the higher dimensions the problem is therefore much more complicated. However, different resonances in multi-dimensional systems generically have different strengths. According to Chirikov's classification (Chirikov 1990), the *guiding resonance* can be chosen arbitrarily; the choice depends on the phase space domain where the motion is considered.

Remaining resonances are considered as driving. The strongest one among the driving resonances is called the *layer resonance*. It drives transport across the chaotic layer of the guiding resonance. This chaotic transport is faster than the Arnold diffusion driven by the remaining resonances, and it can be described by the separatrix map (1.26), see (Chirikov 1979, p. 355). Therefore, one may expect that the generic T_L–T_r relationship, derived above using the separatrix map theory, can be also used in generic dynamical situations in multi-dimensional systems.

It should be noted that there are at least two natural selection biases, mentioned already in Sect. 4.3, that render the emergence of the quadratic relationship (4.21) ubiquitous in numerical simulations. First, the maximum Lyapunov exponent is normally computed on a long enough time interval so that its value were saturated. Consequently, this condition eliminates small recurrence times from consideration. Second, when T_L–T_r relationships are constructed, the computation time of the maximum Lyapunov exponent is normally not greater than T_r; e.g., the computation is stopped if the system decays.

On one hand, imposing the lower bound on the computation time may make the role of sticking regimes prominent, and, on the other hand, imposing the upper bound makes the computed maximum Lyapunov exponent value local, i.e., characterizing a particular sticking event. These two factors facilitate the emergence of the generic relationship (4.21).

Some illustrative numerical-experimental examples are in order. First, let us consider the separatrix map (1.31) itself. We build the T_L–T_r plot, where T_L is the inverse of the maximum Lyapunov exponent computed while the trajectory stays at one side of the chaotic layer, and T_r is the duration of any of these consequent stays. Time is measured in the map iterations.

The staying at one side of the layer means that the w variable in Eqs. (1.31) has a non-changing sign, plus or minus. When the trajectory crosses the unperturbed separatrix (the $w = 0$ axis), the sign of w alternates. Thus, any segment of the trajectory between the crossings of the $w = 0$ axis forms a separate recurrence.

For the separatrix map (1.31) parameters, we choose $\lambda = 3.22$ and $c = 0$. These values are the same as used in Chirikov and Shepelyansky (1984), and they are chosen in such a way that the critical curve has the *golden* winding number, i.e., the winding number equal to the *golden mean*, namely, $(\sqrt{5} - 1)/2$. The golden mean is the irrational number that is farthest from neighbouring rational numbers; see Lichtenberg and Lieberman (1992). The effect of *marginal resonances* (considered in detail in Sects. 5.5 and 5.6) is thus reduced to a minimum, and the generic T_L–T_r relationship should manifest itself most clearly. The recurrences with $T_r < 10$ are eliminated from the analysis, so that the maximum Lyapunov exponent were saturated, as explained above.

In Fig. 4.3, the computed dependence clearly follows the generic theoretical quadratic relationship, as expected.

Another paradigmatic example, presented in Shevchenko (1999b), concerns the standard map (1.23) with the stochasticity parameter chosen to be $K = 2$. In fact, the explored statistical effects can be recovered at any non-zero K that is not too

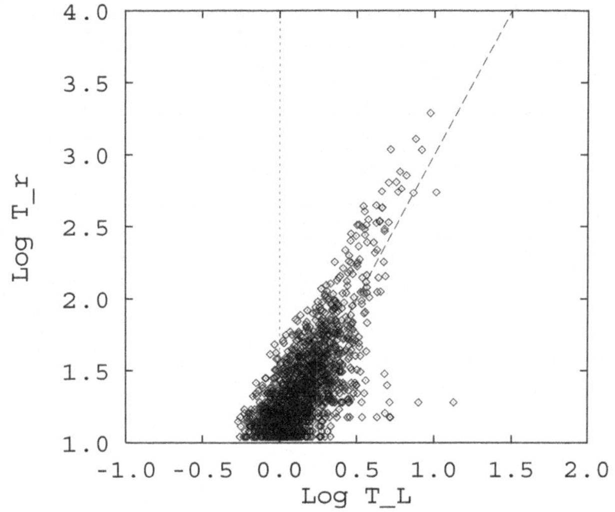

Fig. 4.3 A T_L–T_r statistical relationship for the motion inside the chaotic layer of the separatrix map (1.31). The parameters $\lambda = 3.22$ and $c = 0$ correspond to the "golden" critical curve case. Time is expressed in the map iterations. Straight dashed line: the theoretical quadratic relationship. The logarithmic scales are decimal (Figure 1 from Shevchenko 1998a, with permission from Elsevier)

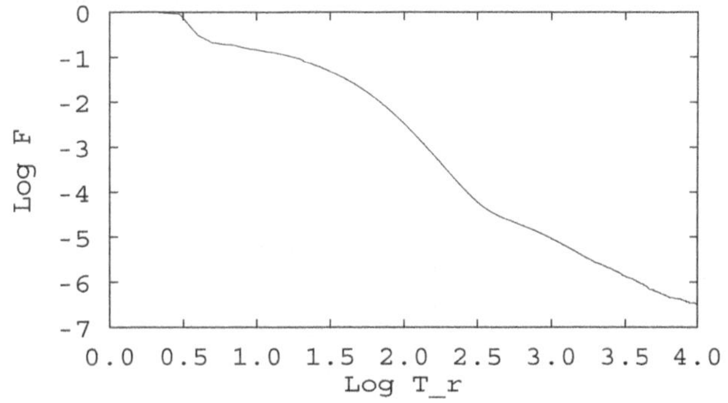

Fig. 4.4 The integral distribution of recurrence times in case of the standard map (1.23) with $K = 2$. The major exponential decay preceded with the initial short-scale drop, as well as the subsequent power-law decay, are prominent. Logarithms are decimal (Figure 1 from Shevchenko 1999b, by permission from Springer Nature, © 1999)

large, say, $K < 4$, for the regular component to be adequately present in the motion phase space.

An integral distribution of recurrence times for a single chaotic orbit is shown in Fig. 4.4. The $F(T_r)$ quantity is the observed relative fraction of recurrences that are longer than T_r. The recurrences are counted at the $y = 0 \bmod 1$ axis. Events of sticking to the island of stability around the integer resonance situated at this axis lead to the initial steep short-scale drop in the distribution. Then, on a limited interval, namely at $0.7 < \log T_r < 1.2$, the distribution is observed to be algebraic with the power-law index equal to -0.56. This is close to the inverse square root law, which is inherent to free diffusion in the central part of a chaotic layer until the trajectory reaches the layer borders (Chirikov and Shepelyansky 1981, p. 9). Here, the borders $y = 0 \bmod 1$ are conventional. At $\log T_r = 1.2$–1.3, the dependence becomes exponential, because the finite width of the layer starts to be important. Indeed, according to (Chirikov and Shepelyansky 1981, p. 10), when the time of diffusion across a stochastic layer is finite, the recurrence distribution decays exponentially due to the diffusion fluctuations. The tail of the distribution in Fig. 4.4 is algebraic, $\propto T_r^{-\alpha}$, with the power-law index $\alpha \approx 1.48$. The over-all structure of the distribution resembles structure of the distributions presented in Fig. 4 of Shevchenko and Scholl (1997) for time intervals between eccentricity jumps of intermittent chaotic trajectories in the 3/1 mean motion Jovian resonance. According to Shevchenko and Scholl (1997), the power-law indices α in the tails of the latter distributions are close to 1.5; this value was theoretically derived in Chirikov (1990, 1996) as imminent for the critical dynamics. Figure 4.4 shows that all the main features (the major exponential decay preceded by the initial steep short-scale drop, as well as the subsequent power-law decay) of the distribution are present already in the much less sophisticated case of the paradigmatic standard map.

In Fig. 4.5, the dependence T_L–T_r in the same case of the standard map is shown. The maximum Lyapunov exponent is measured for a recurrence. As adopted, the

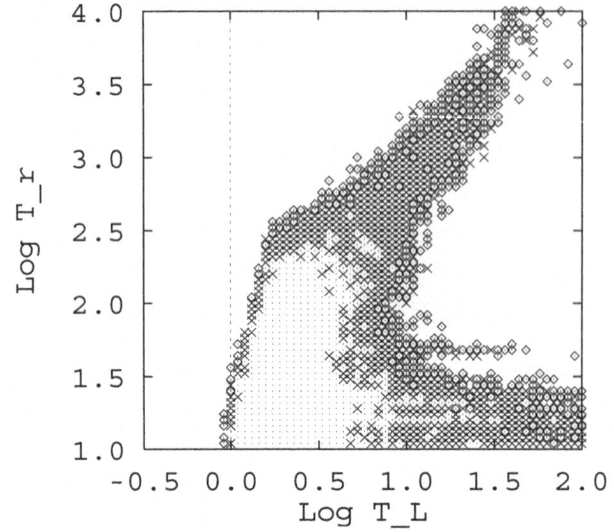

Fig. 4.5 The statistical T_L–T_r relationship in case of the standard map (1.23) with $K = 2$. Dots: the data obtained setting the number of iterations $N_{it} = 10^6$. Dots plus crosses: $N_{it} = 10^8$. Dots plus crosses and diamonds: $N_{it} = 10^9$. Logarithms are decimal (Figure 2 from Shevchenko 1999b, by permission from Springer Nature, © 1999)

recurrences are counted at the line $y = 0 \bmod 1$. For convenience of handling large data arrays, the plot field in Fig. 4.5 is partitioned in pixels; thus, the figure represents a kind of a density plot. Any pixel is turned black, if it contains at least one observed T_L, T_r pair. Increasing the number of iterations N_{it} allows one to recover recurrences with less and less frequent values of the T_L, T_r pair. The plot areas corresponding to N_{it} from 10^6 to 10^8 and then up to 10^9 are covered, respectively, with crosses and diamonds. One may see that the recurrences with $\log T_r > 2.5$ are rare if $N_{it} < 10^6$, and, for them, the diffusion is normal, because T_L does not depend, on average, on T_r. At $\log T_r > 2.5$ the dependence is algebraic with the power-law index $\beta \approx 1.5$–2.

Consider now an analytically more complicated problem, concerning the asteroidal motion in vicinities of the 3/1 mean motion resonance with Jupiter. Let us see how the T_L–T_r relationship looks like in this asteroidal problem, as considered in this statistical respect in Shevchenko (1998a). The analysis was made in the framework of the planar elliptic restricted three-body problem and was limited to asteroidal orbits with eccentricities less than 0.4. The computations were performed using the symplectic map developed in Wisdom (1983) especially to describe the dynamical vicinities of the 3/1 mean motion resonance.

We set Jupiter's perihelion to be at the origin of longitudes, i.e., its longitude of perihelion $\varpi_J = 0$. Jupiter's orbital semimajor axis is set to unity.

Let l be the mean longitudes of an asteroid, ϖ its longitude of perihelion, a and e its semimajor axis and eccentricity. The starting values for the asteroidal orbit are regularly distributed on the rectangular grid $0.48025 \le a_0 \le 0.48200$, $0.005 \le e_0 \le 0.050$, with the step in a_0 equal to 0.00005 and the step in e_0 equal to 0.005. For Jupiter, the eccentricity $e_J = 0.048$, and the initial value of its mean longitude l_J set to be zero. For the asteroid, $l_0 = \pi$ and $\varpi_0 = 0$.

This choice of l_0, ϖ_0 makes the set of starting values of the asteroidal motion representative (Wisdom 1983): almost every orbit in the phase space in the neighbourhood of the 3/1 Jovian resonance intersects the defined plane. The chosen rectangular grid totally covers the chaotic domain at $e_0 \le 0.05$ and includes neighbouring domains of regular motion. In the statistical analysis, the regular orbits are excluded.

The maximum Lyapunov exponents were computed on the time interval $n_{it} = 10^7$ each (in iterations of Wisdom's map), or less, if a jump in eccentricity was encountered. The map's iteration equals one Jupiter period. The jump was fixed if the asteroidal eccentricity exceeded 0.2. The resulting T_L–T_r relationship is shown in Fig. 4.6. As in the above considered cases of the separatrix and standard maps, the observed T_L–T_r statistical dependence is obviously in accord with the generic quadratic relationship expected for the motion near the separatrix of non-linear resonance.

An intriguing feature of the observed relationship consists in the existence of a group of chaotic orbits with unbounded recurrence time; these orbits do not exhibit eccentricity jumps, at least during the adopted time interval of computation. In Fig. 4.6, these are the points with $\log T_r = 7$. These orbits mostly have $\log T_L = 4.0$–4.2; therefore, they are definitely chaotic.

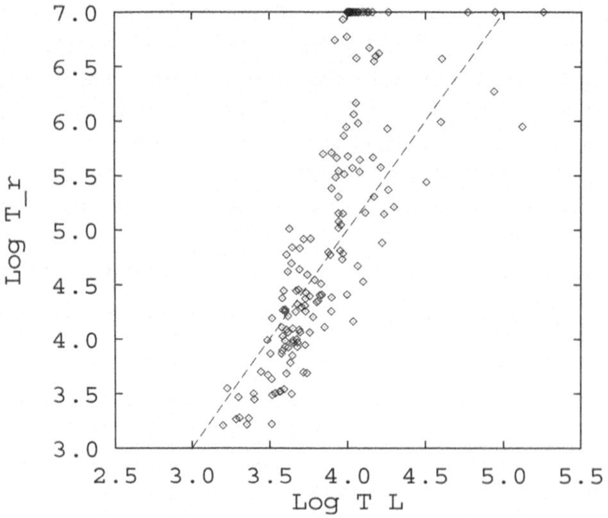

Fig. 4.6 A statistical T_L–T_r relationship for the chaotic asteroidal trajectories in the 3/1 Jovian resonance, as described in the text. Time is in Jupiter orbital periods. Straight dashed line: the generic quadratic relationship, shown for reference. The logarithmic scales are decimal (Figure 2 from Shevchenko 1998a, with permission from Elsevier)

Milani and Nobili (1992) identified chaotic asteroidal orbits without any prominent orbital changes (such as eccentricity jumps) in a study of long-term dynamics of the asteroid Helga. At a first glance, such a phenomenon seems to obviously contradict the theoretical T_L–T_r relationship. In reality, there is no contradiction; the matter is how the "sudden orbital change" is defined in any given model. One and the same definition should not be used when trajectories in the analysed statistical set belong to disconnected chaotic domains in the phase space of motion.

Another statistical manifestation of the critical behaviour was numerically observed by Shevchenko and Scholl (1996, 1997) in the statistical distributions of time intervals between the eccentricity jumps of chaotic asteroidal orbits in the 3/1 Jovian resonance. The distributions in the tails followed the power-law decay. Such kind of decay, according to Chirikov (1990), is inherent to the sticking regime of motion at chaos borders in the phase space. Thus, there are two known long-term effects in the statistics of sudden changes of chaotic asteroidal orbits. These are the power-law character of the T_L–T_r relationship and the power-law decay of the T_r distributions, with specific power-law indices. The both effects are plausibly explained as critical phenomena, arising due to sticking of orbits to chaos borders in the phase space. The presented above results of numerical simulations in the standard and separatrix map models unambiguously recover similar dynamical phenomena.

We have seen that the power-law indices of the observed algebraic dependences are not usually determined precisely, due to large scatter of the numerical results. An immanent distortion of any statistically evaluated power-law β index of the

T_L–T_r relationship is due to dominance of short recurrence times. In any accurate calculations of β, therefore, longer recurrences should be taken with greater weights (Shevchenko 1998a).

On the other hand, notwithstanding the large immanent scatter, the T_L–T_r relationship can be indeed usually used to statistically predict sudden orbital changes, if the initial small-T_r part of the relationship has been recovered numerically.

4.4 Two Kinds of Hamiltonian Intermittency

As shown in the previous Section, the close-to-quadratic relationship (4.21) between the recurrence and Lyapunov times in general Hamiltonian systems with divided phase space has a straightforward explanation, if one takes into account the stickiness phenomenon. The correlation arises because the trajectories sticking to chaos border have large recurrence times, i.e., the times of return to a specified domain of the phase space of motion, and, since they mimic regular orbits, they have also large computed Lyapunov times, which are local in nature.

Speaking in a more general framework, the emergence of the algebraic T_L–T_r relationship, Eq. (4.21), is due to the phenomenon of intermittency, i.e., chaotic behaviour intermittently interchanged with close-to-regular one. This kind of phenomena is well known in studies of dissipative dynamical systems (Lichtenberg and Lieberman 1992). Much less it is discussed in relation with Hamiltonian dynamics. There exist two kinds of Hamiltonian intermittency known to date.

The first one takes place in case of adiabatic chaos, when the motion is "stochastized" at encounters of a trajectory with the separatrix, while most of the time, spent far from the separatrix, the motion is regular-like (Zaslavsky et al. 1991). The separatrix in the simple model by Zaslavsky et al. (1991), as well as in the map models (1.45 and 1.46), is just the singular axis $y = 0$ separating bound and unbound types of motion. The first kind Hamiltonian intermittency is inherent to adiabatic chaos; see Shevchenko (2008a,b) and references therein.

The second kind Hamiltonian intermittency (Shevchenko 1998b) takes place in case of non-adiabatic chaos, when a fractal "chaos–order" boundary is present in the phase space of motion. Sticking of orbits to the chaos border results in the emergence of long time segments of a close-to-regular behaviour; these segments are separated by prominently chaotic dynamical events, arising when the trajectory leaves the border neighbourhood (Shevchenko and Scholl 1996, 1997; Shevchenko 1998a,b).

Therefore, the physical cause for the first kind Hamiltonian intermittency consists in encounters of a trajectory with the slowly pulsating separatrix, whereas the physical cause for the second kind Hamiltonian intermittency consists in the phenomenon of sticking of orbits to chaos border in the phase space of motion.

Statistical properties of the two kinds of Hamiltonian intermittency are very different, as demonstrated further on in this book. This is just the second kind Hamiltonian intermittency that is responsible for the emergence of the algebraic T_L–T_r relationship (4.21). In Sect. 11.1.8 we shall see that, in certain circumstances, a different T_L–T_r relationship can arise, namely, a quasilinear one, and the emergence of the quasilinear dependence is due to the first kind Hamiltonian intermittency.

Chapter 5
Extents of Chaotic Domains

Methods for estimating extents of chaotic domains in the phase space of Hamiltonian systems are potentially broadly applicable in problems of physics and dynamical astronomy. In this Chapter, both analytical and numerical approaches to measuring the widths of chaotic layers are described and discussed. We start with considering the separatrix split phenomenon; then proceed to a discussion of early heuristic estimates of the chaotic layer width. Then, modern methods for estimating the layer width in cases of non-adiabatic chaos and adiabatic chaos are described. Presence of marginal resonances may drastically affect the width in the both cases; therefore, the marginal resonance phenomenon is especially and separately discussed. The presentation is partially based (in Sections 5.4, 5.5, and 5.6) on the papers by Shevchenko (1998a, 2008a) (by permission from Elsevier) and Shevchenko (2012).

Analytical and numerical approaches to measuring the widths of chaotic layers have different merits and different demerits. An immanent shortcoming of any analytical approach consists in that it implies an idealization of the phenomenon under study, and, therefore, the obtained estimates are inherently approximate. The accuracy of the estimates is hard to evaluate, due to a number of approximations involved. On the other hand, numerical approaches to the problem are applicable in a rather narrow range of values of the controlling parameters: the numerical approaches usually cannot be used in case of low relative frequencies of perturbation, i.e., at the adiabaticity parameter values $\lambda \ll 1$ (due to limitations on the computation time), as well as in case of high relative frequencies of perturbation, i.e., at $\lambda \gg 1$ (because the chaotic layer widths are exponentially thin with the perturbation frequency), and also in case of tiny amplitudes of perturbation, i.e., at $\varepsilon \ll 1$, due to limitations on the arithmetic precision. Therefore, only analytical methods may provide the global picture. Their another advantage is that the analytical estimating is easy and rapid to accomplish, as soon as any theoretical model is shown to be valid. Finally, their most important advantage, perhaps,

I. I. Shevchenko, *Dynamical Chaos in Planetary Systems*, Astrophysics and Space Science Library 463, https://doi.org/10.1007/978-3-030-52144-8_5

consists in the physical insight that the analytical methods provide, making the role of each parameter clearly visible.

The widths of chaotic layers of interacting non-linear resonances are mostly controlled by the adiabaticity parameter λ (the ratio of the perturbation frequency to the frequency of small-amplitude phase oscillations on resonance). Chirikov (1979) derived formulas for the widths in case of $\lambda \gtrsim 1$ (non-adiabatic chaos). As follows from these expressions (which are discussed in the following), the chaotic layers are exponentially thin with increasing $\lambda \gg 1$. This means that the cases of intermediate and low relative frequencies of perturbation are most actual in applications.

As an example demonstrating the complexity of the problem of estimating extents of chaotic domains, let us consider the orbital dynamics of asteroids in vicinities of the 3/1 mean motion resonance with Jupiter (Shevchenko 2002b). The trajectories are computed by means of numerical integration of equations of motion in the planar elliptic restricted three-body problem Sun–Jupiter–asteroid. Wisdom's map (Wisdom 1983) is used as an integrator; Jupiter's eccentricity serves as the controlling parameter. The maximum Lyapunov exponent is used as an indicator of chaos. It is computed by means of the shadow trajectory method, illustrated in Fig. 2.1.

As usually designated, a, e, ϖ, l, and n are, respectively, an asteroid's semimajor axis, eccentricity, longitude of perihelion, mean longitude, and mean motion. Jupiter's orbital elements are marked with index "J." We set $a_J = 1$, and the starting mean longitude $l_{J0} = 0$. Jupiter's mass in Solar units is set equal to $1/1047.355$.

By means of massive computations of the maximum Lyapunov exponents, the chaotic domains are identified on a representative plane of starting values. The plane is defined, following Wisdom (1983), as that satisfying the conditions $\varpi = \varpi_J = 0$ and $\varphi = \pi$, where $\varphi = l - 3l_J$ is the resonance phase. Wisdom (1983) argued that almost each zero inclination orbit, that has initial conditions close to the 3/1 mean motion resonance, crosses this plane. In the plots, the starting values a_0 and e_0 on the representative plane are denoted simply by a and e.

Approximate charts of chaotic domains in the 3/1 resonance problem were obtained also in Wisdom (1983) and Murray and Fox (1984). Here the charts are computed with much higher resolution, allowing one to compare the extents of chaotic domains at various values of Jupiter's eccentricity. The resolution of the charts in Figs. 5.1, 5.2, and 5.3 is $5 \cdot 10^{-6}$ in the semimajor axis a and $2.5 \cdot 10^{-4}$ in the eccentricity e. The representative plane of starting values is limited from above, $e \leq 0.1$, because Wisdom's map is applicable at $e < 0.3$–0.4 (Wisdom 1983).

The chaotic orbits are statistically distinguished by the method of "movable peaks" (in histograms of computed finite-time Lyapunov exponents), described above in Sect. 2.6. The identified chaotic domains are presented in Figs. 5.1, 5.2, and 5.3. The three plots correspond, respectively, to three fixed Jovian eccentricities (the minimum, current, and maximum ones). The starting data for the orbits demonstrating jumps of eccentricity (e exceeds 0.2) are marked in black. Those without jumps are marked in grey.

A comparative inspection of extents of the chaotic domains in Figs. 5.1, 5.2, and 5.3 allows one to conclude that, contrary to any common sense expectations,

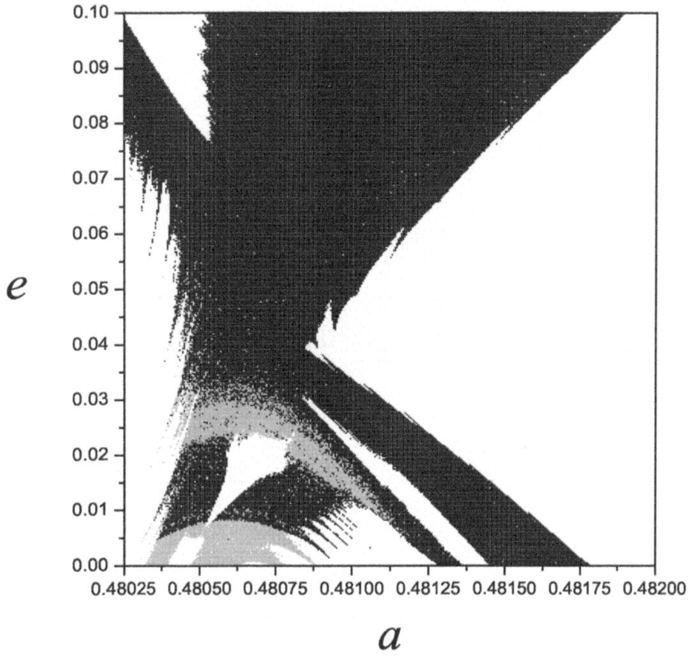

Fig. 5.1 Vicinities of the 3/1 mean motion resonance with Jupiter at Jovian eccentricity fixed at $e_J = 0.030$; a and e are the starting values of asteroidal semimajor axis (in Jovian units) and eccentricity. Chaotic orbits with jumps of eccentricity are marked in black; chaotic orbits without such jumps ($e < 0.2$) are marked in grey (Figure 1a from Shevchenko 2002b)

the maximum chaotic extents are not at all observed when the Jovian eccentricity is maximal. What is more, the violently chaotic asteroidal orbits (those exhibiting eccentricity jumps) are also not at all most abundant when the Jovian eccentricity is maximal.

From Figs. 5.1, 5.2, and 5.3, we see that, during the complete period of slow variation of Jupiter's eccentricity, the extents of the asteroidal chaotic domain in the given example of 3/1 resonance vary substantially, by the order of two. This demonstrates why any ability to analytically predict extents of chaotic domains in applications can be important and insightful.

5.1 The Separatrix Split

The separatrix split was first ever revealed and described in (Poincaré 1899). Poincaré outlined the arising complicated scheme verbally, but did not try to depict it as a graph or a sketch. He wrote: "When one tries to depict the figure formed by these two curves and their infinity of intersections, each of which corresponds

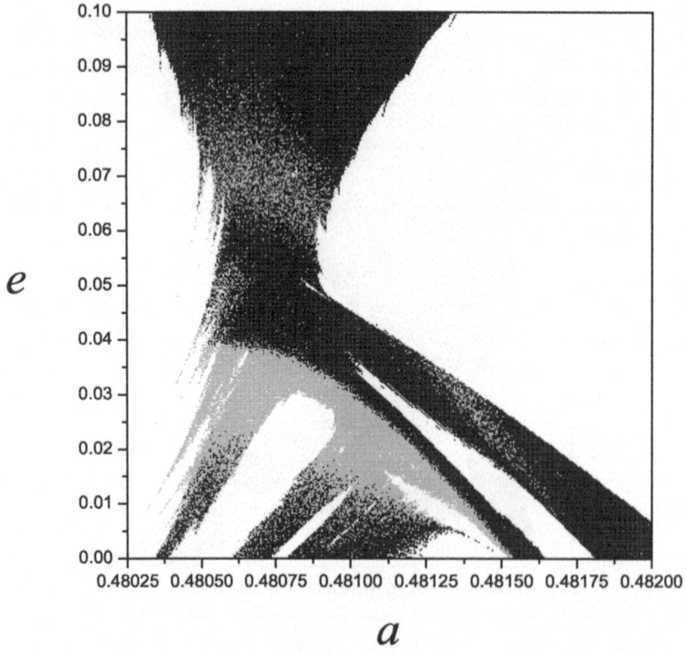

Fig. 5.2 The same as Fig. 5.1, but at $e_J = 0.048$ (Figure 1b from Shevchenko 2002b)

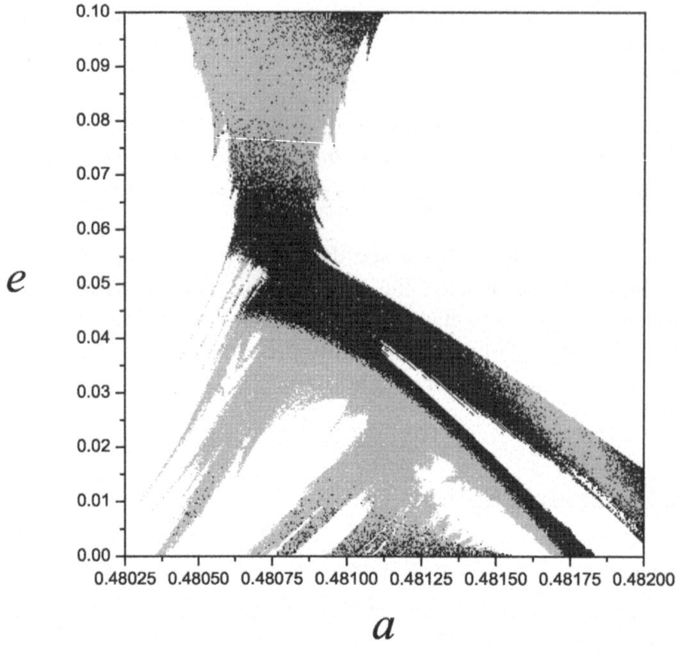

Fig. 5.3 The same as Fig. 5.1, but at $e_J = 0.060$ (Figure 1c from Shevchenko 2002b)

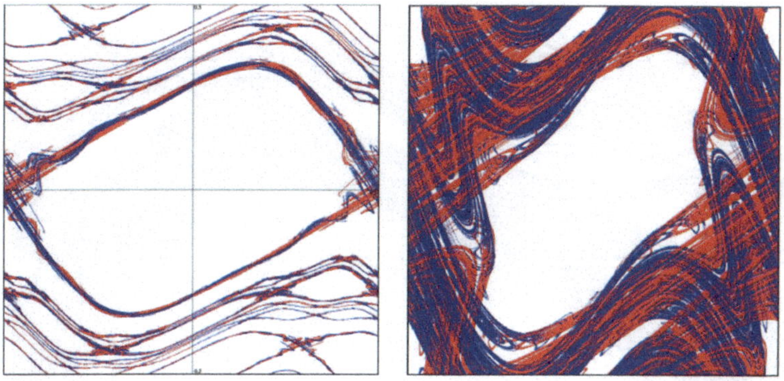

Fig. 5.4 Stable and unstable manifolds for the standard map at the stochasticity parameter $K = 1$ (left panel) and $K = 2$ (right panel) (Figure 7 from Meiss 2008, by permission from Springer Nature, © 2008)

to a doubly asymptotic solution, these intersections form a kind of net, web or infinitely tight mesh; neither of the two curves can ever cross itself, but must fold back on itself in a very complex way in order to cross the links of the web infinitely many times. One is struck by the complexity of this figure that I am not even attempting to draw" (Poincaré 1899). (The English translation is presented as given in Meiss (2008).) The splitting structure, envisaged by Poincaré (1899), was first ever drawn as a sketch much later on by Melnikov (1963). Nowadays, it can be straightforwardly constructed using relevant computer codes; see resulting examples is Fig. 5.4, where splitted separatrices (stable and unstable manifolds) of resonances of the standard map at the stochasticity parameter values $K = 1$ and $K = 2$ are shown. The left panel of Fig. 5.4 presents manifolds for the integer, half-integer, one-third-integer, etc., resonances, whereas the right panel solely the manifolds for the integer resonance.

In the article "Visual explorations of dynamics: The standard map," Meiss (2008) gives a nice example of a phase portrait of the standard map (1.23). This phase portrait is reproduced here in Fig. 5.5. It is obtained by continuous iterations of the standard map (1.23), with randomly selected starting conditions. The stochasticity parameter K is set to $K = 0.971635406$, representing an approximation of its exact critical value K_{G}. In the portrait, the chaotic layers corresponding to integer, half-integer, and one-fifth-integer resonances are shown in different colours and are straightforwardly identifiable.

Comparing Fig. 5.4, left panel, and Fig. 5.5 makes it obvious that the separatrix split is much less than the corresponding chaotic layer's physical width, no matter which resonance one considers. Indeed, the layer width at $\lambda \gg 1$ is $\sim \lambda$ times greater than the split δ; see, e.g., Zaslavsky (1998, 2005).

The $\lambda \gg 1$ case just corresponds to Figs. 5.4 and 5.5, because $\lambda = 2\pi/\sqrt{K}$, see Eq. (1.25). On the other hand, at $\lambda \ll 1$, the layer width is $\sim 1/\lambda$ times greater than the separatrix split, as explained below.

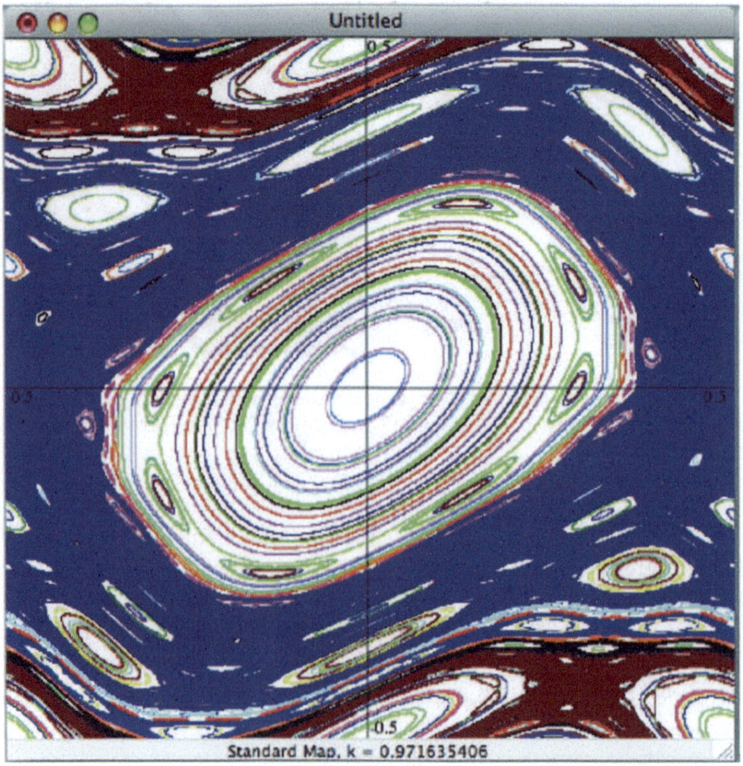

Fig. 5.5 The phase portrait of the standard map at $K = 0.971635406$ (Figure 1 from Meiss 2008, by permission from Springer Nature, © 2008)

5.2 Early Estimates of the Chaotic Layer Width

Extents of chaotic domains, and, in particular, widths of chaotic layers, are among the most important and valuable characteristics of the chaotic motion of Hamiltonian systems. Various aspects of the problem of analytical estimating widths of chaotic layers were considered in Zaslavsky and Filonenko (1968), Chirikov (1978), Chirikov (1979), Zaslavsky (1984), Zaslavsky and Sagdeev (1988), Neishtadt (1986), Sagdeev et al. (1991), Elskens and Eskande (1991), Lazutkin (1991), Ahn et al. (1996), Yamagishi (1995), Treshchev (1998), Shevchenko (1998b), Vecheslavov (2004), Shevchenko (2008a), Soskin and Mannella (2009).

Already at the very beginning of studies on this subject, it became clear that the $\lambda \ll 1$ and $\lambda \gg 1$ cases are dynamically very different and, therefore, require different and separate kinds of analysis.

Usually the width is measured in terms of the unperturbed energy of the motion on the separatrix, i.e., the width $w_\mathrm{b} \equiv \Delta E/E_{\mathrm{s}x}$. In the perturbed pendulum model, an exponential decrease $\Delta E/E_{\mathrm{s}x} \sim \exp(-A\lambda)$ (where $A = \mathrm{const}$) with λ at $\lambda \gg 1$

was advocated in Zaslavsky and Filonenko (1968) and derived rigorously later on in Chirikov (1979).

In the adiabatic $\lambda \ll 1$ and intermediate $\lambda \sim 1$ cases of the paradigmatic perturbed pendulum model (1.6), several analytical estimates were proposed:

- If $\lambda \lesssim 1$, then $w_{\mathrm{b}} \sim \lambda\varepsilon$ (Zaslavsky and Filonenko 1968; Zaslavsky 1984; Zaslavsky and Sagdeev 1988; Sagdeev et al. 1991).
- If $\lambda \ll 1$, then $w_{\mathrm{b}} \sim \lambda\varepsilon$ (Zaslavsky et al. 1991).
- If $\lambda \sim 1$, then $w_{\mathrm{b}} \sim \varepsilon$ (Zaslavsky 1998, 2005).
- If $\lambda \ll 1$, then $w_{\mathrm{b}} \sim \varepsilon$ (Neishtadt 1986; Elskens and Eskande 1991).

Here ε is the relative strength of perturbation, see Eq. (1.6).

The second estimate $w_{\mathrm{b}} \sim \lambda\varepsilon$ at $\lambda \ll 1$ is inadequate. It was obtained in (Zaslavsky et al. 1991, Equation (3.33)) in the same assumption as in Zaslavsky and Filonenko (1968); Zaslavsky (1984); Zaslavsky and Sagdeev (1988); Sagdeev et al. (1991), namely, in the assumption that the layer width is, by the order of magnitude, is equal to the amplitude of variation of the model pendulum energy on the period of rotation (or, equivalently, half-period of libration) of the model pendulum. This period is just the separatrix map period, and the amplitude is just the parameter W of the separatrix map (1.26). In the adiabatic limit $\lambda \to 0$ this assumption fails: it underestimates the width by a factor of λ. The physical reason why the W amplitude (the Melnikov–Arnold integral) underestimates the layer width will be clear from the general analysis given below in the subsequent Sections.

If the "$\lesssim$" symbol is interpreted as "any number less than approximately," i.e., it characterizes the values less than "~ 1," then the first estimate is inadequate, because it is thus reduced to the second estimate.

Note that a general formula for the layer width given in (Zaslavsky et al. 1991, Equation (3.37)) underestimates the width at $\lambda \ll 1$ even by a larger factor (by λ^2), because it represents a generalization of the formula $w_{\mathrm{b}} \approx \lambda W$, valid solely at $\lambda \gg 1$, to the whole range of perturbation frequencies, from zero to infinity.

The third estimate is just an equation that follows from the first two estimates if one sets $\lambda \sim 1$.

The fourth (valid) estimate is a trivial conclusion following from considering the slowly pulsating separatrix in the original coordinate–momentum phase space of the perturbed pendulum model of non-linear resonance.

5.3 The Layer Width in Case of Non-adiabatic Chaos

Let us consider the case of non-adiabatic chaos, i.e., the case $\lambda \gg 1$. First we explore the problem numerically, in numerical experiments with the separatrix map (1.31).

The chaotic layer border y_{b}, corresponding to the maximum energy deviation (from the unperturbed separatrix) of a chaotic trajectory inside the layer, can be obtained as the maximum of $|y_i|$ achieved by a single chaotic trajectory, if the number of iterations of the map is large enough. The λ dependence of the chaotic

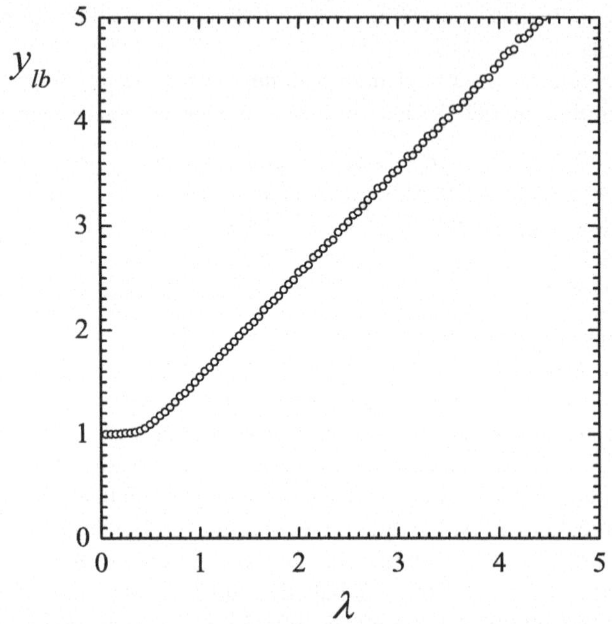

Fig. 5.6 The λ dependence of the chaotic layer half-width $y_{\rm lb}$ in case of the least perturbed border (Figure 1 from Shevchenko 2008a, with permission from Elsevier)

layer half-width $y_{\rm lb}$ is, therefore, constructed in the following way (Shevchenko 2008a).

At each step in λ, the value of c corresponding to the minimum $y_{\rm b}$ is found (on a fine grid exposed over the $0 \leq c < 2\pi$ interval) and the minimum $y_{\rm b}$ is plotted against the current λ value. This $y_{\rm b}$ value is denoted henceforth as $y_{\rm lb}$. It corresponds to the case of the least perturbed border, because, by the given procedure, the role of marginal resonances is minimized.

An initial part of the λ–$y_{\rm lb}$ relationship, computed in this way, is given in Fig. 5.6. The observed dependence clearly follows the piecewise linear law

$$y_{\rm lb} \approx \begin{cases} 1, & \text{if } 0 \leq \lambda \leq 1 - a, \\ \lambda + a, & \text{if } \lambda > 1 - a, \end{cases} \tag{5.1}$$

where $a \approx 1/2$. Indeed, at $\lambda \in [0.5, 10]$, by linear fitting the numerical data in Fig. 5.6 one finds $y_{\rm lb}(\lambda) = a + b\lambda$ with $a = 0.5351 \pm 0.0041$ and $b = 1.0059 \pm 0.0007$ (Shevchenko 2008a).

We see that, at $\lambda > 1/2$, the λ dependence of $y_{\rm lb}$ is a sum of two addends. The first addend, a, is the half-amplitude of the last rotational invariant curve of the standard map at the critical value of the stochasticity parameter $K = K_{\rm G} = 0.971635406\ldots$, because, at $\lambda \gg 1$, the separatrix map can be linearized in y and locally approximated by the standard map. The half-amplitude of the last rotational invariant curve of the standard map can be numerically estimated to be ≈ 0.508, in accord with the observed $a \approx 0.535$. The second addend, $b\lambda$, is the border y value averaged over the phase $x \in [0, 2\pi]$. The chaotic layer half-width at $\lambda \gg 1$ approximately equals λ, see Chirikov (1979); therefore the second addend is also in accord with the existing theory.

The constant a component, conditioned by the border curve bending, can be eliminated by subtracting the bending half-amplitude. The resulting quantity is designated henceforth by $\bar{y}_{\text{lb}}$. At $\lambda \geq \frac{1}{2}$, one has $\bar{y}_{\text{lb}} \approx \lambda$. At $\lambda \leq \frac{1}{2}$, $\bar{y}_{\text{lb}}$ equals to the maximum y divided by two, as follows from the border curve shape revealed in the next Section. Therefore, for the time-averaged half-width one has

$$\bar{y}_{\text{lb}} \approx \begin{cases} \frac{1}{2}, & \text{if } 0 \leq \lambda \leq \frac{1}{2}, \\ \lambda, & \text{if } \lambda > \frac{1}{2}. \end{cases} \tag{5.2}$$

The qualitative change in character of the y_{lb}–λ relationship at $\lambda \approx 1/2$ manifests a sharp distinction between two different types of dynamics, namely, *adiabatic chaos* and *non-adiabatic chaos*. Adiabatic chaos can be interpreted as due to sporadic encounters of the orbit with the $y = 0$ singular line (this mechanism of chaotization was originally evoked in studies of the so-called "relativistic" map in Zaslavsky et al. 1991). Non-adiabatic chaos is naturally explained as due to resonance overlap phenomenon (Chirikov 1979).

However note that the border $\lambda \approx 1/2$ (Shevchenko 2008a) between adiabatic and non-adiabatic chaos does not coincide with the borderline between the cases of overlapping and non-overlapping of resonances: the latter borderline lies much higher in λ. For example, in the standard map phase space the integer resonances start to overlap (on decreasing λ) at $\lambda \approx 2\pi/0.97 \approx 6.5$ (Chirikov 1979).

As already mentioned above, the linear λ dependence for the chaotic layer width at $\lambda \gg 1$ was derived in Chirikov (1979) by means of approximating the separatrix map by the standard map locally in y. In a rigourous mathematical study by Ahn et al. (1996), a lower bound for the chaotic layer width was established; see Equation (5.8) in Ahn et al. (1996). In our notations, this lower bound can be represented as $y_{\text{b}} > \frac{3}{4}\lambda$, in accord with Chirikov's estimate and Eq. (5.2).

The time-averaged half-width can be expressed through the original energy variable w; this gives

$$\overline{w}_{\text{lb}} = |W|\bar{y}_{\text{lb}} = \frac{4\pi|\varepsilon|\lambda^2\bar{y}_{\text{lb}}}{\sinh\frac{\pi\lambda}{2}}, \tag{5.3}$$

where $\bar{y}_{\text{lb}}$ is given by Eq. (5.2), and $\lambda \gtrsim 1$.

5.4 The Layer Width in Case of Adiabatic Chaos

Formula (5.3) is valid for any frequency of perturbation in system (1.6) with $k = 1$, $a = b$, provided that the separatrix map correctly describes the behaviour of the original system. At high relative frequencies of perturbation, $\lambda \gg 1$, expression (1.29) for the W parameter needs correction, see Sect. 3.4. At low relative frequencies of perturbation, $\lambda \ll 1$, formula (5.3) demonstrates that the chaotic layer width, expressed in w, decreases linearly with λ (as soon as the case of the least perturbed border is considered and, therefore, ε is not fixed).

If one fixes ε, the low frequency limit of the width is a non-zero constant. This is a trivial consequence of the *slowly pulsating separatrix* description for adiabatic chaos; see Neishtadt (1986), Bruhwiler and Cary (1989). Indeed, the Hamiltonian (1.6) with $k = 1$, $a = b$ can be naturally rendered in the form of a pendulum with modulated frequency of small-amplitude oscillations:

$$H = \frac{\mathcal{G}p^2}{2} - (\mathcal{F} - 2a\cos\tau)\cos\varphi, \tag{5.4}$$

see Chirikov (1979). Considering the relative full energy $w_H = \frac{H}{\mathcal{F}} - 1$ instead of usual $w = \frac{H_0}{\mathcal{F}} - 1$, a simple heuristic estimate of the chaotic layer width in the limit $\lambda \to 0$ can be given. Indeed, as representation (5.4) implies, the w_H energy on the slowly pulsating separatrix varies in the limits from -2ε to 2ε; therefore, the layer half-width is equal to 2ε. This quantity includes the amplitude of the chaotic layer bending, described above in Sect. 1.8.

In case of $\lambda \ll 1$, the diffusion across the layer is slow, and on a short time interval the phase point follows close to some current curve, called the *guiding curve*, and using equations for this curve, formulas for the layer width can be derived as described above in Sect. 3.2. The half-widths y_{b} and w_{b}, given by formulas (3.16) and (3.17), represent the *maximum* relative energy deviations inside the layer. The *time-averaged* half-widths $\bar{y}_{\mathrm{b}}$ and $\overline{w}_{\mathrm{b}}$, as follows from the geometry of the boundary curve, Eq. (3.15), are two times less.

In Vecheslavov (2004), a different approach was used for reducing the separatrix map (1.26) to a differential equation, when chaos is adiabatic; it was assumed that the increments of the map (1.26) variables were small at *each* iteration. Our representation (1.31) of the separatrix map makes clear that this can be satisfied only if $y \gg 1$, therefore, $y_{\mathrm{b}} \gg 1$. Then it follows from Eq. (3.16) that c should be close to $\approx 0 \bmod 2\pi$.

Let us assess the conditions at which our approximation (3.15) is valid. In deriving the increment in x, the term $\lambda \ln |y_{i+1}|$ in the second equation of Eqs. (1.31) was neglected, therefore, the inequality $c \bmod 2\pi \gg \lambda \ln |y_{\mathrm{b}}|$ should hold. One has

$$c \bmod 2\pi \gg \lambda \ln \left|\operatorname{cosec}\frac{c}{2}\right|. \tag{5.5}$$

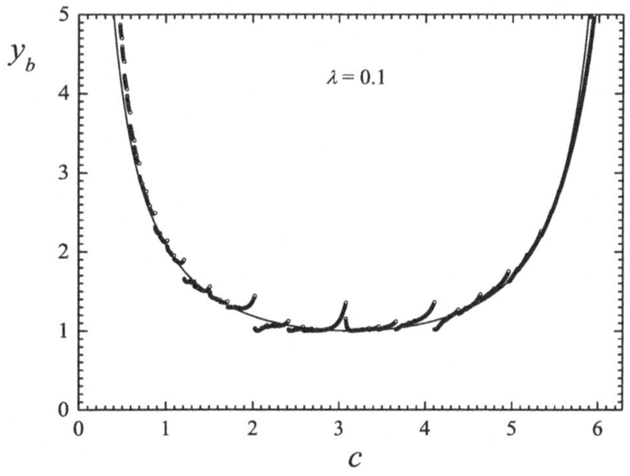

Fig. 5.7 Dependence of y_b on c at $\lambda = 0.1$. Circles: numerical-experimental data. Solid curve: theory given by Eq. (3.16) (Figure 2 from Shevchenko 2008a, with permission from Elsevier)

This means that at $\lambda \ll 1$, the value of c should be far enough from $c \approx 0 \bmod 2\pi$. What is more, c should not correspond to other resonances, the role of which, however, is much less than the main one corresponding to zero c.

Theoretical width (3.16) is confronted with available numerical-experimental data in Fig. 5.7. Numerically, the width was found by measuring maximum relative energy deviations for orbits of the map (1.31). The theoretical curve closely follows the numerical data at all values of c except resonant ones, where narrow discontinuities are observed.

The latter discontinuities arise due to emergence of regular islands inside the chaotic layer at resonant values of c. These disturbances are analogous to those observed in the behaviour of the standard map. In the latter case, analogous local disturbances emerge in the K dependence of the maximum Lyapunov exponent, and they are conditioned by local depressions in the measure of the phase space chaotic component, also due to appearance of regular islands, see Fig. 3.1.

The theory deviations from numerics are most prominent near the main resonance, i.e., at values of c close to $0 \bmod 2\pi$. If λ and c are both close to zero, the relative increments of w and τ in Eqs. (1.26) are both small. Then, the already mentioned above approach by Vecheslavov (2004) becomes applicable. Equations (1.26) reduce to the differential equation

$$\frac{dw}{d\tau} = -\frac{W \sin \tau}{\lambda \ln \frac{32}{|w|}}, \tag{5.6}$$

analogous to a similar one derived in Vecheslavov (2004), except that homogeneous variables are used here and the condition on c is taken into account.

Analogously to Eq. (3.15), the given Eq. (5.6) describes a guiding curve with an arbitrary constant of integration $\mathcal{C}$:

$$w \ln \frac{32e}{|w|} = \frac{W}{\lambda}(\cos \tau + \mathcal{C}). \tag{5.7}$$

As in deriving Eq. (3.16), the expression for chaos border follows from the condition for curve (5.7) to cross the $w = 0$ axis. Then, the constant of integration for the chaotic layer boundary curves is $\mathcal{C} = \pm 1$. The formula for the layer half-width is consequently given by

$$w_b \ln \frac{32e}{w_b} = \frac{2|W|}{\lambda}. \tag{5.8}$$

An expression for the W parameter, which can be used here in the given case of $k = 1$, follows from Eq. (1.29). If $\lambda \ll 1$, then $W \approx 8\varepsilon\lambda$, and

$$w_b \ln \frac{32e}{w_b} = 16|\varepsilon|. \tag{5.9}$$

Hence, at $\lambda \ll 1$, w_b depends solely on ε, i.e., the λ dependence has expired. For different k values, formulas for W would be different, as presented in Shevchenko (2000a), Vecheslavov (2004); however, the λ dependence in Eq. (5.8) expires all the same.

Formulas (3.16) and (3.17) for the chaotic layer half-width are applicable in case of generic values of the c parameter (excluding a vicinity of 0 mod 2π); and Eqs. (5.8) and (5.9) are applicable, conversely, in the vicinity of $c \approx 0 \bmod 2\pi$. In the first case, the layer width depends on both λ and ε, whereas in the second case, the λ dependence expires and the width depends solely on ε. When used in any application, Eqs. (5.9) and (5.8) can be easily numerically solved by iterations.

The performance of the analytical theory versus numerical data is illustrated in Figs. 5.8 and 5.9.

Numerically, the chaotic layer half-width was measured by two methods (Shevchenko 2008a). The first one was proposed in Chirikov (1978, 1979) and developed and extensively used in Vecheslavov (2004). The numerical procedure consists in calculating the minimum period T_{min} of motion inside the layer; then, the layer half-width is estimated by the formula

$$w_b = 32 \exp(-\omega_0 T_{min}) \tag{5.10}$$

(Chirikov 1978, 1979; Vecheslavov 2004). The maximum deviation in energy from the unperturbed separatrix is thus obtained, because it just corresponds to the minimum period of motion. Formula (5.10) directly follows from the second line of the map (1.26).

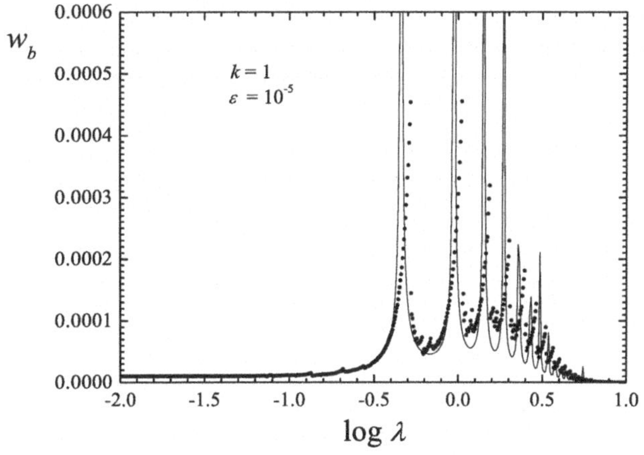

Fig. 5.8 The chaotic layer half-width w_b, in dependence on λ, for system (1.6) with $k = 1$ and $a = b$. Dots: results of direct computation by the first method. Solid curve: theory given by Eq. (3.17). Logarithms are decimal (Figure 3 from Shevchenko 2008a, with permission from Elsevier)

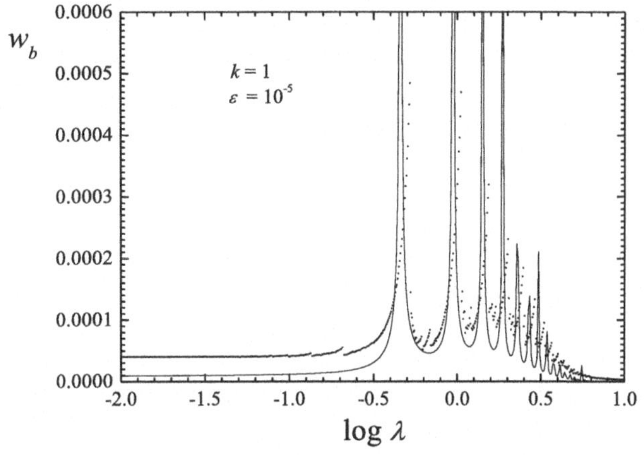

Fig. 5.9 The same as Fig. 5.3, but the half-width w_b is computed by the second method, i.e., as the maximum energy deviation. Logarithms are decimal (Figure 4 from Shevchenko 2008a, with permission from Elsevier)

A more precise method consists in direct continuous measuring of the relative energy deviation from the unperturbed separatrix $w = \frac{H_0}{\mathcal{F}} - 1$ in the course of numerical integration. By fixing the maximum deviation, one finds the layer size.

The direct integrations of the equations of motion were performed using the Dormand–Prince integrator (Hairer et al. 1987). For $k = 1$, $a = b$, and $\varepsilon = 10^{-5}$, the results of w_b estimating by the first method are graphically presented in Fig. 5.8, and the results obtained by the second method are given in Fig. 5.9.

The theory and the numerical-experimental data are generally in accord, even sharp variations are qualitatively reproduced. Theoretical dependence (3.17) approximately reproduces the numerical-experimental one even at $\lambda > 1$. However, Eq. (3.17) is applicable at low and intermediate relative frequencies of perturbation, i.e., at $\lambda < 1$ and $\lambda \sim 1$. At high relative frequencies of perturbation, i.e., at $\lambda > 1$, a different formula, namely Eq. (5.3), which describes the width averaged over c, should be used.

A notable feature of the constructed λ dependences consists in sharp narrow peaks. They are conditioned by encountering resonances $c \approx 2\pi m$, $m = 1, 2, \ldots$, on varying λ. Equations (3.16) and (3.17) predict that at such values of c the width goes to infinity; in reality, the width is of course finite. The λ locations of the peaks are approximately specified by the equation

$$c = \lambda \ln \frac{4}{|\varepsilon|\lambda} = 2\pi m. \tag{5.11}$$

The locations are practically insensitive to the k value. If $|\varepsilon| \ll \lambda$, they can be approximately found as $\lambda_m \approx -2\pi m / \ln \varepsilon$. On decreasing ε, the peaks move slowly to the left.

The abscissas λ_m of the peaks in Figs. 5.8 and 5.9 all exceed 0.5; this means that they are situated in the domain of non-adiabatic chaos. Therefore, the locations of the peaks are expected to be better described by the *tangency condition* (Shevchenko 1998b, 2012) for emergence of integer *marginal resonances*, i.e., integer resonances at the borders of the layer.

At $\lambda \gtrsim 1$, the emergence of prominent (integer and half-integer) marginal resonances cause sporadic strong variations of the relative energy w of individual chaotic orbits (Chirikov 1979; Shevchenko 1998b, 2012). Such violent behaviour becomes possible, when, with variation of a parameter, say, λ or W, an outer border (any of the two outer borders) of the main chaotic layer starts to overlap with the narrow near-separatrix chaotic layer of a prominent (say, integer) resonance; i.e., a *heteroclinic connection* (Meiss 1992) emerges between the main layer and the latter layer. Since the latter layer is relatively narrow, the connection condition is approximately equivalent to a "tangency" of the marginal resonance's unperturbed separatrix and the main layer border. The tangency condition is

$$W = W_{\mathrm{t}}^{(m)}(\lambda) \tag{5.12}$$

(Shevchenko 1998b), where W at $k = 1$ and $a = b$ is given by Eq. (1.29) and

$$W_{\mathrm{t}}^{(m)}(\lambda) = \frac{32}{\lambda^3}\left(\left(1+\lambda^2\right)^{1/2} - 1\right)^2 \exp\left(-\frac{2\pi m}{\lambda}\right). \tag{5.13}$$

Equation (5.12) can be solved numerically, and it does provide good accuracy for the λ locations of the first five peaks $\log \lambda_m$ ($m = 1, 2, \ldots, 5$), with deviations not more than 0.02 (Shevchenko 2008a). Conversely, Eq. (5.11), generically valid in the domain of adiabatic chaos, may serve here only as an extrapolation and, no wonder, is less accurate; it provides the deviations up to 0.06.

The tangency condition can be also employed for analytical estimating the height of the peaks; see Equation (10) in Shevchenko (1998b).

Whereas the integer resonances ($m = 1, 2, \ldots$) manifest themselves in the plots of Figs. 5.8 and 5.9 as the peaks, the $m = 0$ resonance produces, at $\lambda \to 0$, an asymptotic horizontal plateau. The cause for the plateau emerging at small λ is clear: if one fixes ε, then, no matter how small this fixed value is, with λ decreasing the value of $c = \lambda \ln(32/|W|)$ (Eq. (1.32)) also decreases and inevitably approaches $c \approx 0$. The point of transition to the main resonance domain corresponds to intersection between the curve given by formula (3.17) (with $\lambda \to 0$ this curve asymptotically goes down to zero) and the horizontal line given by Eq. (5.8).

The plateau was first ever identified and discussed in Vecheslavov (2004), and an approximate heuristic formula for its asymptotic height in energy was proposed:

$$w_{\mathrm{b}}/\varepsilon \approx 0.22 \cdot 8 = 1.76 \tag{5.14}$$

in case of $k = 1$, see Equation (14) in Vecheslavov (2004); a misprint (missing ε) is corrected here. This estimate is in approximate accord with the plateau heights in Figs. 5.8 and 5.9, with an accuracy of a factor of 2. Equation (5.9) is potentially much more precise: in case of the considered plots, it gives $w_{\mathrm{b}}/\varepsilon = 1.00142$, perfectly close to the numerical-experimental height, which is equal to 1.00585.

Formula (3.17) gives $w_{\mathrm{b}}/\varepsilon = 0.915$, also in quite good accord with the numerics, although c is close to zero.

In Fig. 5.9, the λ–w_{b} relationship, constructed by the second method, at $\lambda \lesssim 0.4$ goes notably higher than the theoretical curve. This is due to the chaotic layer bending phenomenon, considered in Sect. 1.8. This geometrical phenomenon is absent in Fig. 5.8, because, when the first method is used, the bending averages out. The relative energetic amplitude of bending at $k = 1$ in the limit $\lambda \to 0$ at the section of phase space $\varphi = 0 \bmod 2\pi$ is equal to 4ε, see Sect. 1.8. The numerically found value of $w_{\mathrm{b}}/\varepsilon$ at $\lambda = 0.01$ is equal to 4.00158, in perfect agreement with the theoretical prediction; the deviation is only 0.04%.

The plots in Figs. 5.8 and 5.9 concern the $k = 1$ case. Systems with different k can be considered in a similar way. To obtain a λ–w_{b} theoretical dependence, one should simply use a relevant expression for W in Eq. (3.17). For example, in case of $k = 2$ and $a = b$,

$$W = \frac{8\pi\varepsilon\lambda^2(\lambda^2 - 2)}{3\sinh\frac{\pi\lambda}{2}} \tag{5.15}$$

(Shevchenko 2000a, Vecheslavov 2004).

The role of resonances can be illustrated by means of constructing phase space sections. Integer resonances $m = 0, 1, 2, \ldots$ result in stretching the layer in y. The border motion in these cases is quite simple: it follows the guiding curves (5.7). The role of fractional resonances is more intricate. The phase portrait of the separatrix map (1.31) in case of the fractional resonance with winding number $Q = 4/5$ is

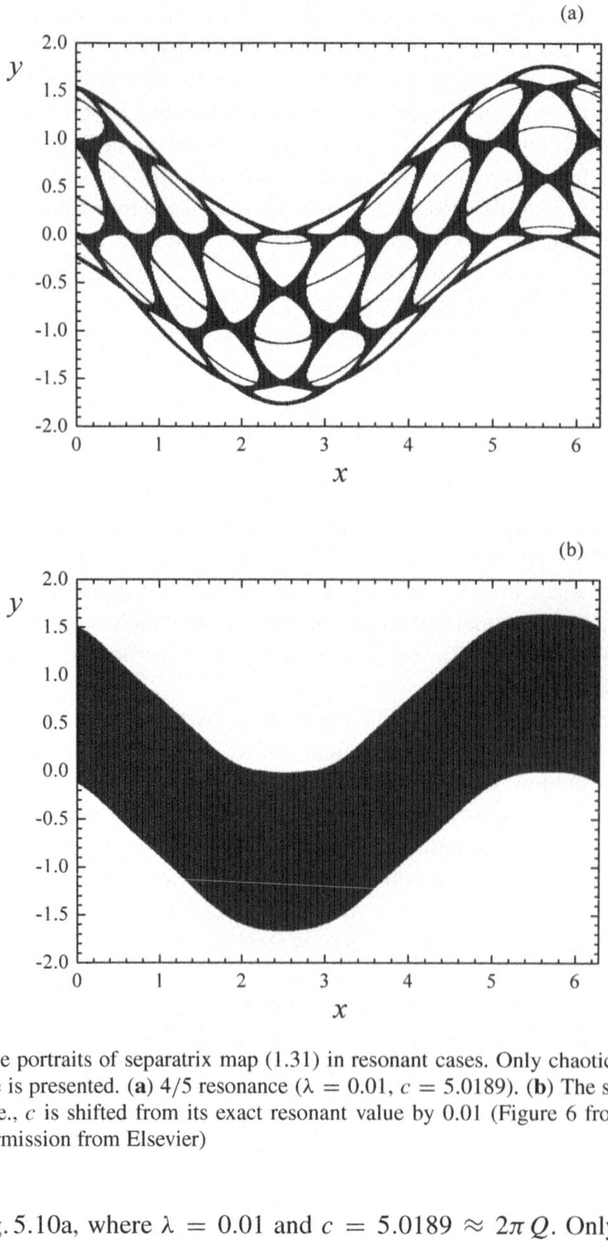

Fig. 5.10 Phase portraits of separatrix map (1.31) in resonant cases. Only chaotic component of the phase space is presented. (**a**) 4/5 resonance ($\lambda = 0.01$, $c = 5.0189$). (**b**) The same as (**a**), but $c = 5.0289$, i.e., c is shifted from its exact resonant value by 0.01 (Figure 6 from Shevchenko 2008a, with permission from Elsevier)

shown in Fig. 5.10a, where $\lambda = 0.01$ and $c = 5.0189 \approx 2\pi Q$. Only the chaotic component is shown. The choice of c corresponds to the minimum measure of the chaotic component inside the layer.

In Fig. 5.10a, a "porous" structure of the layer is evident. The main pattern is formed by 5 curves of sinusoidal form, embedded in narrow bands of generic chaos. These curves are nothing but the singular curve $y = -\sin x$ and its four consecutively iterated images. The plot illustrates how resonant structures with large amount of inner regular component are formed: in case of resonance with the winding number equal to rational number p/q, the $q+1$th image of the singular curve $y = -\sin x$ coincides with the initial singular curve exactly; therefore, the bands of generic chaos in the neighbourhood of the singular curves are not broadened; this results in the presence of large inner regular component.

At the given values of parameters, any small ($\approx$0.002 is enough) positive or negative shift in c leads to complete visual ergodicity of the motion inside the layer; therefore, in Fig. 5.10b the layer turns totally black. However, the proximity to resonance still determines the form of the outer borders. On further shifting c away from the resonance, this influence decays, and the borders become more and more close in form to the guiding curve given by Eq. (3.15).

In Fig. 5.10b, the value of the c parameter is shifted away from the resonant one by 0.01. No regular islands are seen inside the layer; however, this visual impression is deceptive: tiny islands are always present and can be revealed by implementing especial numerical techniques (Shevchenko 2008a), such as massive computation of Lyapunov exponents on fine grids of initial data.

Tiny peaks in numerical data in Figs. 5.8 and 5.9 are due to fractional resonances. On decreasing λ, no matter what the ε value is, one reaches the plateau corresponding to the $m = 0$ resonance.

Instead, if λ is fixed and ε is decreased, one finds much more intricate behaviour. According to formula (1.32), no matter how small the λ value is, one can achieve any value of c by diminishing ε. In other words, the complete set of resonances is traversed once and once again, if ε is steadily decreased. At $\lambda \ll 1$ (in fact, already at $\lambda = 0.01$) the encounters with prominent resonances take place at microscopic ε values. Setting $c = 2\pi(Q + m)$ (where $m = 0, 1, 2, \ldots$ and Q is taken modulo 1) and rearranging Eq. (1.32), one has for the resonant value of ε:

$$\varepsilon_{\text{res}} = \frac{4}{\lambda} \exp\left(-\frac{2\pi(Q+m)}{\lambda}\right). \tag{5.16}$$

According to this formula, if $\lambda = 0.01$, the 4/5 (mod 1) resonance considered above is located at $\varepsilon_{\text{res}} \approx 2.004 \cdot 10^{-216}$ ($m = 0$), $\varepsilon_{\text{res}} \approx 2.670 \cdot 10^{-489}$ ($m = 1$), $\varepsilon_{\text{res}} \approx 3.559 \cdot 10^{-762}$ ($m = 2$), ... That is why the resonant structure of the separatrix map phase space portrait, presented in Fig. 5.10a, is impossible to reveal by means of numerical computing sections of phase space of original system (1.6): the relative magnitude of perturbation is microscopic.

Such situation is typical; this explains why chaotic layers in phase space sections of slowly perturbed Hamiltonian systems usually do not show any sign of inner regular component (see, e.g., Neishtadt et al. 1997): a very fine tuning of the values of system parameters is necessary to achieve its visibility, and, besides, the relative perturbation magnitudes, at which significant resonant structures emerge, are microscopic.

In applications, phase space sections are usually constructed not in the time and energy variables (in which the separatrix map is rendered, see illustrative portraits in Figs. 1.8 and 1.11), but in "resonant phase–conjugate momentum" variables (see, e.g., Figs. 1.6, 1.5, 1.4, and 5.5). Therefore, it is important to be able to transform the width rendered in energy to the width rendered in momentum. We have already considered this problem above in Sect. 1.6, in connection with the regular projection algorithm. In case of relatively thin layers, the layer width in momentum can be assessed in a more straightforward way, using simple analytical relations. Let us demonstrate how this can be done. We assess the total maximal width of the chaotic domain around a resonance, in momentum $p = \dot{\varphi}$; for the model with Hamiltonian (1.6), this position corresponds to $\varphi = 0 \mod 2\pi$; see, e.g., Fig. 1.5.

As follows from Hamiltonian (1.6), the width Δp of a thin chaotic layer in momentum p is related to the width Δw in the normalized energy w by the formula

$$\Delta p = \frac{\omega_0}{4} \Delta w \tag{5.17}$$

(Chirikov 1990, Shevchenko 2008b). Therefore, the total width of the chaotic domain in momentum is given by

$$\Delta \dot{\varphi} = \Delta \dot{\varphi}_{\text{res}} + \frac{\omega_0}{2} \left(\overline{w}_{\text{lb}}(\lambda, \varepsilon_1) + \overline{w}_{\text{lb}}(\lambda, \varepsilon_2) \right), \tag{5.18}$$

where the time-averaged half-width $\overline{w}_{\text{lb}}$ in energy (in case of the least perturbed border of the chaotic layer) is given by Eq. (5.3).

5.5 Marginal Resonances in Case of Non-adiabatic Chaos

As already discussed in part above in Sect. 4.4, Hamiltonian intermittency, manifesting itself in energy sporadic jumps, separated in time by relatively long periods of its quiet behaviour, is prominent in the separatrix map dynamics when the map's parameters have specific fine-tuned values. The jumps become possible when, upon variation of parameters, a heteroclinic connection is formed between the main chaotic layer and the narrow chaotic layer of an integer resonance, i.e., they start to overlap; thus a marginal integer resonance emerges (Chirikov 1979; Shevchenko 1998b, 2012). This phenomenon can be described by an approximate scheme: the unperturbed separatrix of an integer resonance starts to be tangent to the border of the main chaotic layer.

The chaotic layer width is maximal, or close to maximal, when a prominent marginal resonance is present at the border of the chaotic layer, and its separatrix chaotic layer is in heteroclinic connection with the main layer. The width is maximal when the two layers are on the brink of heteroclinic disconnection. Relevant formulas for the maximal width and for the critical values of parameters were derived in Shevchenko (1998b, 2012) in the perturbed pendulum model of nonlinear resonance.

Any theory for estimating the layer width should be developed separately for adiabatic and non-adiabatic cases of perturbation. Here we consider Hamiltonian intermittency in the adiabatic case. As shown above in Sect. 5.3, the border line between non-adiabatic and adiabatic chaos is rather sharp and is located at the adiabaticity parameter value $\lambda \approx 1/2$ for any system described by the separatrix map; see Fig. 5.6 for an illustration.

An approximate condition for the tangency of the unperturbed separatrix of a marginal integer resonance to the border of the main chaotic layer (equivalently, the approximate condition for the energy jumps in the near-separatrix motion) is given by Eq. (5.12) (Shevchenko 1998b), where W and $W_{\mathrm{t}}^{(m)}$ are expressed as follows. The formula for W is determined by the choice of a system under study. For the Hamiltonian (1.6) with $k = 1$ and $a = b$ it is given by Eq. (1.29). The expression for $W_{\mathrm{t}}^{(m)}$ does not depend on the choice of the system. In the pendulum model of the marginal resonance it has universal form (5.13).

The maximum value of the relative energy during the energy jump is

$$w_{\mathrm{extr}} = \pm \left(64 \exp\left(-\frac{2\pi m}{\lambda} \right) - \lambda W_{\mathrm{t}}^{(m)}(\lambda) \right), \tag{5.19}$$

where m is the order of the marginal resonance (Shevchenko 1998b). The λ location of the mth peak, $\lambda = \lambda_m$, at any value of ε can be found by numerically solving the functional equation (5.12) with respect to λ at any m. In particular, this can be easily accomplished for any m in the Maple computer algebra system (Char et al. 1993). Then the height of the mth can be found by setting $\lambda = \lambda_m$ in Eq. (5.19).

In the considered case of non-adiabatic chaos, the approximation of marginal resonance in the pendulum model perfectly works (Shevchenko 2012). In Fig. 5.11, a phase portrait of the separatrix map (1.31) with $\lambda = 3$ and $c = 5.55 \bmod 2\pi$ is presented. Solely the upper part of the layer ($y \geq 0$) and solely the chaotic component of phase space is shown in Fig. 5.11. The chosen values of λ and c correspond to the brink of heteroclinic disconnection between the primary chaotic layer (shown in black) and the chaotic layer of the marginal resonance (the secondary layer; shown in grey). A slightest increase in c disconnects the layers, and the width momentarily drops to that of the primary layer.

In Fig. 5.11, it is evident that no regular approximation for the marginal resonance can describe precisely the critical parameter values that would provide the maximal width in case of non-adiabatic chaos, because the secondary layer has substantial width. This width must be taken into account in any high-precision theory for estimating the conditions for the critical heteroclinic connection.

It is also evident from Fig. 5.11, that the marginal resonance separatrix cell is perfectly described by the theoretical pendulum cell. The borders of the theoretical cell are depicted by solid curves, given by the equation

$$y = y^{(m)} \pm 2(y^{(m)}/\lambda)^{1/2} \cos\frac{x}{2}, \tag{5.20}$$

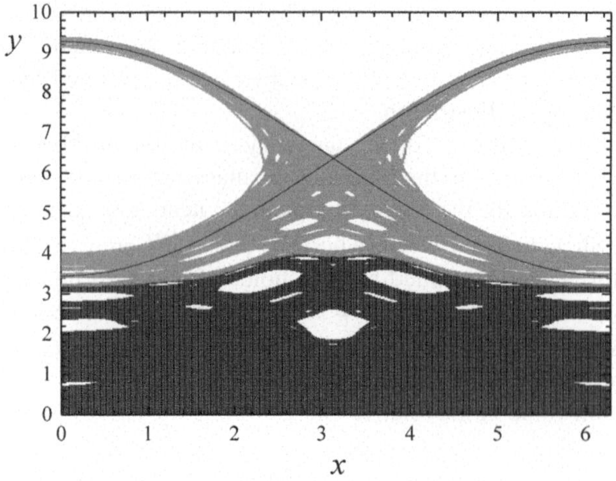

Fig. 5.11 The phase portrait of the separatrix map at $\lambda = 3$ and $c = 5.55 \bmod 2\pi$. The theoretical pendulum cell for the marginal resonance is shown by solid curves. (Reprinted Figure 1 with permission from Shevchenko 2012. © 2012 by the American Physical Society)

where $y^{(m)}$ is the location of the centre of the marginal resonance of order m:

$$y^{(m)} = \exp \frac{c - 2\pi m}{\lambda}, \tag{5.21}$$

as can be straightforwardly derived from Eqs. (1.31) (Shevchenko 2010).

The phase portrait here is synchronized: the pairs x_{i-1}, $(y_i + y_{i-1})/2$ are drawn instead of x_i, y_i, so that the portrait corresponds to a unified surface of section of phase space.

On decreasing λ, the marginal resonance separatrix cell deforms more and more. In case of adiabatic chaos, when $\lambda \ll 1$, the pendulum model is not anymore applicable for its description. This is evident from the phase portrait in Fig. 5.12, where $\lambda = 0.001$ and $c = 0.0076008 \bmod 2\pi$. The separatrix cell does not anymore have the well-known lenticular form, specific for the pendulum case.

5.6 Marginal Resonances in Case of Adiabatic Chaos

Soskin and Mannella (2009) presented a theoretical method for the calculating maximal width of the separatrix chaotic layer, which is suitable for a wide class of slowly periodically perturbed one-degree of freedom Hamiltonian systems. The theory by Soskin and Mannella (2009) describes the shape of the resulting peak in the "frequency of perturbation–relative energy" coordinates, the peak's location on the perturbation frequency axis and its height in the relative energy.

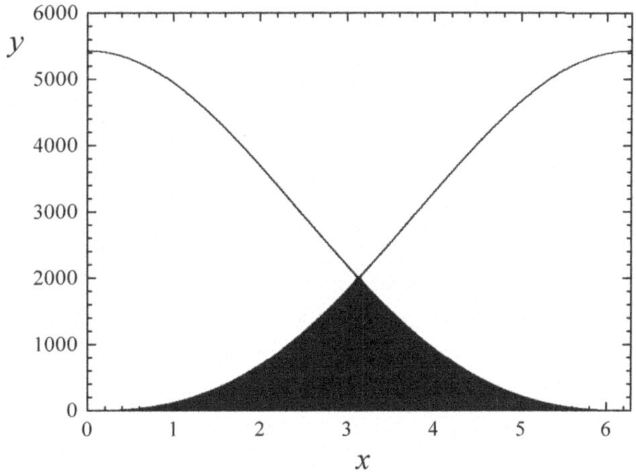

Fig. 5.12 A phase portrait of the separatrix map; $\lambda = 0.001$ and $c = 0.0076008 \bmod 2\pi$. (Reprinted Figure 2 with permission from Shevchenko 2012. © 2012 by the American Physical Society)

In the framework of a classification, given by Soskin and Mannella (2009) for periodically perturbed one-degree of freedom Hamiltonian systems, it provides a general approach for description of Hamiltonian intermittency in such kind of systems.

In the considered case of adiabatic chaos, the λ location of the mth peak λ_m, according to Soskin and Mannella (2009), is given by

$$\lambda_m \approx -\frac{2\pi m}{\ln \frac{h}{4}} = -\frac{2\pi m}{\ln \frac{\varepsilon}{2}} \tag{5.22}$$

(Equation 63 in Soskin and Mannella 2009), and the corresponding theoretical value of the chaotic layer maximal half-width is

$$w_{\text{extr}} \approx 2(4e+1)h = 4(4e+1)\varepsilon \approx 23.75h \tag{5.23}$$

(Equation 72 in Soskin and Mannella 2009). Formulas (5.22) and (5.23) are exact in the limit $h \to 0$, otherwise they are approximate.

The layer width definition depends on whether you take into account the layer bending effect or not. (The bending effect is described above in Sect. 1.8.) The emergence of this effect in numerical experiments depends on which method is used for measuring the layer width. In case of the method based on calculating the minimum period of the motion in the chaotic layer, the effect is averaged out; see Sect. 5.4. The theoretical value of the maximum half-width of the chaotic layer is then (when the bending is absent) given by

$$w_{\mathrm{extr}} \approx 8eh = 16e\varepsilon \approx 21.75h, \tag{5.24}$$

instead of formula (5.23).

One may see that the relative difference between (5.23) and (5.24) is rather small: about 10% $\approx 1/(4e)$. The bending effect seems to be also small in estimating the maximum width in the domain of non-adiabatic chaos also, because the good accord of the theoretical estimates and numerical-experimental data in the domain of applicability of the theory exists, as described above. Comparing Figs. 5.8 and 5.9, we see that the height of the first peak in Fig. 5.8 is lower by about 10% ($\approx 1/(4e)$) than that in Fig. 5.9. The cause of this difference is that the layer bending in the first case is averaged out, while in the second case it is present.

The geometrical form of the separatrix cell in Fig. 5.12 (the adiabatic case) can be quite easily described analytically, if one uses the separatrix map representation (1.31). Using this representation, formulas (5.22) and (5.24) can be also derived.

This is accomplished as follows. Let $y \geq 0$ and assume that the increments of x and y per iteration in Eqs. (1.31) are small compared to their total magnitudes of variation. This approximation is analogous to that used in Vecheslavov (2004) in case of the separatrix map in classical form. It is valid in presence of integer marginal resonances, because the amplitude of variation of y is then much greater than 1, while the increment of y per iteration is less than 1.[1] Therefore, the map is reduced to the differential equation

$$\frac{dx}{dy} = \frac{-\lambda \ln y + c - 2\pi m}{\sin x}. \tag{5.25}$$

Here c may take any values, $-\infty < c < +\infty$, whereas the m value is chosen in such a way that $0 \leq c - 2\pi m < 2\pi$. Integrating, one finds the guiding curve in the form

$$y(\lambda - \lambda \ln y + c - 2\pi m) = -\cos x + 1, \tag{5.26}$$

where the integration constant is set equal to 1, so that the curve is tangent to the $y = 0$ axis. This tangency is critical, because a slightest change of the map parameters may disconnect the curve from the $y = 0$ axis, and then the motion becomes regular.

It is evident that the geometrical form of the separatrix cell of the marginal resonance in Fig. 5.12 is perfectly described by Eq. (5.26): the analytical curve visually coincides with the cell's computed borders.

The critical value of c is found as follows. The unstable fixed point of the marginal resonance m of map (1.31) is situated at $x = \pi$, $y = \exp((c - 2\pi m)/\lambda)$,

[1] Note that in a general situation, when there are no marginal resonances, such an approximation is invalid; see discussion in Shevchenko (2008a).

whereas the stable fixed point (centre) of the same resonance is situated at $x = 0 \bmod 2\pi$, $y = \exp((c - 2\pi m)/\lambda)$; see Eq. (5.21). By evaluating the coordinates of the unstable fixed point in Eq. (5.26) and solving the resulting equation with respect to c, one finds the critical value of c:

$$c_m = 2\pi m - \lambda \ln \frac{\lambda}{2}. \tag{5.27}$$

Substituting $c = c_m$ in Eq. (5.26) and solving the resulting equation with respect to y, one finds that, at $x = 0$ and $x = 2\pi$, there exist two solutions of Eq. (5.26): $y = 0$ and $y = 2e/\lambda$. As easily analytically checked, they correspond to two different extrema of the $y(x)$ function. Therefore, the maximal value of y is

$$y_{\text{extr}} = \frac{2e}{\lambda}. \tag{5.28}$$

To relate the obtained parameter values with the original Hamiltonian parameters, recall that $y = w/W$, and c is given by formula (1.32), where, in the considered Hamiltonian model,

$$W \approx 8\varepsilon\lambda, \tag{5.29}$$

if $\lambda \ll 1$. Hence, $w_{\text{extr}} \approx 8eh$ and $\lambda_m \approx -2\pi m/\ln(\varepsilon/2)$, in accord with Eqs. (5.24) and (5.22).

Formula (5.28) combined with the equality of the y coordinate of the unstable fixed point to $2/\lambda$ (at $x = \pi$, there exists only one solution, $y = 2/\lambda$) gives that the relative amplitude of the motion at the layer's outermost border is equal, in the adiabatic limit, to $e \approx 2.718$.

Let us underline that, concerning theoretical descriptions of marginal resonances, the case of adiabatic chaos, in comparison with the cases of intermediary chaos and non-adiabatic chaos, are much less actual, at least to date, for any physical or technical applications. The cause is that even the first ($m = 1$) marginal resonances appear at $\lambda < 1/2$ only if the perturbation strength ε is microscopic; indeed, Eq. (5.22) implies that, for λ_1 to be less than 1/2, the ε value should be less than $2e^{-4\pi} \approx 10^{-5}$. In typical applications, perturbation amplitudes are usually much greater. For the second and higher order resonances ($m \geq 2$) to appear in the domain of adiabatic chaos, ε should be drastically less, supermicroscopic. All the resonances with $m > 1$ emerge at $\lambda_m > \lambda_1$; thus, when the perturbation is strong enough ($h \gtrsim 10^{-5}$), all the peaks are situated in the domain of non-adiabatic chaos.

Part II
Resonances and Chaos in the Solar System

In the second part of the book, most remarkable manifestations of chaotic behaviour in the dynamics of the Solar system bodies are considered, described and analysed. These manifestations include: chaos in rotational dynamics of various bodies, in orbital dynamics of satellite systems, asteroids and comets, and, finally, in orbital dynamics of planets. At present, only several objects or classes of objects in the Solar system demonstrate a directly observable strongly chaotic behaviour. However, if one considers the long-term dynamical evolution of any class of objects, it is usually found that the role of prior chaotic stages in determining the present dynamical states is generically great.

Considering the rotational and orbital dynamics of celestial bodies, it is often possible to explain chaotic manifestations by using various models of interaction and overlap of non-linear resonances, thoroughly discussed in the first part of the book. In particular, one may obtain analytical estimates of the Lyapunov timescales (representing the "time horizons" of predictable dynamics) and the chaotic diffusion timescales. Locations and extents of chaotic domains in the phase space of motion are also of great interest, and they can also be characterized in the same models.

Chapter 6
Defying the Orrery Paradigm: Historical Background

... But when the planets
In evil mixture to disorder wander,
What plagues and what portents, what mutiny ...

William Shakespeare,
Troilus and Cressida, I, iii (1602)

In course of many centuries, up to the 20th one, nothing had seemed to any scientist or philosopher to be less prone to chaos and accident than the repetitive measured motion of the Solar system bodies, first of all planets and the Moon.

The idea of harmonious celestial order inspired and led Johannes Kepler in his revolutionary studies of planetary motions. In 1593, 50 years after publication of *De Revolutionibus Orbium Coelestium*[1] by Nicolaus Copernicus, he found and was inspired by the fact that the circle inscribed in an equilateral triangle and the circle described around the same triangle (Fig. 6.1) have approximately the same ratio of radii as the orbits of Jupiter and Saturn. In the following years, Kepler developed a more sophisticated model of the Solar system in the form of a concentric sequence of five principal polyhedra (Fig. 6.2). In this model, afterwards named *Kepler's goblet*, the Solar system obeyed a static geometrical order.

In 1784, 5 years before the Great French revolution, Pierre Simon Laplace paid attention to another, not geometrical but dynamical, relationship between the orbits of Jupiter and Saturn: the orbital periods of these planets are close to the integer commensurability (resonance) 2/5. Taking this relationship into account in a newly developed analytical theory, based on Newton's law of gravitation, he succeeded to explain observed anomalies in the orbital motion of Jupiter and Saturn and revealed a periodic character of these anomalies, thus showing that they do not lead to any monotonous changes.

[1] *On the Revolutions of the Heavenly Spheres.*

I. I. Shevchenko, *Dynamical Chaos in Planetary Systems*, Astrophysics and Space Science Library 463, https://doi.org/10.1007/978-3-030-52144-8_6

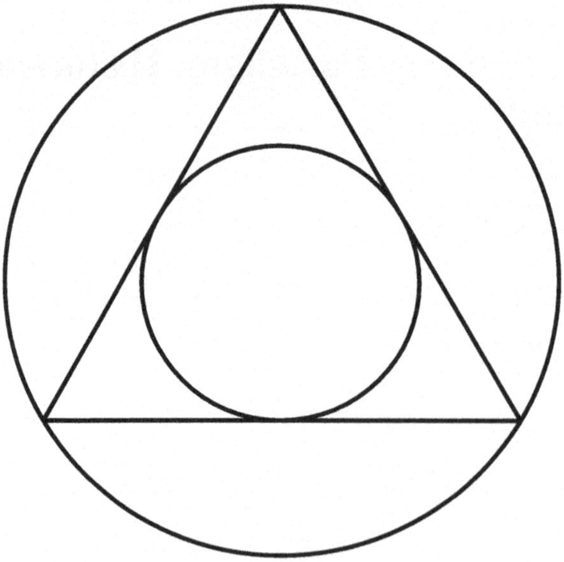

Fig. 6.1 The orbits of Jupiter and Saturn, and Kepler's triangle

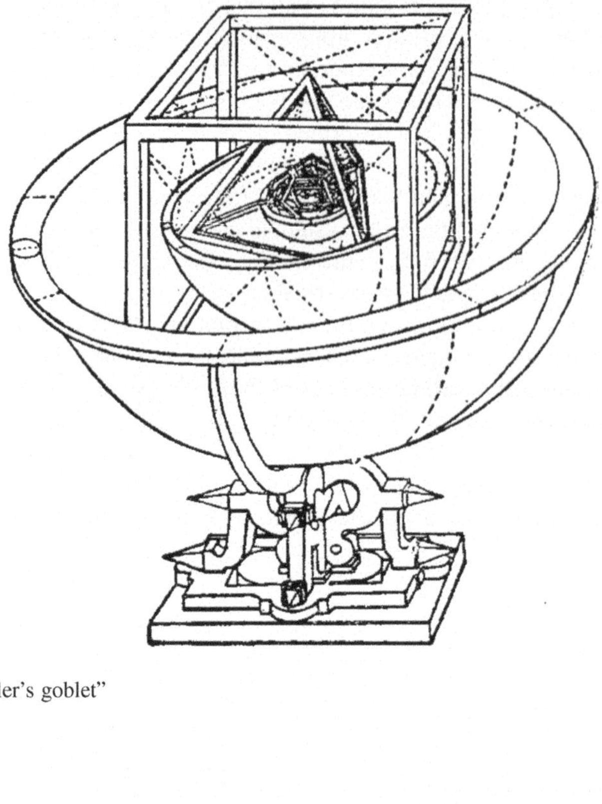

Fig. 6.2 "Kepler's goblet"

This triumph of celestial mechanics was preceded in 1759 by another outstanding achievement: on the basis of Newton's theory of gravitation, Alexis Claude Clairaut accurately predicted the time of return of Comet Halley (the time of passage of perihelion by the comet), and observations of the comet confirmed this prediction triumphantly.

These and other remarkable successes of Newton's theory in describing the orbits of celestial bodies inspired Laplace to put forward a philosophical concept that all motions in the Universe are absolutely predetermined: if, at any moment of time, the initial positions and velocities of all the world-comprising particles are set, then the world's subsequent history is also set. Pannekoek (1961) gives a clear-cut short description of this scientific-philosophical platform: "The Solar system was thought to be a huge mechanism brought and pushed in the motion solely by the force of universal gravitation. It was a fully cognizable and calculable clockwork which kept the motion forever."

This deterministic concept is graphically embodied in a mechanical device, called *orrery* (Fig. 6.3). The orrery displays the motions of the Solar system planets and their satellites by means of a clockwork, demonstrating the perfect predictability of the motions.

Since the work of Laplace, the deterministic nature of the motions of large and minor bodies of the Solar system had never been put in question for about 200 years, until the middle of the twentieth century. Observed rational commensurabilities between the orbital periods of some Solar system bodies, such as Galilean moons of Jupiter, were regarded as a graphical manifestation of the reigning order and harmony.

Fig. 6.3 An orrery

From our modern dynamical viewpoint, it can be stated that the deterministic idea originally stemmed from the consideration, by Laplace, of a specific commensurability (resonance), namely, the 2/5 one between the motions between Jupiter and Saturn.

It may seem paradoxical (although historically logical) that, two centuries after Laplace, studies of *interactions of resonances* by Boris Chirikov, Jack Wisdom, and other scientists led to abandoning the deterministic concept in celestial mechanics. The process of adopting a new concept, that of *dynamical chaos*, was promoted, in particular, by Comet Halley's return in 1986. This was the third one after the return accurately predicted by Clairaut in 1759. In 1986, in the works of Petrosky (1986) and Chirikov and Vecheslavov (1986), Comet Halley's orbit was not considered anymore a triumphant example of the predetermined motion, but was regarded as a prominent manifestation of dynamical chaos.

Thus, the Laplace determinism reigned in celestial mechanics within the time span of about two centuries, covered by three orbital revolutions of Comet Halley. It is curious that even during these two centuries, the educated community as a whole did hardly perceive the cosmic world as an ideal predictable mechanism: for example, the apparitions of comets were generally perceived as sudden and dangerous; see Fig. 6.4 and a discussion in Boime (1990).

Fig. 6.4 *Starry night* by Vincent Van Gogh (a sketch, 1889). Credit: https://commons.wikimedia.org/wiki/File:Van_Gogh_Starry_Night_Drawing.jpg

Chapter 7
Rotational Dynamics

Considering the rotational dynamics of celestial bodies, it is often possible to explain chaotic manifestations by using various models of interaction and overlap of non-linear resonances, addressed in the first part of the book. In particular, analytical estimates of the Lyapunov timescales and the chaotic diffusion timescales can be obtained. Locations and extents of chaotic domains in the phase space of motion are also of great interest, and they are characterized in this Chapter in the same models. We concentrate on chaos phenomena in the rotational dynamics of planetary satellites (first of all, in the rotation of Hyperion, the seventh satellite of Saturn) and on chaotic obliquities of planets. The presentation partially follows (in Section 7.1) the papers by Kouprianov and Shevchenko (2005) (by permission from Elsevier) and Melnikov and Shevchenko (1998).

7.1 Rotational Dynamics of Satellites

Rotational dynamics of planetary satellites presents a variety of intriguing dynamical behaviours, including the historically first ever case of predicted and directly observed dynamical chaos, that in the rotation of Hyperion, the seventh satellite of Saturn.

7.1.1 Hyperion and Other Chaotic Rotators

Wisdom et al. (1984) predicted, on a basis of numerical experiments and theoretical estimates, that the seventh satellite of Saturn, Hyperion, rotates chaotically with respect to its centre of masses: the orientation of the satellite and the velocity of its rotation change in time chaotically.

I. I. Shevchenko, *Dynamical Chaos in Planetary Systems*, Astrophysics and Space Science Library 463, https://doi.org/10.1007/978-3-030-52144-8_7

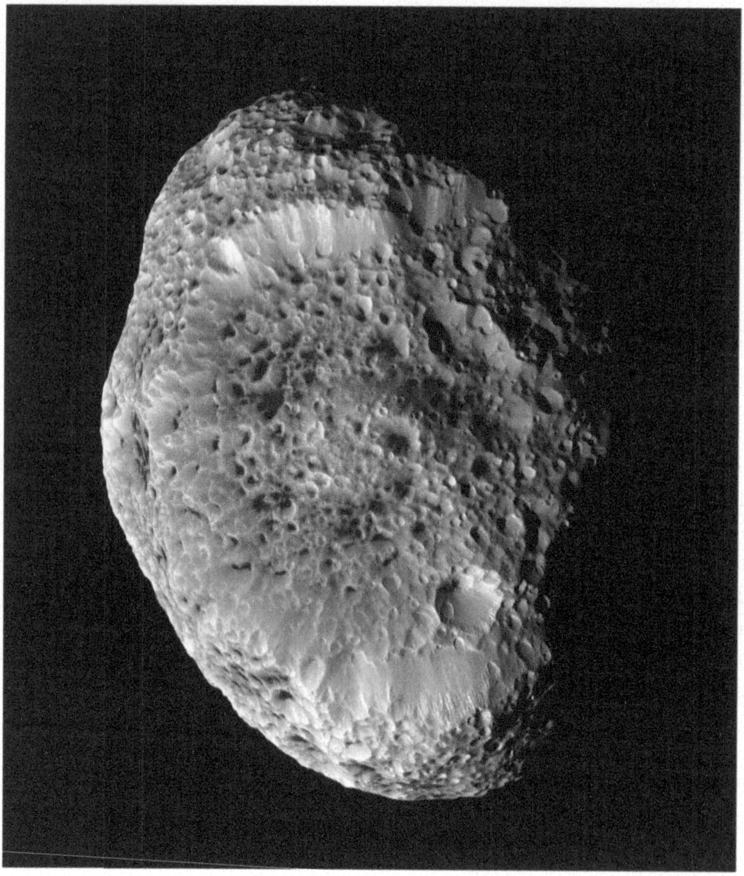

Fig. 7.1 Hyperion. An image taken by the *Cassini* space probe (Figure 1a from Thomas et al. 2007, by permission from Springer Nature, © 2007)

The cause of chaos in the rotational dynamics of satellites consists in interaction of *spin-orbit resonances* (resonances between the rotational and orbital motions). It turned out that in case of Hyperion this interaction is especially strong, due to a prominent non-sphericity of its figure (see Fig. 7.1) and its appreciable orbital eccentricity ($e \approx 0.1$).

The primary source of information on the rotational dynamics of minor Solar system bodies is provided by analysing their *lightcurves*. The lightcurve is nothing but a time sequence of values of flux of light from any observed astronomical object. Klavetter (1989b) constructed precise lightcurves of Hyperion with high resolution in time, and carried out their analysis and modelling. He concluded that this satellite was, most likely, indeed in the chaotic rotation state. Later on, the "tumbling" character of Hyperion's rotation was directly observed during the flyby of *Voyager*-2 near Saturn (Black et al. 1995). To date, the rotation of Hyperion is the only known

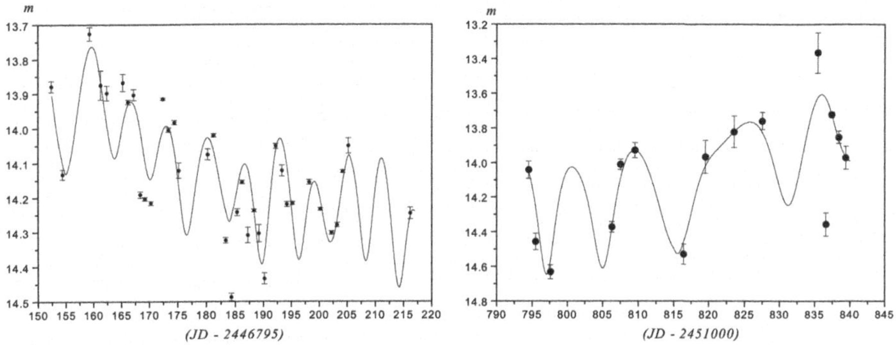

Fig. 7.2 Model lightcurves of Hyperion (solid curves), for observational data by Klavetter (1989a) (left panel) and for observational data by Devyatkin et al. (2002) (right panel). The observed lightcurves are shown by dots. JD is for Julian date, *m* is for stellar magnitude (Figure 6 from Devyatkin et al. 2002. With permission from Pleiades Publishing Inc.)

example of observable chaos in rotational dynamics of planetary satellites (Wisdom et al. 1984; Wisdom 1987a; Klavetter 1989b; Black et al. 1995; Devyatkin et al. 2002; Melnikov 2002; Harbison et al. 2011; Tarnopolski 2015).

Melnikov (2002) modelled Hyperion's lightcurves, using the observational data by Klavetter (1989a) and Devyatkin et al. (2002). By means of computation of the Lyapunov exponents of the motion he arrived to an observationally rigourous conclusion that Hyperion was indeed in the chaotic mode of rotation. In Fig. 7.2, the model lightcurves of Hyperion are presented.

Planar (in the orbit plane) oscillations and rotations of any satellite near synchronous spin-orbit resonance (the resonance at which the period of rotation of a satellite is equal to its orbital period, as in case of the Moon) are described by the perturbed pendulum Hamiltonian (1.6) with specific parameters. The role of the deviation angle of the pendulum is played by the angle specifying the satellite orientation with respect to the direction to the planet. Therefore, methods of analytical estimating the maximum Lyapunov exponent, based on the separatrix map theory (Chap. 3), are applicable. For Hyperion, the resulting theoretical estimates of the Lyapunov time ($\approx$30 d) turn out to be in accord with numerical-experimental ones (Shevchenko and Kouprianov 2002; Kouprianov and Shevchenko 2005).

Do there in the Solar system exist any other (except Hyperion) satellites that are chaotically tumbling? For many satellites the character of rotation is yet unknown. Most of the satellites, for which it is determined, rotate synchronously with the orbital motion. Similar to the Moon, they permanently face their planet by one side. However, in the course of the long-term dynamical evolution, any satellite at some moment of time enters a state of chaotic rotation, especially violent in a neighbourhood of the separatrices of synchronous resonance.

The theoretical research by Kouprianov and Shevchenko (2005) showed that, apart from Hyperion, the 16th and 17th Saturnian satellites Prometheus and Pandora may also rotate chaotically. If they do, the Lyapunov times of their rotation are very small, $\sim$1 d.

The following problem is important: whether in the course of the chaotic tumbling there exists a preferable orientation of the satellite, or all its orientations in this case are equiprobable? Calculations by Melnikov and Shevchenko (2008) show that Prometheus and Pandora in case of chaotic rotation do have preferable orientations of the largest axis of the figure in direction to Saturn. This complicates making conclusions about the character of rotation of these satellites from observations, because the chaotic mode is observationally similar to the usual synchronous rotation. According to numerical-experimental and analytical estimates by Shevchenko and Kouprianov (2002), Kouprianov and Shevchenko (2005), the Lyapunov times of chaotic rotation of minor planetary satellites can be very small, as small as $\sim$1 day (as, theoretically, in case of Prometheus and Pandora), i.e., chaos in case of its presence is observable.

Up to now, Hyperion remains the sole proved example of a Solar System body observed to rotate chaotically. Another candidate for chaotic rotation was suggested to be Nereid, the second satellite of Neptune (Dobrovolskis 1995), mainly on the grounds of the large orbital eccentricity ($\approx$0.75). The large eccentricity, on one hand, broadens chaotic domains in the rotation phase space and, on the other hand, as found by Dobrovolskis (1995), calls forth shortening the time of satellite's tidal despinning. However, observations by Grav et al. (2003) proved that Nereid rotates regularly; its rotation is too fast (too far from synchronous resonance) to be chaotic (Hesselbrock et al. 2013; Terai and Itoh 2013).

Four circumbinary satellites Nix, Hydra, Styx, and Kerberos in the Pluto–Charon system were supposed to rotate chaotically (Showalter and Hamilton 2015; Correia et al. 2015), but this was not confirmed by imaging data from the *New Horizons* space mission (Weaver 2016); with a possible exception of Styx, see Quillen et al. (2017).

Namaka, one of two known moons of the dwarf planet Haumea, may rotate chaotically (Hastings et al. 2016), due to its large orbital eccentricity, but the chaos requires further observational confirmation.

For all latter six moons (Nereid, Nix, Hydra, Styx, Kerberos, Namaka), the rotation rates are actually (or possibly) too large for a satellite to reside in the rotation phase space chaotic domain around synchronous resonance; i.e., the satellites have not been enough tidally despun, in the course of their long-term dynamical evolution.

By means of analysis of stability of possible rotation states of planetary satellites (judging from their location in the "inertial parameter–orbital eccentricity" plane), Melnikov and Shevchenko (2010) showed, that although the majority of moons with known rotation states rotate synchronously, more than two thirds of all satellites with unknown rotation states are in other (fast or, less probably, chaotic) modes of rotation.

About 200 planetary satellites have been discovered to date. The rotation state has been observationally determined for not more than a quarter of them. Amongst those with identified rotation states, most are in synchronous resonance, whereas the remaining minority has periods of rotation much less than orbital periods. This is a selection effect: small moons, which usually rotate fast (being not tidally despun), are more difficult to observe than big ones, which have been mostly despun down to entering the synchronous state.

Small and big satellites differ in the character of chaotic rotation. For the minor (irregularly-shaped) satellites, the rotation represents a three-dimensional tumbling; for majority of the large satellites (which are close to spherical in shape), the rotational motion in the chaotic domain at the separatrix of synchronous resonance preserves planarity: the rotation axis keeps the approximate orthogonality to the orbit plane (Kouprianov and Shevchenko 2005).

And what about the Moon? As it is well-known to everyone, the Moon always faces the Earth by one and the same side, i.e., it is in synchronous resonance. Not so widely known is that it is subject to small oscillations with respect to the exact synchronous state. This phenomenon carries the name of *physical libration*. How much regular and predictable these small oscillations of our Moon can be, on long time timescales? This problem has not been considered till now, maybe due to its complexity.

7.1.2 Spin-Orbit Resonances

Assume that a satellite of negligible mass moves in a fixed ellipse around a gravitating point (planet). The satellite's vector of the angular momentum coincides with the axis of its maximum moment of inertia and is orthogonal to the orbit plane; i.e., the rotational motion is planar.

The motion is then described by the *Beletsky equation* (Beletsky 1965) for the planar librations/rotations of a satellite moving in an elliptic orbit. If the orbital eccentricity is small ($e \ll 1$), then the Hamiltonian, which is derived by neglecting all terms beyond the first order in the eccentricity, is given by

$$H = \frac{y^2}{2} - \frac{\omega_0^2}{4}\cos(2x - 2t) - \frac{7e\omega_0^2}{8}\cos(2x - 3t) + \frac{e\omega_0^2}{8}\cos(2x - t) \tag{7.1}$$

(Wisdom et al. 1984, Celletti 1990), where x is the satellite orientation angle, i.e., the angle between the axis of the minimum moment of inertia and the line of apsides; $y = \dot{x}$, t is time, and "$\dot{x}$" designates the time derivative of x. The model parameters comprise the eccentricity e and the inertial parameter

$$\omega_0 = \left(\frac{3(B - A)}{C}\right)^{1/2}, \tag{7.2}$$

where $A < B < C$ are the principal moments of inertia of the satellite. The time unit is equal to $1/(2\pi)$ times the orbital period. The ω_0 parameter characterizes the satellite's dynamical asymmetry.

By means of the canonical transformation $p = \frac{y-1}{2}$, $\varphi = 2(x - t)$, the Hamiltonian (7.1) is readily reducible to the perturbed pendulum paradigm (1.6):

$$H = 2p^2 - \frac{\omega_0^2}{4}\cos\varphi - \frac{7e\omega_0^2}{8}\cos(\varphi - t) + \frac{e\omega_0^2}{8}\cos(\varphi + t). \tag{7.3}$$

The motion in a vicinity of the separatrix of system (7.3) can be described by the separatrix algorithmic map (1.33). Comparing Eqs. (1.6) and (7.3), one finds that $k = 1$, $\mathcal{F} = \omega_0^2/4$, $\mathcal{G} = 4$, $\Omega = 1$, $a = -7e\omega_0^2/8$, and $b = e\omega_0^2/8$.

Hence the separatrix algorithmic map parameters $\lambda = 1/\omega_0$, W^+ and W^- are determined by ω_0 and e; see Eqs. (1.54) and (1.55). For the η parameter, $\eta = b/a = -1/7$; thus, the perturbation asymmetry is intermediate. If $\lambda > 1$, then $W^+ \approx -7W^-$.

The separatrix algorithmic map can therefore be straightforwardly used to describe the rotational motion of satellites in the vicinity of the separatrix of synchronous spin-orbit resonance. Usually, this vicinity is not at all small (Shevchenko 1999a); rather often, it is large enough to engulf, apart from the synchronous resonance, the neighbouring important spin-orbit resonances.

Data on the inertial and orbital parameters for several minor satellites, along with calculated values of the separatrix algorithmic map parameters, are given in Table 7.1.

In Fig. 7.3, a phase portrait of a model near-synchronous planar rotation of Phobos is shown, as obtained by using the separatrix algorithmic map (1.33) and the regular projection algorithm, described in Sect. 1.6. The map's phase portrait corresponds to the phase space section of system (7.3), taken at $t = 0 \pmod{2\pi}$. Note that the transformation from (φ, p) to (x, y) at this plane is: $x = \frac{\varphi}{2}$, $y = 1 + 2p$.

The corresponding section of the phase space of system (7.3) can be constructed by means of a direct numerical integration of the equations of motion of the original system (7.1). This section is given in (Wisdom 1987a, Figure 1). It closely agrees with Fig. 7.3, resulting from the separatrix map application.

Table 7.1 Inertial, orbital and separatrix algorithmic map parameters of several minor satellites (Shevchenko 1999a)

Satellite	ω_0	e	Ref.	λ	W^+	W^-
Phobos	0.86	0.015	Wisdom (1987a)	1.163	−0.286	0.0336
Deimos	0.81	0.0005	Wisdom (1987a)	1.235	−0.00962	0.00118
Amalthea	1.14	0.003	Wisdom (1987a)	0.877	−0.0509	0.00408
Janus	0.37	0.009	Goździewski (1997)	2.673	−0.0850	0.0121
Epimetheus	0.87	0.007	Goździewski (1997)	1.149	−0.133	0.0155
Pandora	0.93	0.004	Goździewski and Maciejewski (1995)	1.075	−0.0749	0.00818
Prometheus	1.17	0.004	Goździewski and Maciejewski (1995)	0.855	−0.0668	0.00503
Bifurcation case	0.50	0.01	–	2	−0.152	0.0214

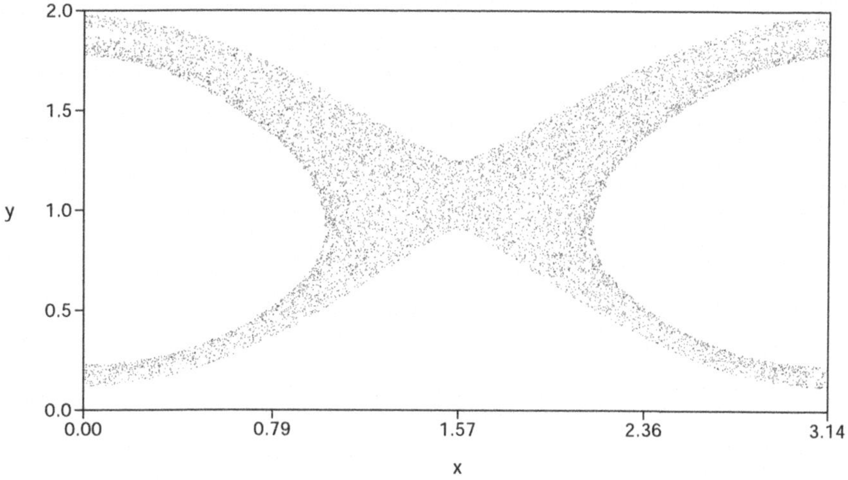

Fig. 7.3 A model for rotating Phobos. The phase portrait of the near-synchronous rotation state, obtained with the separatrix algorithmic map and the regular projection algorithm (Figure 1 from Shevchenko 1999a, by permission from Springer Nature, © 1999)

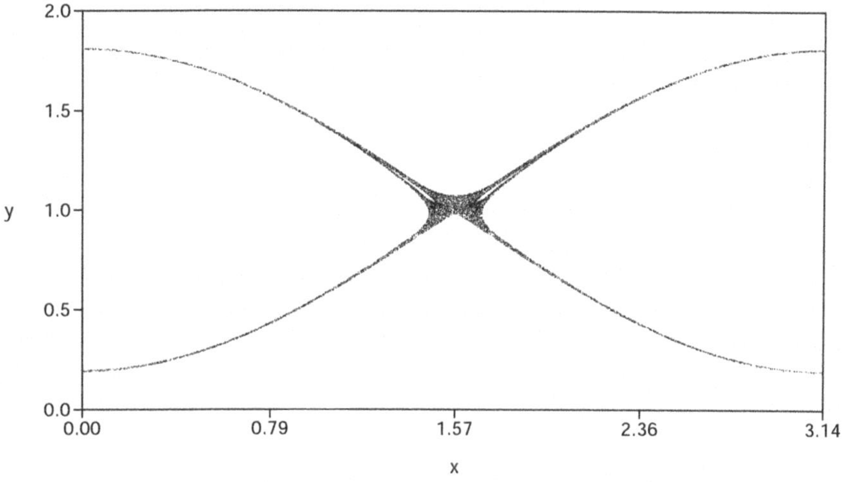

Fig. 7.4 Same as Fig. 7.3, but for model Deimos (Figure 2 from Shevchenko 1999a, by permission from Springer Nature, © 1999)

In Figs. 7.4 and 7.5, analogous phase portraits are presented for Deimos and Janus models. Corresponding phase space sections, obtained by direct integrations, can be found in Wisdom (1987a) and Goździewski (1997); see Figure 3 in Wisdom (1987a) (Deimos) and Figure 5a in Goździewski (1997) (Janus). They all closely agree with the phase portraits obtained by the separatrix map techniques. Generally, the separatrix algorithmic map, in concert with the algorithm of regular

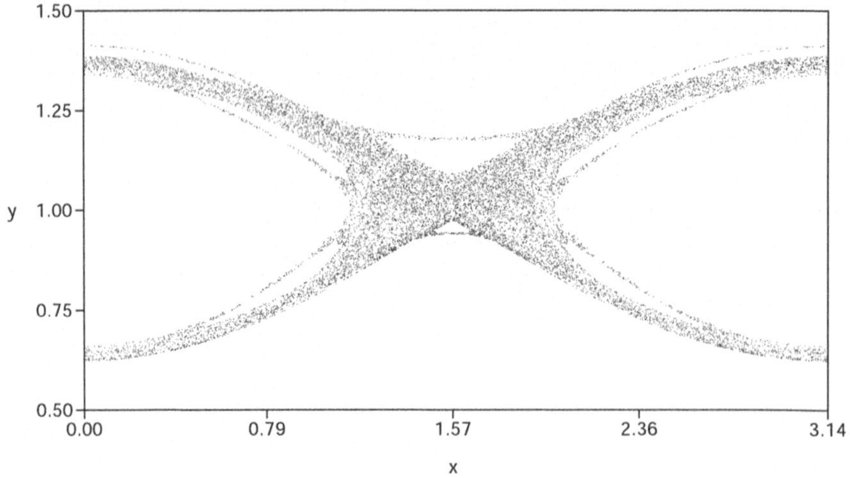

Fig. 7.5 Same as Fig. 7.3, but for model Janus (Figure 3 from Shevchenko 1999a, by permission from Springer Nature, © 1999)

projection, reproduces phase portraits of the rotational motion in close agreement with the known phase space sections (Shevchenko 1999a). This demonstrates the separatrix map theory opportunities, and verifies its validity for further applications in assessing major properties of chaos in the given problem, such as widths of chaotic domains, Lyapunov and diffusion timescales of the rotational dynamics.

The phase space structure can be also readily analysed using the separatrix map techniques. For example, let us see how locations of the major 1/2 and 3/2 spin-orbit resonances, neighbouring to the synchronous one, can be found.

The time-averaged derivative of the resonance argument is $\langle\dot{\varphi}\rangle = 2\langle y\rangle - 2$; therefore, these two resonances both correspond to an integer resonance of the separatrix map. This resonance has the winding number $Q = 1/\langle\dot{\varphi}\rangle = \pm 1$. If, say, $\langle y\rangle = 5/2$, then $Q = 1/3$. The elliptic modulus $k^{(Q)}$ of the motion with the winding number Q can be found by numerically solving the equation

$$\lambda k^{(Q)} K\left(k^{(Q)}\right) = \pi |Q|, \tag{7.4}$$

following from Eq. (1.35). The equation can be efficiently solved by iterations.

The energy deviation is $w^{(Q)} = 2\left(\left(k^{(Q)}\right)^{-2} - 1\right)$. On the other hand, the expression for H_0 (cf. Eq. (1.6)) is reducible to

$$y = 1 \pm \frac{\omega_0}{2^{1/2}}(1 + w + \cos\varphi)^{1/2}. \tag{7.5}$$

The plus and minus signs correspond here to the prograde and retrograde motions, respectively. In the x axis, the resonances are centred at $x = \pi/2 \bmod \pi$ (resonance

1/2) and at $x = 0 \bmod \pi$ (resonance 3/2). Since $\varphi = 2x$ at $t = 0$, Eq. (7.5) gives the following y locations for the centres:

$$y_{1/2} = 1 - \frac{\omega_0}{2^{1/2}} \left(w^{(1)} \right)^{1/2},$$

$$y_{3/2} = 1 + \frac{\omega_0}{2^{1/2}} \left(2 + w^{(1)} \right)^{1/2}. \tag{7.6}$$

The separatrix-map estimates of $y_{1/2}$ and $y_{3/2}$, given by formulas (7.6), are in accord with data in (Wisdom et al. 1984, Figure 3), obtained by localization of the resonances in direct numerical integrations.

7.1.3 Lyapunov Timescales of Chaotic Rotation

A theoretical analysis of rotation stability accomplished in Kouprianov and Shevchenko (2005) showed that the 16th and 17th Saturnian satellites Prometheus and Pandora may rotate chaotically. Let us estimate analytically the Lyapunov times of the rotation of Prometheus and Pandora, if it were chaotic. Here we adopt the planar rotation model, in which the rotation axis is set to coincide with that of the satellite's maximum moment of inertia and to be orthogonal to the orbit plane. If the satellite's orbital eccentricity is small ($e \ll 1$), then the equations of motion are specified by Hamiltonian (1.6).

According to Sect. 7.1.2 and Eq. (7.3), in the pendulum paradigm (1.6) one has: $\varepsilon_1 = -7e/2$, $\varepsilon_2 = e/2$, $\lambda = 1/\omega_0$, where ω_0 is given by formula (7.2). Since $|\varepsilon_1| \gg |\varepsilon_2|$, one can neglect ε_2, setting it to zero. Thus, one of the two perturbing resonances is set to be non-existent, and the resonance triplet is reduced to a doublet, and henceforth we denote ε_1 by ε. The adiabaticity parameter λ, relative perturbation strength ε, and perturbation period $T_{\text{pert}} = 2\pi/\Omega$ for any satellite can be evaluated based on observational data, compiled, e.g., in Shevchenko and Kouprianov (2002). This gives $\lambda = 0.85$, $\varepsilon = -0.014$, and $T_{\text{pert}} = T_{\text{orb}} = 0.61$ days for Prometheus; and $\lambda = 1.07$, $\varepsilon = -0.014$, and $T_{\text{pert}} = T_{\text{orb}} = 0.63$ days for Pandora; T_{orb} is the orbital period.

We see that $\lambda > 1/2$ in the both cases. Therefore, chaos is non-adiabatic, and, according to Sect. 3.3 and Eq. (3.28), the Lyapunov time in the chaotic resonance duad is given by

$$T_{\text{L}} \approx \frac{5T_{\text{pert}}}{2\pi} \cdot \left(\frac{4f(2\lambda)}{g(2\lambda, W(\lambda))} + \frac{f(\lambda)}{g(\lambda, W(\lambda))} \right)^{-1}, \tag{7.7}$$

where

$$W(\lambda) = 4\pi\varepsilon \frac{\lambda^2}{\sinh \frac{\pi\lambda}{2}} \tag{7.8}$$

and

$$f(x) = C_{\rm h}\frac{2x}{1+2x}, \quad g(x, y) = x \ln \frac{32e}{x|y|}, \tag{7.9}$$

where $C_{\rm h} \approx 0.80$ is Chirikov's constant given by formula (3.8), here e is the base of natural logarithms and x and y are "dummy" variables.

Equation (7.7) gives $T_{\rm L} = 1.6$ days for Prometheus and $T_{\rm L} = 1.9$ days for Pandora. Comparing these theoretical values with numerical-experimental ones, namely, $T_{\rm L} = 1.5$ days for Prometheus and $T_{\rm L} = 1.8$ days for Pandora, obtained in Shevchenko and Kouprianov (2002) for the planar chaotic rotation of these moons by means of direct numerical integrations of equations of motion, we see that the separatrix map techniques are perfectly accurate. If the chaotic rotation is non-planar (i.e., it represents a spatial tumbling), the Lyapunov times computed in Shevchenko and Kouprianov (2002) are somewhat smaller: $T_{\rm L} = 0.8$ days for Prometheus, and $T_{\rm L} = 1.1$ days for Pandora.

In the spatial case, the dynamical system has 3 and 1/2 degrees of freedom, instead of 1 and 1/2 in the planar case. Notwithstanding this difference, the Lyapunov time analytical prediction is, by an order of magnitude, still in accord to the estimates found in numerical integrations. In fact, this is no wander, because the synchronous resonance still plays the dominant role in the rotational phase space.

The Lyapunov times of the hypothetical chaotic rotation of Prometheus and Pandora are about thirty times less than the Lyapunov time of the observed chaotic rotation of Hyperion. This means that the chaotic nature of rotation of these two satellites, if their rotation were indeed chaotic, can be established on time intervals of observation much shorter than needed in case of Hyperion. However, there exists a dynamical effect that may impede the observational identification of chaos (Melnikov and Shevchenko 2008); this is the effect of preferential orientation of satellite's figure in the course of chaotic rotation.

7.1.4 Widths of Chaotic Layers

Wisdom et al. (1984) and Wisdom (1987a) estimated widths of the near-separatrix chaotic layers of spin-orbit resonances in the dynamics of planetary satellites and Mercury, using Chirikov's approach (Chirikov 1979) based on the separatrix map theory. Let us see how the widths can be estimated.

Here we designate the satellite's orientation by θ. The total width of the chaotic domain in momentum is given by Eq. (5.18). Comparing Hamiltonians (1.6), (7.1), and (7.3), one has $\Delta\dot{\varphi} = 2\Delta\dot{\theta} = 2\Delta\dot{x}$.

According to Hamiltonian (7.1), for the satellite's orientation one has $\theta = x = \varphi/2$; and, according to Eq. (5.18),

$$\Delta\dot{\theta} = \Delta\dot{\theta}_{\rm res} + \frac{\omega_0}{4}\left(\overline{w}_{\rm lb}(\lambda, \varepsilon_1) + \overline{w}_{\rm lb}(\lambda, \varepsilon_2)\right), \tag{7.10}$$

where $\overline{w}_{\rm lb}$ is given by Eq. (5.3). Therefore,

$$\Delta\dot{\theta} = \omega_0 \left(2 + \frac{1}{4}(\overline{w}_{\rm lb}(\lambda, \varepsilon_1) + \overline{w}_{\rm lb}(\lambda, \varepsilon_2))\right). \tag{7.11}$$

For Hyperion, the parameters are: $\omega_0 = \left(3\frac{B-A}{C}\right)^{1/2} \approx 0.89$, $\lambda = \omega_0^{-1} = 1.13$, $e = 0.1$, $\varepsilon_1 = -\frac{7e}{2} = -0.35$, and $\varepsilon_2 = \frac{e}{2} = 0.05$ (Melnikov and Shevchenko 1998). Since $\Delta\dot{\theta}_{\rm res} = 2\omega_0$, one arrives at

$$\Delta\dot{\theta}_{\rm res} = 1.77, \quad \overline{w}_{\rm lb}(\lambda, \varepsilon_1) = 2.31, \quad \overline{w}_{\rm lb}(\lambda, \varepsilon_2) = 0.32, \tag{7.12}$$

and

$$\Delta\dot{\theta} = 2.33, \tag{7.13}$$

in perfect agreement with the rotational phase space section obtained by direct numerical integrations, see Fig. 7.6.

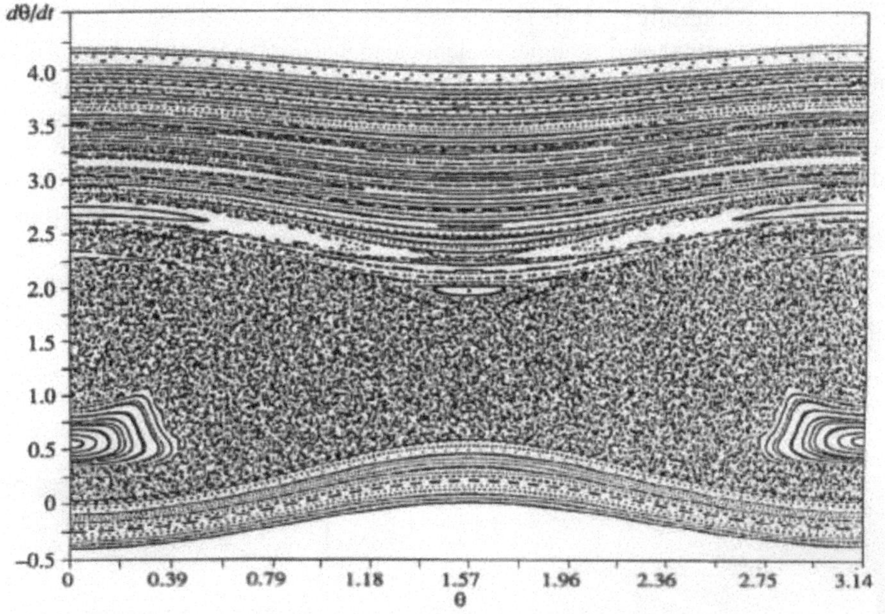

Fig. 7.6 Section of the rotational phase space in case of Hyperion (Figure 3 from Melnikov and Shevchenko 1998. With permission from Pleiades Publishing Inc.)

7.1.5 Chaotic Planar Rotation and Chaotic Tumbling

Assume that a triaxial satellite moves in an eccentric orbit with fixed eccentricity around a gravitating point mass (planet). The inertial parameters comprise the A/C and B/C ratios of the principal central moments of inertia $A \leq B \leq C$ of the satellite. The sole orbital parameter is the eccentricity e. The size of the satellite is supposed to be negligibly small compared with the distance to the planet: $r \ll R$.

Let the initial orientation of the satellite in space be arbitrary. An xyz inertial reference frame (Fig. 7.7) is defined at the pericentre of the satellite's orbit as follows. The x axis is directed along the planet–satellite radius vector, the y axis is directed along the satellite's orbital velocity vector, and the z axis is orthogonal to the satellite's orbital plane and completes the reference system to the right-handed one. Let a right-handed set of axes a, b, and c, directed along the satellite's principal axes of inertia with the moments A, B, and C, be "frozen" inside the satellite and coincide with the xyz axes set. The spatial orientation of the satellite in the xyz frame is defined by a sequence of imaginary rotations of the satellite by the Euler angles θ, ϕ, and ψ from the given position until the satellite reaches its actual orientation. The imaginary rotations are performed in the following order: (1) rotation by θ about c; (2) rotation by ϕ about a; (3) rotation by ψ about b. These rotations are graphically presented in Fig. 7.7.

This definition of the Euler angles is identical to that used in Wisdom et al. (1984) and is different from the standard one. The motivation for using the non-standard definition is that the standard one has a coordinate singularity when the c axis (the axis of the maximum moment of inertia) is orthogonal to the orbital plane. In the adopted system, this situation corresponds to $\phi = 0$, and the singularity is moved to $\phi = \pm\pi/2$, which corresponds to the satellite's axis of rotation lying in the orbital plane.

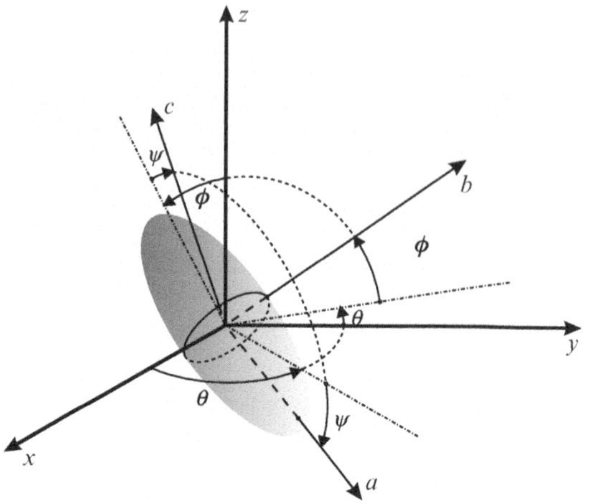

Fig. 7.7 Definition of the Euler angles in the xyz inertial reference frame (Figure 1 from Kouprianov and Shevchenko 2005, with permission from Elsevier)

The adopted set is related to the standard ($\psi', \phi' \in [0, 2\pi)$, $\theta' \in [0, \pi]$) set as follows:

$$
\begin{aligned}
\sin\theta' &= \sqrt{1-\cos^2\phi\cos^2\psi},\\
\cos\theta' &= \cos\phi\cos\psi,\\
\sin\phi' &= -\cos\phi\sin\psi/\sin\theta',\\
\cos\phi' &= \sin\phi/\sin\theta',\\
\sin\psi' &= (\cos\theta\sin\psi+\sin\theta\sin\phi\cos\psi)/\sin\theta',\\
\cos\psi' &= (-\sin\theta\sin\psi+\cos\theta\sin\phi\cos\psi)/\sin\theta'.
\end{aligned}
\tag{7.14}
$$

In the adopted coordinate frame, the Euler equations read:

$$
\begin{aligned}
\dot{\omega}_a &= \tfrac{B-C}{A}\left(\omega_b\omega_c - \tfrac{3}{R^3}\beta\gamma\right),\\
\dot{\omega}_b &= \tfrac{C-A}{B}\left(\omega_c\omega_a - \tfrac{3}{R^3}\gamma\alpha\right),\\
\dot{\omega}_c &= \tfrac{A-B}{C}\left(\omega_a\omega_b - \tfrac{3}{R^3}\alpha\beta\right),\\
\dot{\theta} &= (\omega_c\cos\psi-\omega_a\sin\psi)/\cos\phi,\\
\dot{\phi} &= \omega_a\cos\psi+\omega_c\sin\psi,\\
\dot{\psi} &= \omega_b-(\omega_c\cos\psi-\omega_a\sin\psi)\tan\phi,
\end{aligned}
\tag{7.15}
$$

where

$$R = 1 - e\cos E$$

is the satellite–planet distance, and the eccentric anomaly E can be calculated by solving the Kepler equation

$$E - e\sin E = t. \tag{7.16}$$

Time t is measured in units of $T_{\mathrm{orb}}/2\pi$, where T_{orb} is the satellite's orbital period. The direction cosines of the planet–satellite radius vector on the axes a, b, and c are

$$
\begin{aligned}
\alpha &= \cos\psi\cos(\theta-f)-\sin\psi\sin\phi\sin(\theta-f),\\
\beta &= -\cos\phi\sin(\theta-f),\\
\gamma &= \sin\psi\cos(\theta-f)+\cos\psi\sin\phi\sin(\theta-f),
\end{aligned}
\tag{7.17}
$$

where the true anomaly f is determined by the equation

$$\tan\frac{f}{2} = \left(\frac{1+e}{1-e}\right)^{1/2}\tan\frac{E}{2}. \tag{7.18}$$

The current state of the system is characterized by the vector

$$x = (\omega_a, \omega_b, \omega_c, \theta, \phi, \psi), \tag{7.19}$$

where ω_a, ω_b, ω_c represent projections of the angular velocity vector $\boldsymbol{\omega}$ on the a, b, c axes, and $\theta \in [0, 2\pi)$, $\phi\ [-\pi/2, \pi/2]$, $\psi \in [0, 2\pi)$ are the Euler angles.

The Lyapunov spectrum of system (7.15) consists of six elements: $L^{(1)} \geq L^{(2)} \geq L^{(3)} \geq L^{(4)} \geq L^{(5)} \geq L^{(6)}$, with $L^{(1)} = -L^{(6)}$, $L^{(2)} = -L^{(5)}$, and $L^{(3)} = -L^{(4)}$. Expressed in the original time units, the Lyapunov time is $T_{\mathrm{L}} = T_{\mathrm{orb}}/(\pi(L^{(1)} - L^{(6)})) = T_{\mathrm{orb}}/2\pi L^{(1)}$ days.

In the model of a uniform-density ellipsoidal satellite, the inertial parameters are defined by the satellite's shape parameters, namely, by the semi-axes $a \geq b \geq c$ of the triaxial ellipsoid that approximates the satellite shape:

$$A/C = (b^2 + c^2)/(a^2 + b^2), \quad B/C = (a^2 + c^2)/(a^2 + b^2). \tag{7.20}$$

The shape parameters of many satellites can be found in Seidelmann et al. (2002).

If the near-separatrix chaotic layer is thin, then the half-width of the synchronous resonance cell, bounded by the separatrix, can be calculated as equal to the amplitude of variation of the angular velocity on the separatrix:

$$\Delta\omega = \omega_0 = \left(\frac{3(B-A)}{C}\right)^{1/2} \tag{7.21}$$

(Beletsky 1965; Wisdom et al. 1984). The half-width is equal to the frequency ω_0 of small-amplitude oscillations on the synchronous resonance; ω_0 is given by formula (7.2).

In case of Deimos, the chaotic layer is thin, see Fig. 7.4 and also Figure 3 in Wisdom (1987a) and Figure 5 in Melnikov and Shevchenko (1998). Then, the resonance size is described by the given formula. In case of Hyperion, the chaotic layer is broad, see Fig. 7.6 and also Figure 2 in Wisdom et al. (1984) and Figure 4 in Klavetter (1989b); then, the maximal total variation of the angular velocity on the resonance can be calculated only if the chaotic layer width is taken into account. In case of Nereid, whose orbit is extremely eccentric (~0.75), even a moderate non-sphericity of the satellite's shape may lead to emergence of a vast chaotic domain formed by overlap of a great number of integer spin-orbit resonances, see Figures 8 and 9 in Dobrovolskis (1995).

In case of quasi-spherical (big) planetary satellites moving in close-to-circular orbits, the chaotic layers can be so thin that they cannot be discerned in any sections of the rotational phase space. Nevertheless, the chaos presence can be identified by calculating the Lyapunov exponents, if the initial conditions reside in the thin chaotic layer. For big satellites, the chaotic rotational motion, initially set close to planar one, remains to be almost planar, i.e., it is stable with respect to tilting the rotation axis. In case of, e.g., a model satellite with the parameters of Io, $|\phi|$ and $|\psi|$ both do not exceed 0.0025 radians, if one sets $\phi_0 = \psi_0 = 0.001$ (Kouprianov

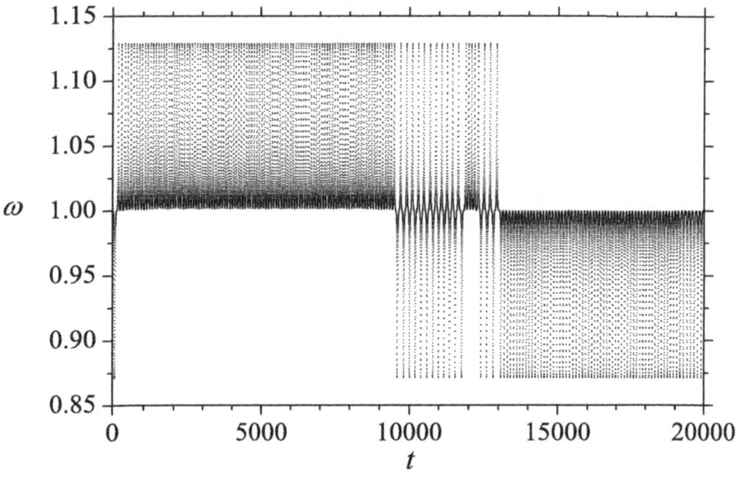

Fig. 7.8 How chaotic rotation of a big planetary satellite may look like: a time history of the angular velocity ω of a model Io, if the initial conditions are set inside the near-separatrix chaotic domain of the synchronous resonance (Figure 2 from Kouprianov and Shevchenko 2005, with permission from Elsevier)

and Shevchenko 2005). This stability is unlike the behaviour numerically observed by Wisdom (1987a) in case of small irregularly-shaped moons, which tumbled chaotically before being captured into the synchronous resonance.

We see that, for the big satellites, entering the chaotic layer of the synchronous resonance does not necessarily lead to any chaotic tumbling. However, a thorough shake-up is still inevitable: on entering the chaotic layer, the more or less uniform rotation is replaced by librations with extremal amplitudes. This is illustrated in Fig. 7.8, where a time history of the spin rate of Io, with the initial conditions set inside the chaotic layer. The second kind Hamiltonian intermittency pattern, considered in Sect. 4.4, is readily recognizable.

How the overall statistics of satellite rotation states in the Solar system can be characterized? By analysing the stability of the possible rotational states of planetary satellites, Melnikov and Shevchenko (2010) showed that, although most satellites with known rotation states rotate synchronously, a significant part (at least two thirds) of all satellites with unknown rotation states reside in other (fast or chaotic) rotation modes. To demonstrate this, a stability diagram in the (ω_0, e) plane was constructed, where ω_0 is the satellite dynamical asymmetry parameter, given by formula (7.2), and e is the orbital eccentricity (Fig. 7.9).

In the constructed diagram, theoretical borders of the zones of existence of synchronous resonance modes are drawn according to Melnikov (2001). In the given ranges of ω_0 and e, the marked domains are delineated by four curves. At the $(\omega_0 = 1,\ e = 0)$ point, a branching curve stems. To the left of the curve, one odd 2π–periodic solution of the Beletsky equation can exist; this solution is called *α-resonance*. To the right of the curve, one or two stable solutions can exist; these

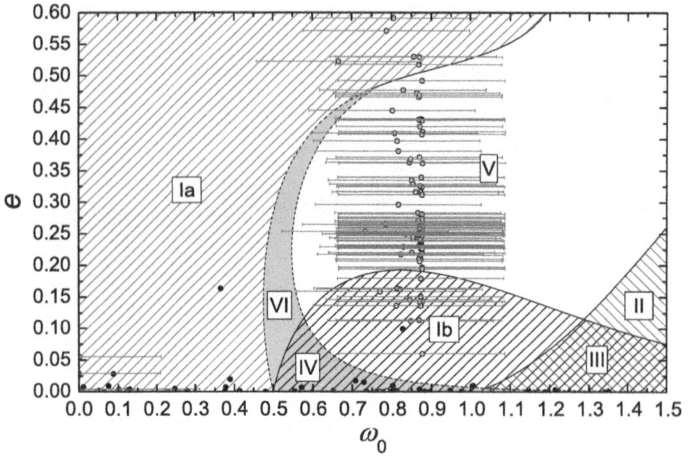

Fig. 7.9 Distribution of moons with known radii in the (ω_0, e) plane (Figure 3 from Melnikov and Shevchenko 2010, with permission from Elsevier)

are, respectively, *β-resonance* or both α-resonance and β-resonance. α-resonance exists both at $\omega_0 \geq 1$ and at $\omega_0 < 1$; β-resonance exists only at $\omega_0 \geq 1$. The synchronous resonance modes are illustrated further on in phase space sections in Figs. 7.12 and 7.13.

At the ($\omega_0 = 1/2$, $e = 0$) point, the zone of parametric resonance emerges. The left border of this zone corresponds to the loss of stability of the α-resonance in domain Ia, through the period-doubling bifurcation. In the plot, this is the dashed curve on the left. The dashed curve on the right corresponds to the loss of stability of the bifurcated mode, through the second consecutive bifurcation. The borders of domains IV and VI are therefore formed by the first doubling bifurcation curve (on the left) and by the second doubling bifurcation curve (on the right). For a detailed classification of synchronous resonance modes, see Melnikov and Shevchenko (1998), Melnikov (2001), Melnikov and Shevchenko (2008).

In sum, the zones can be characterized as follows. Zones Ia and Ib represent the domains of existence of solely α-resonance. Zone II is the domain of existence of solely β-resonance. Zone III is the domain of coexistence of α and β-resonances. Zone IV is the domain of coexistence of α-resonance and period-doubling bifurcation modes of α-resonance. Zone V is the domain of non-existence of any 1/1 synchronous resonance. Zone VI is the domain of existence of solely period-doubling bifurcation modes of α-resonance.

To place actual satellites in the diagram, one should first estimate their ω_0 and e parameters. For 34 satellites, the ω_0 and e data are directly available; see compilation and references in Kouprianov and Shevchenko (2005). For the remaining objects in the sample of 145 satellites, ω_0 was estimated statistically, and e were taken from satellite databases; see Melnikov and Shevchenko (2010). According to Fig. 7.10, satellites with physical radii less than 300 km are irregularly-shaped and, consequently, have large ω_0, whose values can be statistically inferred

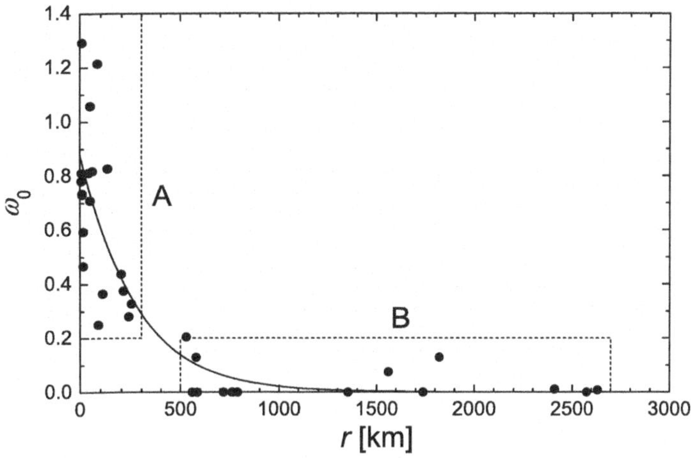

Fig. 7.10 The dynamical asymmetry parameter ω_0 dependence on the physical radius r. Boxes A and B delineate two types of satellites: small irregularly-shaped and big spherical, respectively. Solid curve: exponential approximation $\omega_0(r) = A_0 \exp(-r/r_0)$, where $A_0 = 0.88$ and $r_0 = 270$ km (Figure 1 from Melnikov and Shevchenko 2010, with permission from Elsevier)

from sizes of the satellites. In Fig. 7.9, the open circles show the objects with the ω_0 parameter determined statistically in Melnikov and Shevchenko (2010). Three-sigma errors in estimating ω_0 are shown by horizontal bars. The solid circles show the objects with ω_0 known from observations.

No synchronous rotation states exist in domain V. For most of satellites in domain Ib, the synchronous state is highly probable to be attitude unstable. Therefore, all satellites in domain V and most of satellites in domain Ib rotate either regularly (if they are tidally unevolved) or chaotically (if they are tidally evolved). In sum, at least two thirds of all presented satellites cannot rotate synchronously (Melnikov and Shevchenko 2010).

In Fig. 7.11, a representative example of section of the phase space of the planar rotation of a satellite residing in domain V is shown. The adopted $e = 0.25$ and $\omega_0 = 0.9$ values correspond to the domain V centre. The phase space section is defined at the pericentre of the orbit, i.e., the motion is mapped each orbital period. Obviously, no synchronous state is present in the phase space section; there is a broad chaotic band instead.

Representative examples of the phase space sections, corresponding to other domains, namely, Ia, Ib, III, and IV, are given in Melnikov and Shevchenko (2010).

With the gradual emergence of new observational ground-based and space-based capabilities, the number of discovered planetary satellites is increasing. Since all newly discovered satellites are small, they are all strongly non-spherical; see Fig. 7.10. Besides, the orbits of newly discovered objects are typically highly eccentric. Therefore, all newly discovered satellites are expected to reside mainly in domain V of the (ω_0, e) plane, where the 1/1 synchronous state of rotation is impossible.

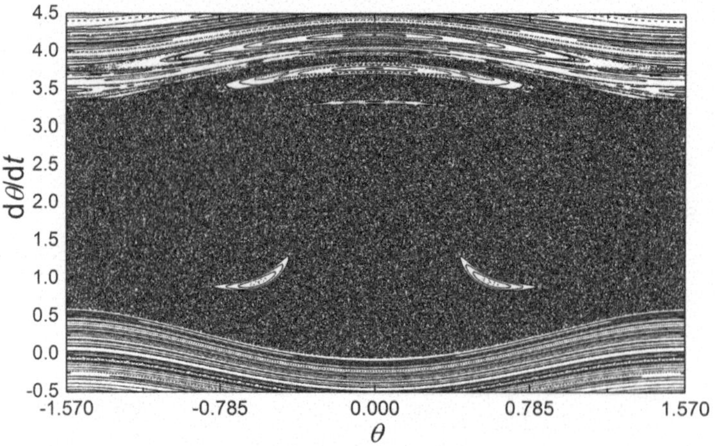

Fig. 7.11 A section of the phase space section of the planar rotation of a satellite in domain V of the (ω_0, e) diagram in Fig. 7.10 (Figure 6 from Melnikov and Shevchenko 2010, with permission from Elsevier)

7.1.6 Stability with Respect to Tilting the Axis of Rotation

Even in absence of external torques, an isolated rotating non-spherical body generically suffers permanent deformations by centrifugal forces, depending on the position of the rotation axis inside the body (Burns and Safronov 1973). The rotation energy dissipates, but the total angular momentum is conserved, while the rotation axis is slowly turning inside the body until the state with the minimally possible, for the given total angular momentum, rotational energy is achieved (Peale 1977; Wisdom 1987b). In this state, the angular momentum vector coincides in direction with the axis of the largest moment of inertia.

In case of a planetary satellite, the rotation is subject to external torques. The gravitational tides stretch the satellite approximately along the planet–satellite line, but the tidal bump is displaced in the direction of rotation, because the material reaction to the tidal force is not instantaneous.

In the beginning of the long-term tidal evolution, satellites have rotation periods much smaller then their typical orbital periods (Peale 1977). The planet's gravitation, acting on the tidal bump, first, tends to reduce to zero the angle between the rotation axis and the normal to the satellite's orbit plane, and, second, creates a torque, which tends to slow the rotation down, typically, to the synchronous state (Peale 1977; Wisdom 1987b). Therefore, the natural final state of the long-term tidal evolution is the planar (the rotation axis being orthogonal to the orbit plane) rotation of the satellite in synchronous resonance with its orbital motion. In this state, the planar rotation axis coincides with the axis of the maximum moment of inertia of the satellite.

Depending on initial conditions and tidal parameters, the way to the final synchronous "calm harbour" may turn out to be not smooth at all. The despinning satellite successively enters near-separatrix chaotic layers of various spin-orbit resonances, where the rotation can be unstable with respect to tilting the rotation axis; then, the satellite starts to chaotically tumble. The major obstacle is the near-separatrix chaotic layer of the synchronous resonance, where such instability was revealed by means of direct numerical integrations by Wisdom (1987a). Whether the satellite starts to tumble or eludes this fate, depends on its tidal evolution speed and on its shape and orbit parameters. For the final synchronous state to be possible, the motion in the centre of synchronous resonance should itself be stable with respect to tilting the rotation axis.

In the present Section, we discuss the stability with respect to tilting the rotation axis in the synchronous and other spin-orbit resonances. The orbital and inertial parameters are taken for the cases of Hyperion, Phobos, Deimos, and Amalthea. To separate the rotation trajectories into regular and chaotic ones, the modal structure of the finite-time Lyapunov exponent distributions is analysed, as described in Sect. 2.6.

In the course of the long-term tidal evolution, the satellite passes through various resonant spin-orbit states and crosses corresponding near-separatrix chaotic layers. Therefore, the stability analysis should be accomplished not only for the synchronous resonance centre, but also for a large-enough phase-space domain near the synchronous resonance.

Consider the satellites listed in Table 7.2. Since the orbital eccentricity of Hyperion is subject to long-periodic oscillations (caused by perturbation from Titan; see Klavetter 1989b), in the e range from 0.08 to 0.12, three cases were considered in Melnikov and Shevchenko (1998): $e = 0.08$, $e = 0.1$, and $e = 0.1236$.

The model Amalthea is illustrated in Figs. 7.12 and 7.13. Both kinds of trajectories (chaotic and regular) are shown in Fig. 7.12, whereas only those unstable with respect to tilting the rotation axis are shown in Fig. 7.13. This doubled representation allows one to class the trajectories into stable and unstable ones with respect to tilting the rotation axis, irrespective of the planar trajectory kind.

Therefore, the phase space sections in Figs. 7.12 and 7.13 contain information on the stability of the main spin-orbit resonances (as well as non-resonant spin-orbit states) with respect to tilting the rotation axis. Major spin-orbit states are easily

Table 7.2 The orbital eccentricity e and the inertial parameters A/C and B/C (Klavetter 1989a; Black et al. 1995; Thomas et al. 1995; Wisdom 1987a; Melnikov and Shevchenko 1998)

Satellite	e	A/C	B/C
Hyperion	0.08–0.12	0.622	0.884
Phobos	0.015	0.723	0.850
Deimos	0.0005	0.711	0.914
Amalthea	0.003	0.473	0.957

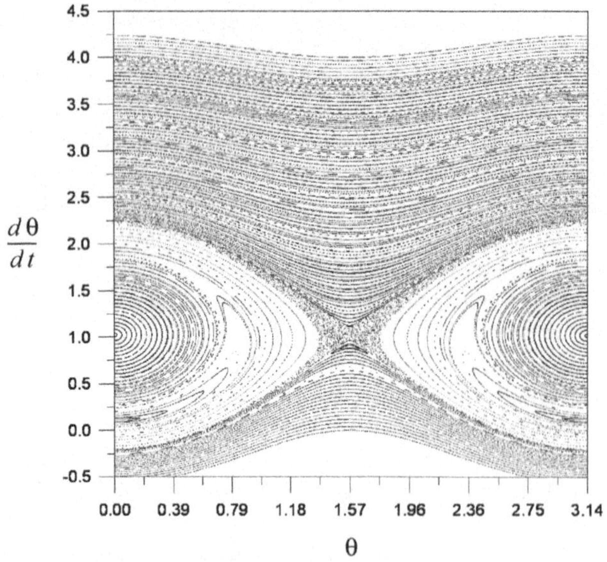

Fig. 7.12 The phase space section defined at the orbit pericentre; $e = 0.003$, $A/C = 0.473$, and $B/C = 0.957$ (model Amalthea) (Figure 3a from Melnikov and Shevchenko 1998. With permission from Pleiades Publishing Inc.)

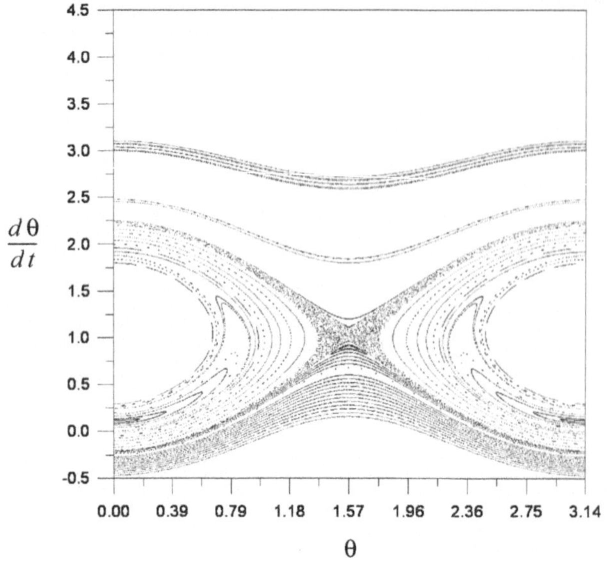

Fig. 7.13 Same as Fig. 7.12, but solely the trajectories unstable with respect to tilting the rotation axis are shown (Figure 3b from Melnikov and Shevchenko 1998. With permission from Pleiades Publishing Inc.)

identifiable, in which the satellite may or may not reside in the course of its long-term tidal evolution, if the orbital eccentricity is kept close to their present-day value.

The stability/instability in spin-orbit resonances 1/2, 1/1 (unstable), 2/1, and 9/4 (stable) in case of Hyperion with $e = 0.1$ was numerically investigated in Wisdom et al. (1984), and that in resonances 1/2 (unstable), 1/1, and 3/2 (stable) in case of Phobos were studied in Wisdom (1987a). The data following from the presented phase space sections in Figs. 7.12 and 7.13 perfectly agree with the results of Wisdom et al. (1984), Wisdom (1987a).

In case of Amalthea, there are two centres of synchronous resonance in the phase space section; see Figs. 7.12 and 7.13. Comparing Figs. 7.12 and 7.13, we see that the motion in one of the centres of synchronous resonance is stable, and in the second one it is unstable. These two centres of the synchronous resonance correspond to the α *mode* and β *mode*, considered above in Sect. 7.1.5.

7.2 Chaotic Obliquities of Planets

Laskar et al. (1993) and Néron de Surgy and Laskar (1997) argued that the current obliquity of the Earth is long-term stable due to the presence of the Moon: the Moon-caused precession of Earth's spin axis is rapid enough to prevent the chaotic diffusion between relevant spin-orbit resonances, as they are more widely separated in the phase space, due to this precession. This phenomenon is important to take into account when assessing habitability properties of rocky exoplanets with and without big moons (Shevchenko 2017b).

With the Moon, Earth's obliquity stays within the range from 22.1° to 24.5° (Lissauer et al. 2012). If the Moon were absent, the Earth would suffer large variations of its obliquity (between 0° and 85°), and this would cause catastrophic variations of climate (Laskar et al. 1993; Néron de Surgy and Laskar 1997). Conversely, the obliquity of Mars, which does not have a large satellite, varies in the range 0–60° (Laskar and Robutel 1993). This must be at least one of the causes of its long-term non-habitability, although Mars is marginally inside the Solar habitability zone.

On the other hand, Lissauer et al. (2012), Li and Batygin (2014) performed series of simulations of the long-term Earth rotation and concluded that in the Moon absence the chaotic diffusion rate in the obliquity would be low enough for the development of life to be successful, and, therefore, the long-term habitability would not be precluded.

7.2.1 Relevant Spin-Orbit Resonances

In Fig. 7.14, real and possible dynamical causes for the long-term variations in obliquity of the Earth are illustrated, as outlined in Laskar and Robutel (1993). In

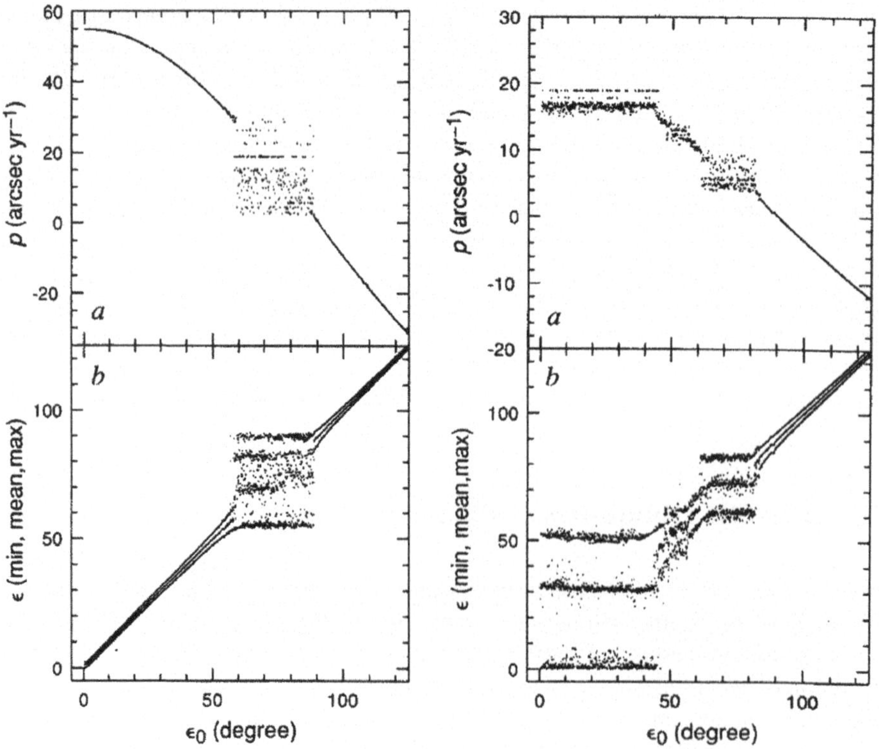

Fig. 7.14 The precession frequency and the maximum, average and minimum obliquities of the Earth's rotation axis, as functions of the initial obliquity. Left panel: the Moon present; right panel: the Moon absent (Figures 2 and 3 from Laskar and Robutel 1993, by permission from Springer Nature, © 1993)

the upper panels, the precession frequency p of the Earth's rotational axis is given, and, in the lower panels, the maximum, average and minimum obliquities ϵ of the Earth's rotation axis are presented. All the quantities are given as functions of the initial obliquity ϵ_0. In the left panel, the plots for the model taking into account the Lunar torque are shown. In the right panel, conversely, the plots for the single Earth model (the Moon absent) are shown. One may see that the Moon stabilizes the Earth's obliquity.

Conversely, a Moonless Mars may have large chaotic variations of the obliquity, causing catastrophic variations in climate; this may be the reason of its sterility (Laskar and Robutel 1993).

Tantalizingly, the favourable range of the Earth's surface temperature variations is conditioned by the suitable obliquity of the Earth, most probably caused by a giant impact (Williams 1993). Conversely, the generic tilt of any planet in a relatively low orbit around its host star is equal to zero, as in case of Mercury and Venus, because this is a natural outcome of the tidal spin-orbit evolution of orbiting bodies; see, e.g., Wisdom et al. (1984).

Chapter 8
Orbital Dynamics of Minor Bodies

The Moon upon her fluent Route
Defiant of a Road

Emily Dickinson, Poem 1528

In studies of orbital dynamics of celestial bodies it is also usually possible to explain chaotic manifestations by using paradigmatic models of interaction and overlap of non-linear resonances, addressed in the first part of the book. In this chapter, we consider major classes of the Solar system minor bodies on the subject of their actual or potential chaotic behaviour. These classes include: planetary satellite systems (in particular, the Miranda–Umbriel, Mimas–Tethys, and Prometheus–Pandora systems); near-Earth asteroids (NEAs), main-belt asteroids and Kuiper belt objects; and comets, which are typically highly chaotic. In connection with the considered dynamical populations, concepts of two-body and three-body resonances and secular resonances are considered. Dynamical environments of small bodies, and, in particular, chaotic zones around contact binaries (such as famous KBO 2014 MU69), are analytically characterized. The presentation in this chapter is partially based (in Sects. 8.1, 8.2, and 8.3) on the papers by Melnikov and Shevchenko (2005), Shevchenko (2008b) (by permission from © Oxford University Press), Shevchenko (2007a) (by permission from Cambridge University Press), Smirnov and Shevchenko (2013) (by permission from Elsevier), and Shevchenko (2010).

8.1 Dynamics of Satellite Systems

If the orbital frequencies (mean motions) of two planets in a planetary system, or two satellites in a satellite system, are approximately commensurable, i.e., their ratio

I. I. Shevchenko, *Dynamical Chaos in Planetary Systems*, Astrophysics and Space Science Library 463, https://doi.org/10.1007/978-3-030-52144-8_8

is approximately equal to the ratio of two integers, the system is, from a common sense viewpoint, either close to orbital resonance,[1] or resides in it.

In the Solar system, many planetary satellites form resonant or close-to-resonant configurations. In the Jovian satellite system, the Galilean satellites Io and Europa, as well as Europa and Ganymede, reside in the 2/1 mean motion resonance; thus, the system of these three satellites is involved in the three-body resonance 4:2:1 (called the *Laplace resonance*). In the Saturnian system, Mimas and Tethys, as well as Enceladus and Dione, reside in the 2/1 mean motion resonance, Dione and Rhea are close to the 5/3 resonance, and Titan and Hyperion are in the 4/3 resonance. In the Uranian system, all commensurabilities are approximate: Miranda and Umbriel are close to resonance 3/1, Ariel and Umbriel to 5/3, Umbriel and Titania to 2/1, Titania and Oberon to 3/2.

Captures of satellite systems in orbital resonances represent natural stages of the long-term tidal evolution of these systems (Tittemore and Wisdom 1990). Capture in a resonance may considerably affect the subsequent dynamical history of any satellite system (Malhotra and Dermott 1990; Tittemore and Wisdom 1989) and even internal structure of satellites (Dermott et al. 1988).

For a satellite system to be captured in a mean motion resonance, it should cross the near-separatrix chaotic layer of the corresponding resonance. Inside the layer, the system moves chaotically. Therefore, any system now observed to reside in a resonance, spent some time of its long-term dynamical history in a chaotic regime. When neighbouring resonances overlap, chaos can be wide-spread and the time of residence in the chaotic regime can be rather prolonged.

Chaos in the orbital dynamics of satellite systems (and, generally, many other celestial-mechanical systems, including planetary systems) may emerge due to interaction and overlap of neighbouring mean motion resonances, as well as, inside a separate mean motion resonance, due to interaction and overlap of subresonances corresponding to the given mean motion resonance; see, e.g., Malhotra (1994, 1998). The orientation of orbits in space is subject to variations including secular precession of pericentres and nodes, and this precession is just the reason of splitting of orbital resonances into subresonances. Chaos acting on longer (orbital precession) timescales is due to interaction of secular resonances, see Sect. 8.2.5.

Stages of dynamical chaos played important role in the long-term orbital evolution of planetary satellites in the Solar system. The Miranda–Umbriel system is perhaps the best studied one in this respect. Miranda and Umbriel are the second and fifth satellites of Uranus; at present, the system is close to the 3/1 mean motion resonance. It is rather probable that the system attended this resonance and its chaotic domain in the past in the course of tidal evolution (Malhotra and Dermott 1990; Tittemore and Wisdom 1990). To the accuracy of the second order in eccentricities and inclinations, the 3/1 mean motion resonance is split into six subresonances (Malhotra 1990; Malhotra and Dermott 1990): three inclination-type

[1]The expression "close to resonance" means that the frequencies satisfy the given integer relation only approximately, and, therefore, the corresponding resonant phase rotates, instead of libration.

subresonances and three eccentricity-type subresonances. The resonance dynamics can be considered in this approximation separately in the circular-inclined and planar-eccentric problems. The Hamiltonian of each of these problems can be reduced to that of the non-linear pendulum with periodic perturbations. Analytical and numerical-experimental estimates of the Lyapunov time for several selected stages of evolution of the system are given in Melnikov and Shevchenko (2005). Both the theory and numerical integrations give the Lyapunov time of order of 100 yr. This means that, even if the system were chaotic at present, its chaos would be practically unobservable.

Apart from the Miranda–Umbriel system, chaotic states in past epochs of orbital evolution and their effect on the present orbital states are known to have been prominent in the Enceladus–Dione system (S2 and S4) (Ferraz-Mello and Dvorak 1987; Karch and Dvorak 1988; Callegari and Yokoyama 2007), Miranda–Ariel system (U5 and U1) (Tittemore and Wisdom 1990), Ariel–Umbriel system (U1 and U2) (Tittemore and Wisdom 1988, 1990), Titan–Iapetus system (S6 and S8) (Noyelles and Vienne 2005), Galilean system (J1, J2, J3 and J4) (Tittemore 1990; Noyelles and Vienne 2005, 2007).

Prominent chaos in the orbital dynamics of satellite systems took place not only in the past epochs, but is also immanent to the nowadays dynamics of some systems. Champenois and Vienne (1999a,b) considered dynamics of Mimas and Tethys, the first and third satellites of Saturn, residing at present in the 4/2 orbital inclination-type mean motion resonance. This resonant pair is unique in the Solar system, as it has large-amplitude librations ($\simeq 95°$) of the resonant phase, with a rather small time period (about 70 yr). The averaged Hamiltonian of the problem was reduced in Champenois and Vienne (1999b) to the form of a non-linear pendulum with periodic perturbations. Using this analytical representation of the Hamiltonian, one may readily obtain analytical estimates of the Lyapunov time. Such estimates were obtained in Melnikov and Shevchenko (2005), and were confirmed therein by direct integrations of the equations of motion. They turned out to be 300–600 yr in various models. Thus, chaos in this system does not manifest itself on short enough timescales accessible to observations.

To date, the chaotic orbital behaviour of Prometheus and Pandora is the only known example of directly observable orbital chaos in the Solar system, just as the rotation of the 7th satellite of Saturn, Hyperion, is the only known example of observable chaos in the rotational dynamics of planetary satellites.

Historically, chaos in the orbital motion of Pandora was theoretically envisaged by Borderies et al. (1984), as early as in 1984, in the same year when chaos in the rotation of Hyperion was predicted by Wisdom et al. (1984). However, contrary to the Hyperion case, chaos in the Prometheus–Pandora system was not especially sought for in observations, but was independently discovered later on.

In fact, the Halley comet dynamics, considered further on in Sect. 8.3, provides another example of chaos identified in observations. However, there exists a principal difference. Comet Halley's chaos is not directly observable; it was revealed by Chirikov and Vecheslavov (1986), Vecheslavov and Chirikov (1988), Chirikov and Vecheslavov (1989) by means of analysing data of historical chronicles covering many hundreds of years. Conversely, chaos in the Prometheus–Pandora system can be observed directly anytime on short enough timescales.

Here we are concerned with solely chaotic orbital regimes of satellite systems. For general reviews on the dynamics of planetary satellites see Peale (1976, 1986, 1999). An introductive theory and discussions can be found in Malhotra (1998).

8.1.1 The Miranda–Umbriel System

Chaotic regimes play prominent roles in the long term orbital evolution of the planetary satellite systems. The Miranda–Umbriel system (U5 and U2) is perhaps the best studied one in this respect.

No low-order commensurabilities are observed now in the orbital motion of the satellites of Uranus (Malhotra 1990). Tittemore and Wisdom (1989) found that chaos associated with the 3/1 mean motion resonance of the Uranian satellites Miranda and Umbriel could result in a significant change in the orbital elements of these moons and in an escape from the resonance in the course of their long-term orbital evolution; a passage of the moons through this resonance in the past could have caused the present high inclination ($\simeq 4.3°$) of Miranda's orbit.

In this context, Malhotra and Dermott (1990) considered a role of secondary resonances in the long-term orbital evolution of Miranda and Umbriel. By definition, the secondary resonances represent resonances between phase oscillations on the primary resonance and an external periodic perturbation. Malhotra and Dermott (1990) developed a perturbed pendulum model describing the dynamics of the Miranda–Umbriel system in a secondary orbital resonance. Within the framework of this model, they showed that the capture in the 3/1 mean-motion resonance and the following capture in the 3/1 secondary resonance with the backward final escape from the main resonance resulted in the present-day anomalously high orbital inclination of Miranda.

Malhotra (1990) obtained analytical and numerical estimates of the probability of capture in the 3/1 secondary resonance. The probability turned out to be high enough for this mechanism to work. Therefore, the Miranda–Umbriel system could pass through the near-separatrix chaotic layer of the 3/1 orbital resonance twice during its dynamical history.

The problem of chaotic motion in the 3/1 orbital resonance is of interest, since its study helps to clarify details of the long-term orbital evolution of the Miranda–Umbriel system. The averaged (on the orbital timescale) Hamiltonians of the three-body problem "planet—two satellites" in the vicinities of the 3/1 and 2/1 mean motion resonances can be reduced, in some approximation, to the Hamiltonian of the non-linear pendulum with periodic perturbations (Malhotra 1990; Malhotra and Dermott 1990; Champenois and Vienne 1999b). Due to the precession of lines of apsides of satellite orbits, orbital resonance splits into subresonances. In case of the 3/1 mean motion resonance, in approximation of the second order in eccentricities and inclinations, there exist six subresonances

(Malhotra 1990; Malhotra and Dermott 1990): three *inclination-type subresonances* and three *eccentricity-type subresonances*. In this approximation, the resonant dynamics can be considered separately in frameworks of *inclined-circular* and *planar-elliptic* three-body problems. The system Hamiltonian in each of these cases is reducible to a Hamiltonian with two degrees of freedom.

According to Malhotra and Dermott (1990), approximate equations of the motion in the inclined-circular and planar-eccentric problems have similar analytical structures. If the terms of order higher than two in inclinations are neglected, the Hamiltonian of the inclined-circular three-body problem "planet—two satellites" in the vicinity of the 3/1 mean-motion resonance is reducible, after averaging on the orbital timescale, to the form

$$H = -\frac{GI^2}{2} + F\cos\phi + \varepsilon_1 \sin\left(\frac{\phi}{2} + \tilde{\tau}\right) + \varepsilon_2 \sin\left(\frac{\phi}{2} - \tilde{\tau}\right) \tag{8.1}$$

(Malhotra 1990). In Eq. (8.1), $\phi = 3\lambda_2 - \lambda_1 - 2\Omega_1 - \pi$ is the resonant phase of the so-called inclination-type i_1^2 resonance (Malhotra 1990; Malhotra and Dermott 1990), $\tilde{\tau} = \tilde{\Omega}t$ is the perturbation phase angle, $\tilde{\Omega}$ is the perturbation frequency, equal to the half-frequency of the rotation of the phase angle $3\lambda_2 - \lambda_1 - 2\Omega_2$ of the neighbouring i_2^2 resonance, λ_1 and λ_2 are the mean longitudes of the inner and outer moons, respectively; Ω_1 and Ω_2 are the longitudes of the ascending nodes of the inner and outer moons, respectively; I is the momentum conjugated to ϕ; t is time. The G, F, ε_1, and ε_2 parameters are given by

$$G = \frac{3}{m_1 a_1^2}\left(1 + 9\alpha^2 \frac{m_1}{m_2}\right), \quad F = \frac{\mathcal{G} m_1 m_2}{8a_1}\alpha^2 b_{3/2}^{(2)}(\alpha)(i_1^{\text{res}})^2,$$

$$\varepsilon_1 = -\frac{\mathcal{G} m_1 m_2}{4a_1}\alpha^2 b_{3/2}^{(2)}(\alpha) i_1^{\text{res}} i_2, \quad \varepsilon_2 = \frac{\mathcal{G} m_1 m_2}{4a_1}\alpha^2 b_{3/2}^{(1)}(\alpha) i_1^{\text{res}} i_2. \tag{8.2}$$

Here, $\mathcal{G}$ is the universal gravitational constant; m_1 and m_2 are the masses of the inner and outer moons, respectively; i_1^{res} is the orbital inclination of the inner moon in the exact 3/1 resonance (this inclination is approximately equal to the mean value of i_1 for the motion inside the resonance); i_2 is the outer moon's orbital inclination, assumed to be a model-dependent constant, see Malhotra (1990), Malhotra and Dermott (1990); $b_{3/2}^{(1)}(\alpha)$ and $b_{3/2}^{(2)}(\alpha)$ are Laplace coefficients; and $\alpha = a_1/a_2 \approx 0.481$ is the ratio of the semimajor axes of the inner and outer moons in the 3/1 resonance. It follows from Eq. (8.2) that

$$\frac{\varepsilon_1}{F} = -2\frac{i_2}{i_1^{\text{res}}}, \quad \frac{\varepsilon_2}{F} = 2\frac{i_2}{i_1^{\text{res}}}\frac{b_{3/2}^{(1)}(\alpha)}{b_{3/2}^{(2)}(\alpha)} \simeq -1.718\frac{\varepsilon_1}{F} \tag{8.3}$$

(Malhotra 1990; Shevchenko 2000b). The perturbation frequency $\widetilde{\Omega}$ is determined mostly by the dynamical oblateness of the host planet:

$$\widetilde{\Omega} \simeq -\frac{1}{2}\sqrt{\frac{\mathcal{G}M}{a_1^3}}\left\{\alpha^{3/2}\left[-3\mathcal{J}_2\left(\frac{R}{a_2}\right)^2+\frac{1}{2}\frac{m_1}{M}\alpha b_{3/2}^{(1)}(\alpha)\right]-\right.$$
$$\left.\left[-3\mathcal{J}_2\left(\frac{R}{a_1}\right)^2+\frac{1}{2}\frac{m_2}{M}\alpha^2\left(b_{3/2}^{(1)}(\alpha)-\frac{1}{2}b_{3/2}^{(2)}(\alpha)\right)\right]\right\} \tag{8.4}$$

(Malhotra 1990), where M, R, and $\mathcal{J}_2$ are the mass, equatorial radius and second zonal harmonic of the gravitational field of the host planet, respectively.

For the Miranda–Umbriel system, we set $R = 26{,}200\,\text{km}$, $\mathcal{J}_2 = 3.346 \cdot 10^{-3}$, $m_1/M = 7.9 \cdot 10^{-7}$, $m_2/M = 1.45 \cdot 10^{-5}$, $a_1/R = 4.8626$, and $a_2/R = 10.1145$. According to formula (8.4), $\widetilde{\Omega} \simeq -0.3251\,\text{yr}^{-1}$. Formula (65) in Malhotra (1990) gives nearly the same value: $\widetilde{\Omega} \simeq -0.3306\,\text{yr}^{-1}$. Following Shevchenko (2000b) and Melnikov and Shevchenko (2005), we reduce the Hamiltonian (8.1) to the perturbed pendulum standard form, by performing the canonical transformation $I = -p$, $\phi = \phi$ (with valence equal to -1), the change $\widetilde{\Omega} \to -\widetilde{\Omega}$ (to make the $\widetilde{\Omega}$ frequency positive), and the shift $\tilde{\tau} = \tau + \frac{\pi}{2}$. One gets the paradigmatic Hamiltonian (1.6), where φ is the deviation angle of the model pendulum from its position of equilibrium, p is the momentum, $\tau = \widetilde{\Omega}t + \tau_0$, where $\widetilde{\Omega}$ and τ_0 are the frequency and the initial phase of perturbation, respectively; $k = 1/2$, $a = -\varepsilon_1$, and $b = \varepsilon_2$.

In the inclined-circular approximation, the inclinations of the moons slowly vary in time in the course of the system long-term evolution. Data on the orbital inclinations of Miranda and Umbriel (i_1^{res} and i_2, respectively) are presented in Table 8.1 for four epochs of the past long-term evolution, according to the numerical modelling in Malhotra and Dermott (1990). Each epoch corresponds to a separate dynamical model with individual fixed parameters. In model 1, the system dynamics is distinguished by the presence of the 4/1 secondary resonance. In models 2 and 3, the 3/1 and 2/1 secondary resonances, respectively, dominate. Model 4 corresponds to the time epoch of the system escape from the i_1^2 resonance.

In the course of the long-term evolution, the Miranda–Umbriel system passes, in succession, through secondary resonant states corresponding to the models presented in Table 8.1. If the i_1 initial value is $0.1°$, the inclination of Miranda's orbit reaches $2.2°$ (model 1) when $t \simeq 1.4 \cdot 10^5$ yr. Models 2, 3, and 4 correspond to the time epochs $t \simeq 2 \cdot 10^5$, $4 \cdot 10^5$, and $8 \cdot 10^5$ yr, respectively.

Table 8.1 The Miranda–Umbriel system, four epochs of dynamical history (Melnikov and Shevchenko 2005)

Model	i_1^{res}, °	i_2, °	$\widetilde{\Omega}$, yr^{-1}	$\varepsilon_1/\mathcal{F}$	$\varepsilon_2/\mathcal{F}$
1	2.2	0.20	−0.325	−0.182	0.312
2	2.8	0.20	−0.325	−0.143	0.245
3	4.1	0.20	−0.325	−0.098	0.168
4	5.7	0.15	−0.325	−0.053	0.090

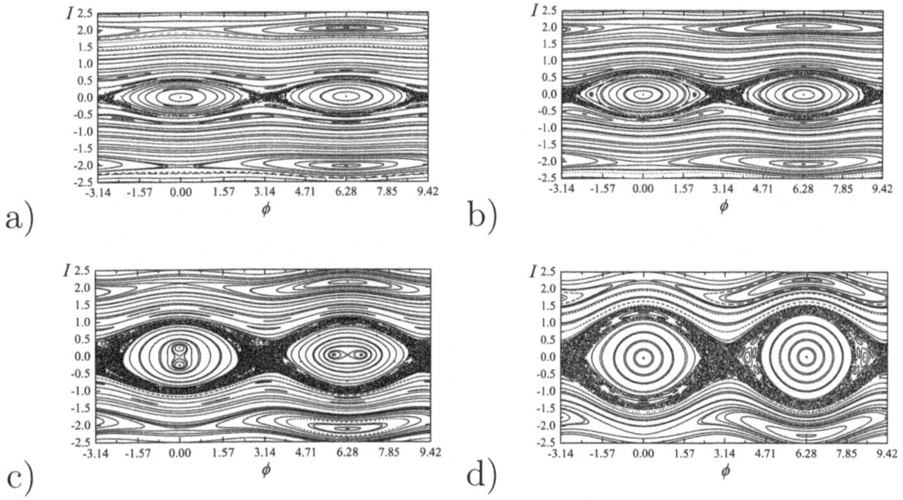

Fig. 8.1 Phase space sections $\tau = 0 \pmod{2\pi}$ for the four models presented in Table 8.1. Panels (**a**, **b**, **c**, **d**) correspond to models 1, 2, 3, 4, respectively (Figure 1 from Melnikov and Shevchenko 2005. With permission from Pleiades Publishing Inc.)

Figure 8.1 shows phase space sections, defined at $\tau = 0 \pmod{2\pi}$, for the system (8.1). The sections were constructed in Melnikov and Shevchenko (2005) by means of direct numerical integrations of the equations of motion defined by the Hamiltonian (8.1).

In the next section, we discuss how such phase space sections can be interpreted in the framework of the separatrix map theory.

8.1.2 Generalization of the Separatrix Algorithmic Map

Recall that Eqs. (1.33) constitute the separatrix algorithmic map for the Hamiltonian (1.6) in case of $k = 1$. In case of $k = 1/2$, the unperturbed pendulum Hamiltonian H_0 is 2π-periodic with respect to φ, but the perturbed Hamiltonian H is 4π-periodic (see Eq. (1.6) or (8.1)). This makes construction of the separatrix algorithmic map in case of $k = 1/2$ somewhat different.

If $-\pi < \varphi < \pi$, the W quantity has a specific sign, and, when φ is in the interval $\pi < \varphi < 3\pi$, the sign of W is opposite. These φ intervals alternate at each iteration of the separatrix map, if the model pendulum rotates; in case of its libration, the trajectory stays inside one particular φ interval. Taking this into account, it is straightforward to write down the separatrix algorithmic map in case of $k = 1/2$:

$$\text{if } w_n > 0 \text{ then } W := -W,$$
$$\text{if } w_n < 0 \text{ and } W = \pm W^{\pm} \text{ then } W := \pm W^{\mp};$$

$$w_{n+1} = w_n - W \sin \tau_n,$$
$$\tau_{n+1} = \tau_n + \Delta_{n+1}\tau \pmod{2\pi} \tag{8.5}$$

(Shevchenko 2000b). Here the first line corresponds to the model pendulum rotation, and the second one to its libration. The expressions for W^+ and W^- in case of any positive integer or half-integer k can be found by using formulas given in Appendix C.

In case of $k = 1$, a complete picture of the motion in the (τ, w) variables is provided by constructing two phase portraits of the separatrix algorithmic map (1.33), one for the prograde ($W = W^+$) motion, and one for the retrograde ($W = W^-$) motion; see Sect. 1.6. In case of $k = 1/2$, a complete picture is provided by constructing four phase portraits of the map (8.5): these are two prograde-retrograde pairs, one for $-\pi \le \varphi < \pi$ and second for $\pi \le \varphi < 3\pi$. If $W = W^{\pm}$, then the motion takes place in the first interval; if $W = -W^{\pm}$, then in the second one.

The separatrix algorithmic map takes the motion of the system (1.6) on the (τ, w) plane at fixed values of the resonant phase angle φ, equal to 0 and $\pm\pi$. When Poincaré sections are constructed numerically in applied problems, it is customary to use another plane, namely the (φ, p) plane, taken at a fixed value of the perturbation phase angle, say, $\tau = 0 \pmod{2\pi}$. In our case, to construct a section in the usual phase-angle–momentum variables, an adequate projection procedure should be employed.

For $k = 1$, a regular projection algorithm is described in Sect. 1.6; it is given by the scheme (1.41). The algorithm is called regular, because it is based on a regular approximation of the chaotic motion on small (much smaller than the Lyapunov time) time intervals. Let us rewrite the original algorithm (1.41), rendering it in the current variables. We introduce designations $w = w_n$, $\hat{\tau} = \tau_n$, $\Delta\tau = \Delta_n\tau = \tau_n - \tau_{n-1}$, and $W = W_{n+1}$.

The projection of a function $f = \varphi$ or $f = p$ to the surface $\tau = 0 \pmod{2\pi}$ is given by the formula

$$f|_{projected} = \begin{cases} f\left(t = -\frac{\hat{\tau}}{\Omega}\right), & \text{if } W = W^+, \\ -f\left(t = -\frac{\hat{\tau}}{\Omega}\right), & \text{if } W = W^-. \end{cases} \tag{8.6}$$

The functions $\varphi(t)$ and $p(t)$ represent the explicit solution (via elliptic functions) of the equations of the unperturbed non-linear pendulum; these functions can be evaluated given the current value of w.

One iteration of the separatrix algorithmic map can produce several (or even many) projected points. The algorithm (1.41) for finding all projected points for a current step of the map can be rewritten as

$$\begin{aligned}
&\text{while } \Delta\tau > \hat{\tau} \text{ do} \\
&\text{evaluate } \varphi,\ p \text{ by Eq. (8.6)} \\
&\hat{\tau} := \hat{\tau} + 2\pi \\
&\text{end do}
\end{aligned} \tag{8.7}$$

Here the $\hat{\tau}$ initial value is taken modulo 2π, while the increment $\Delta\tau$ and consequent values of $\hat{\tau}$ are not.

In more detail, the algorithm can be commented as follows (Shevchenko 1999a, 2000b). First, it is verified whether the condition of intersection of the trajectory with the chosen surface of section is valid for the time span corresponding to the current step of the map, and if yes, the projection is made. Then the $\hat{\tau}$ interval is incremented by 2π, and it is verified whether the intersection condition is still valid. If yes, the projection is made once more with the new value of $\hat{\tau}$, and one more phase point on the (φ, p) plane, at $\tau = 0 \pmod{2\pi}$, is found. The cycle is repeated until the $\Delta\tau \le \hat{\tau}$ condition starts to hold. The procedure is performed at each iteration of the separatrix algorithmic map.

This is the regular projection algorithm for the case $k = 1$. Let us generalize it to the $k = 1/2$ case. In case of $k = 1/2$, Eq. (8.6) changes to

$$f|_{projected} = \begin{cases} f\left(t = -\frac{\hat{\tau}}{\Omega}\right), & \text{if } |W| = |W^+|, \\ -f\left(t = -\frac{\hat{\tau}}{\Omega}\right), & \text{if } |W| = |W^-|. \end{cases} \tag{8.8}$$

Besides, one should attribute each projected point to a specific resonance domain. In case of libration, the trajectory stays inside one particular resonance domain. If $W = W^{\pm}$, then the motion takes place inside the first primary resonance, $-\pi < \varphi < \pi$. If $W = -W^{\pm}$, then it takes place inside the second primary resonance, $\pi < \varphi < 3\pi$. By using Eq. (8.8), one obtains a value of φ modulo 2π, for example, $-\pi < \varphi < \pi$. According to the sign of W, this φ value should be shifted (attributed) to the corresponding domain of the motion. This shift is zero if $W = W^{\pm}$, or is equal to 2π if $W = -W^{\pm}$.

In case of rotation, the trajectory jumps from one φ interval to another one at each iteration of the separatrix algorithmic map; i.e., the intervals alternate. In this situation, one should take into account the property of the separatrix map asynchronism (see Sect. 1.6): τ is mapped with a delay in relation to w. Since the sign of W alternates, this delay affects the attribution of any projected point to the specific primary interval in φ. Namely, for the initial value $-\pi \le \varphi < \pi$ given by Eq. (8.8): if $|W| = |W^+|$ and $\varphi > 0$, then one should set $\varphi := \varphi - 2\pi$; and if $|W| = |W^-|$ and $\varphi < 0$, then one should set $\varphi := \varphi + 2\pi$. Afterwards, as in case of libration, if $W = -W^{\pm}$, then one shifts $\varphi := \varphi + 2\pi$ once more. Finally, φ is taken modulo 4π. This completes the regular projection algorithm in the $k = 1/2$ case.

Let us see how the separatrix algorithmic map in the $k = 1/2$ case and the corresponding regular projection algorithm work in application to the chaotic orbital dynamics of the Miranda–Umbriel system, at a particular stage of its long-term orbital evolution. For an illustration, we choose model 3 in Table 8.1, because this model corresponds to parametric resonance,[2] i.e., the adiabaticity parameter $\lambda = 2$.

Indeed, the value of $\lambda = \frac{\tilde{\Omega}}{\omega_0}$ is given by the approximate relation $\lambda \approx \frac{0.143}{i_M^{\mathrm{res}}} = \frac{8.17}{i_M^{\mathrm{res}\,\circ}}$ (Malhotra 1990). The inclinations of Miranda and Umbriel in model 3 are $i_M^{\mathrm{res}} = 4.1°$, $i_U = 0.20°$; therefore, $\lambda = 2$. From Eqs. (8.3), (1.54), and (1.55) one has:

$$\frac{a}{\mathcal{F}} = 0.0976, \qquad \frac{b}{\mathcal{F}} = 0.168,$$

$$W^+ = 0.106, \qquad W^- = 0.182.$$

In Fig. 8.2, the corresponding phase space section for the system (8.1) in the (φ, p) variables, taken at $\tau = 0 \pmod{2\pi}$, is shown. The section is constructed by applying the separatrix algorithmic map (8.5) and the regular projection algorithm (8.7). In Fig. 8.3, the same section is shown, but obtained by direct numerical integrations of the equations of motion defined by the Hamiltonian (8.1). One may see that the both ways for constructing the section provide almost identical results.

However, as one could have expected, a small zone near the resonance centre, where the period-doubling bifurcation occurs due to the parametric resonance, is out of reach for the separatrix algorithmic map. This pattern is located too far from the separatrix.

Figures 8.2 and 8.3 graphically illustrate a mechanism of destabilization of the 3/1 mean motion resonance due to capture of the Miranda–Umbriel system in the 3/1 secondary resonance (Tittemore and Wisdom 1989; Malhotra and Dermott 1990; Malhotra 1990; Henrard and Moons 1992; Moons and Henrard 1994). The 3/1 secondary resonance represents here the resonance between the circulation and libration frequencies of the neighbouring i_2^2 and i_1^2 resonances, respectively. In the course of the Miranda–Umbriel system long-term dynamical evolution, this secondary resonance moves outward from the centre of the primary 3/1 resonance; in the considered model 3 it is already deep inside the chaotic layer of the primary resonance; therefore, it is already close to complete destabilization. Three small regular islands, corresponding to the secondary resonance, are yet clearly present inside the chaotic layer.

---

[2]For the parametric resonance theory, see Chirikov (1979).

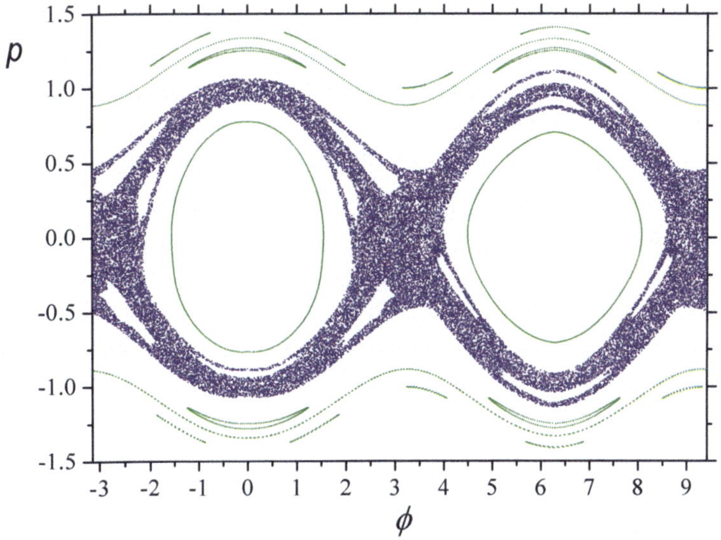

Fig. 8.2 The Miranda–Umbriel system, model 3 of Table 8.1: the phase space section at $\tau = 0 \pmod{2\pi}$. Constructed by applying the separatrix algorithmic map and regular projection algorithm; $\lambda = 2$, $W^+ = 0.106$, $W^- = 0.182$. The chaotic layer is shown in violet, and regular orbits in green (Figure 1 from Shevchenko 2000b, by permission. Copyright © 2000 by World Scientific Publishing Co. Pte. Ltd.)

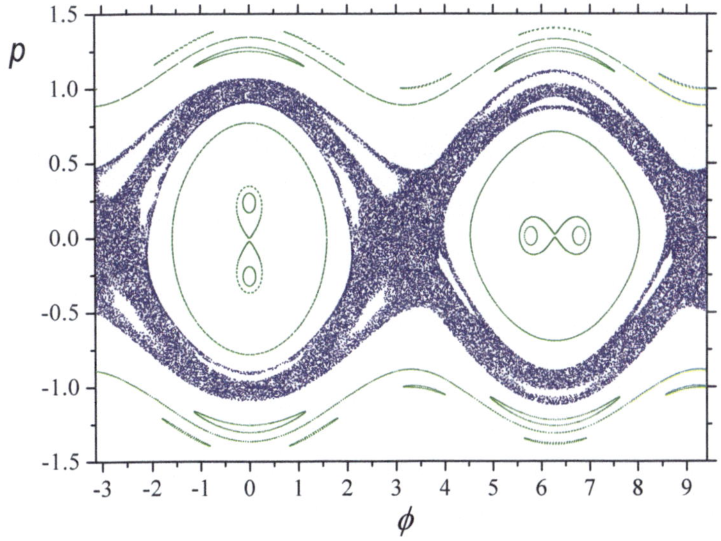

Fig. 8.3 Same as Fig. 8.2, but obtained by direct numerical integrations of the equations of motion. Note the period-doubling bifurcation of the resonance centre. This central pattern is too far from the separatrix, and the separatrix map cannot be used to describe it (Figure 2 from Shevchenko 2000b, by permission. Copyright © 2000 by World Scientific Publishing Co. Pte. Ltd.)

8.1.3 The Mimas–Tethys System

Amongst Saturnian satellites, there exist at present three resonant relations (Champenois and Vienne 1999b), a mean motion resonance in the Mimas–Tethys system among them. Namely, Mimas (inner moon) and Tethys (outer moon) reside now in the inclination-type $i_1 i_2$ resonance 4/2.[3] The subscripts 1 and 2 designate the inner and outer moons, respectively. The resonance phase $\varphi = 2\lambda_1 - 4\lambda_2 + \Omega_1 + \Omega_2$ librates about 0° with the amplitude $\simeq$95° and the period $\approx$70 yr (Champenois and Vienne 1999a,b). Due to the large amplitude of the resonant argument oscillations, the system resides inside the near-separatrix chaotic layer. Before entering the layer, Mimas and Tethys spent some time in the i_1^2 resonance, from which they eventually escaped (Champenois and Vienne 1999b).

In total, there are six subresonances corresponding to the 4/2 mean motion resonance, in a model of the second order of smallness in the eccentricities and inclinations. Champenois and Vienne (1999b) showed that, in the Mimas–Tethys system, resonances of higher orders in inclinations and eccentricities, namely, the $i_1^2 e_2$, $i_1 i_2 e_2$, and $i_2^2 e_2$ resonances, strongly affect the dynamics in the $i_1 i_2$ resonance. Neglecting the terms of order greater than 3 in the inclinations and eccentricities, the Hamiltonian of the "planet—two satellites" three-body problem in the vicinity of the 2/1 mean motion resonance can be represented as

$$H = \frac{\mathcal{G} I^2}{2} - \mathcal{F}\cos\varphi + \varepsilon_1 \cos\left(\frac{\varphi}{2} - \tau\right) + \varepsilon_2 \cos\left(\frac{\varphi}{2} + \tau\right) + \varepsilon_3 \cos\left(\frac{3}{2}\varphi + \tau\right) \tag{8.9}$$

(Champenois and Vienne 1999b), where $\varphi = 2\lambda_1 - 4\lambda_2 + \Omega_1 + \Omega_2$ is the resonance angle of the $i_1 i_2$ resonance, $\tau = \tilde{\Omega} t + \tau_0$ is the perturbation phase angle, $\tilde{\Omega}$ is the perturbation frequency equal to the rotation frequency of the angle $\frac{1}{2}\Omega_1 - \frac{3}{2}\Omega_2 + \varpi_2$, τ_0 is the initial perturbation phase, λ_1 and λ_2 are the mean longitudes of the satellites, Ω_1 and Ω_2 are the longitudes of ascending nodes, I is the momentum conjugated to φ; t is time. The parameters $\mathcal{G}$, $\mathcal{F}$, ε_1, ε_2, and ε_3 are given by

$$\begin{aligned} &\mathcal{G} = -1, \quad \mathcal{F} = f_0(\alpha)\sigma\gamma_1\gamma_2, \\ &\varepsilon_1 = -e_2 f_3(\alpha)\sigma\gamma_2^2, \quad \varepsilon_2 = -e_2 f_2(\alpha)\sigma\gamma_1\gamma_2, \quad \varepsilon_3 = -e_2 f_1(\alpha)\sigma\gamma_1^2, \end{aligned} \tag{8.10}$$

where $\sigma = 12 n_1^2 \alpha m_2 + 48 n_2^2 m_1$, $\gamma_1 = \sin(i_1/2)$, and $\gamma_2 = \sin(i_2/2)$; n_1, n_2 are the mean motions of Mimas and Tethys; m_1 and m_2 are the masses of the satellites in units of Saturn's mass; $\alpha = a_1/a_2$ is the ratio of the semimajor axes of the moons. The functions $f_k(\alpha)$ are expressed in terms of Laplace coefficients and their derivatives; see Champenois and Vienne (1999b).

[3]Here the "4/2 resonance" designation instead of the "2/1 resonance" designation is justified because the mean longitudes of Mimas λ_1 and Tethys λ_2 enter the corresponding resonant terms of the Hamiltonian expansion with the coefficients 2 and 4.

By analogy with (8.3), one has

$$\frac{\varepsilon_1}{\mathcal{F}} = -e_2 \frac{f_3(\alpha)}{f_0(\alpha)} \frac{\gamma_2}{\gamma_1}, \qquad \frac{\varepsilon_2}{\mathcal{F}} = -e_2 \frac{f_2(\alpha)}{f_0(\alpha)}, \qquad \frac{\varepsilon_3}{\mathcal{F}} = -e_2 \frac{f_1(\alpha)}{f_0(\alpha)} \frac{\gamma_1}{\gamma_2}. \quad (8.11)$$

In the Hamiltonian (8.9), the terms with the ε_1 and ε_2 coefficients correspond to the $i_2^2 e_2$ and $i_1 i_2 e_2$ resonances, respectively, whereas the term with the ε_3 coefficient corresponds to the $i_1^2 e_2$ resonance. Following Champenois and Vienne (1999b), we set $n_1 = 2422.44\,\mathrm{yr}^{-1}$, $n_2 = 1213.17\,\mathrm{yr}^{-1}$, $m_1 = 6.34 \cdot 10^{-8}$, and $m_2 = 1.06 \cdot 10^{-6}$. Assuming $\alpha = 0.6306$, one has $f_0 = -1.6509$, $f_1 = 5.2379$, $f_2 = 9.7082$, and $f_3 = 0.2219$.

Model parameters used below are given in Table 8.2. Models 1 and 2 in Table 8.2 approximately correspond to the current observational status of the system, within the observational data accuracy limits. Note that, due to its smallness, the orbital eccentricity of Tethys is known with relatively low accuracy; $e_2 = 0.001$ is maximally possible. Models 3–6 give possible orbital parameters during the capture in the $i_1 i_2$ resonance $\simeq 2 \cdot 10^8$ yr ago.

Sections of the phase space of the Hamiltonian (8.9), obtained by direct numerical integrations at $\tau = 0$ (mod2π), are shown in Fig. 8.4 for all six models given in Table 8.2. The sections graphically illustrate that the chaotic domains may occupy very different volumes in the phase space, depending on the model choice (i.e., the evolutionary stage), from practically non-existent to significant levels.

8.1.4 The Prometheus–Pandora System

To date, a unique case of observable orbital chaos in satellite systems is known to exist; this is chaos in the Prometheus–Pandora system. The Saturnian satellites Prometheus (S16) and Pandora (S17) are the shepherds of Saturn's F ring. The both moons were discovered during the *Voyager* space mission in 1980–1981 (Synnott et al. 1984). Consequent observations from the *Hubble Space Telescope* in 1995 revealed that the mean longitudes of Prometheus and Pandora differed at that time

Table 8.2 The Mimas–Tethys system and the parameters of the Hamiltonian (8.9) (Champenois and Vienne 1999b; Melnikov and Shevchenko 2005)

Model	i_1,°	i_2,°	e_2	$\widetilde{\Omega}$, yr^{-1}	$\varepsilon_1/\mathcal{F}$	$\varepsilon_2/\mathcal{F}$	$\varepsilon_3/\mathcal{F}$
1	1.62	1.093	0.000235	$2\pi/200$	0.000021	0.001382	0.001105
2	1.62	1.093	0.001	$2\pi/200$	0.000091	0.005881	0.004702
3	0.5	1.093	0.0005	$2\pi/185$	0.000147	0.002940	0.000726
4	0.5	1.093	0.002	$2\pi/185$	0.000588	0.011761	0.002904
5	0.5	1.093	0.0028	$2\pi/185$	0.000823	0.016466	0.004064
6	0.5	1.093	0.009	$2\pi/185$	0.002644	0.052926	0.013063

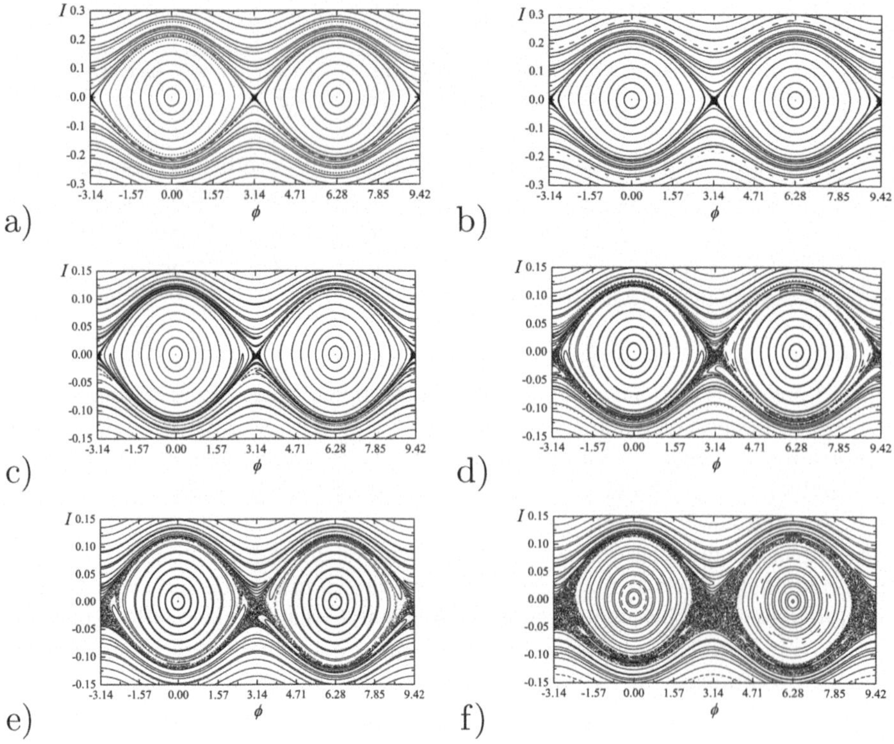

Fig. 8.4 Phase space sections at $\tau = 0 \pmod{2\pi}$, for the Mimas–Tethys system models given in Table 8.2. Panels (**a**, **b**, **c**, **d**, **e**, **f**) correspond to models 1, 2, 3, 4, 5, 6, respectively (Figure 2 from Melnikov and Shevchenko 2005. With permission from Pleiades Publishing Inc.)

by some 20° from the values predicted by the *Voyager* ephemerides (Bosh and Rivkin 1996; Nicholson et al. 1996; McGhee et al. 2001), with Prometheus lagging its predicted longitude, and Pandora leading by the same amount. What is more, abrupt changes ("kinks") in the mean motions were observed (French et al. 2003). Goldreich and Rappaport (2003a) interpreted the kinks as a signature of dynamical chaos naturally arising in mutual gravitational interactions of these almost coorbital satellites. In direct numerical integrations, they found that the system gradually deviated from the Voyager ephemerides due to a slow chaotic diffusion in the space of orbital elements; they attributed the kinks to bursts of gravitational interactions during the apsidal anti-alignments, which occur every 6.2 yr.

Goldreich and Rappaport (2003b) and Renner and Sicardy (2003) interpreted the chaos as due to interaction of subresonances in the 121/118 mean motion resonance multiplet. The resonant librations were identified by surveying the rates of change of all possible critical arguments for the Prometheus–Pandora system (Goldreich and Rappaport 2003b, Figure 1). Chaos manifests itself in numerical simulations of the system dynamics in all observationally possible initial settings (Goldreich and

Table 8.3 The Prometheus–Pandora system parameters (Goldreich and Rappaport 2003b; Farmer and Goldreich 2006)

Satellite	m/M	a, km	e	n, $^\circ s^{-1}$	$\dot{\varpi}$, $^\circ s^{-1}$
Prometheus	$5.80 \cdot 10^{-10}$	139,000	$2.29 \cdot 10^{-3}$	$6.80 \cdot 10^{-3}$	$3.1911 \cdot 10^{-5}$
Pandora	$3.43 \cdot 10^{-10}$	142,000	$4.37 \cdot 10^{-3}$	$6.63 \cdot 10^{-3}$	$3.0082 \cdot 10^{-5}$

Rappaport 2003a,b; Cooper and Murray 2004; Jacobson and French 2004; Renner et al. 2005), with or without including perturbations from other Saturnian satellites. Farmer and Goldreich (2006) studied the dynamics of the Prometheus–Pandora system by analogy with a nearly adiabatic parametric pendulum and explored the separatrix crossing phenomena in the system in this analogy.

The parameters of the satellite system, as adopted in Goldreich and Rappaport (2003b); Farmer and Goldreich (2006), are given in Table 8.3. This Table contains masses m (in units of Saturn's mass M), orbital semimajor axes a and eccentricities e, mean motions n, rates $\dot{\varpi}$ of change of the longitudes of pericentres. Further on, the quantities corresponding to Pandora are primed.

Saturn's oblateness causes rapid apsidal precession of the orbits of Prometheus and Pandora. The influence of satellite–satellite gravitational interactions on the apsidal angles and orbital eccentricities can therefore be neglected, and one may analyse solely changes in mean motions or in the resonance angle $\psi \equiv 121\lambda' - 118\lambda$, where λ and λ' are the mean longitudes of the moons.

As shown by Goldreich and Rappaport (2003b), the equations of motion of the Prometheus–Pandora system can be reduced in this way to the paradigm of the non-linear pendulum with periodic perturbations. The dynamics of the Prometheus–Pandora system in this paradigmatic approximation was studied in detail by Farmer and Goldreich (2006) in terms of adiabatic invariants and separatrix crossing phenomena. Conversely, we use the separatrix map theory.

The system dynamics near the 121/118 resonance is approximately described by the equation

$$\frac{d^2\psi}{dt^2} = 3(121n')^2 \frac{m}{M} \left(1 + \frac{am'}{a'm}\right) \Sigma_{q=1}^{4} C_q \sin(\psi - \delta_q) \tag{8.12}$$

(Goldreich and Rappaport 2003b), where $\psi = 121\lambda' - 118\lambda$ is the general resonant argument, and

$$\delta_1 = 3\varpi, \qquad \delta_2 = 2\varpi + \varpi',$$

$$\delta_3 = \varpi + 2\varpi', \qquad \delta_4 = 3\varpi',$$

where ϖ and ϖ' are the longitudes of pericentres of the moons. The four trigonometric terms correspond to the components of the resonance multiplet. The constant coefficients are: $C_1 = -0.00108$, $C_2 = 0.00626$, $C_3 = -0.0121$, and $C_4 = 0.00782$. In fact, the resonance quartet can be shortened, in a good approximation, to a triplet, because C_1 is negligible in comparison with C_2, C_3, and C_4.

We choose $\varphi = \psi - \delta_3$ as the phase angle of the guiding resonance in the triplet, because the corresponding term is the strongest one. Then, the frequency of small-amplitude phase oscillations on the guiding resonance is

$$\omega_0 = \left| 3(121n')^2 \frac{m}{M} \left(1 + \frac{am'}{a'm} \right) C_3 \right|^{1/2}, \tag{8.13}$$

and the perturbation frequency is

$$\Omega = \dot{\varpi} - \dot{\varpi}'. \tag{8.14}$$

The ratio of $\Omega = 1.01$ rad yr^{-1} and $\omega_0 = 2.55$ rad yr^{-1} gives the adiabaticity parameter: $\lambda = 0.395 \approx 0.4$.

Since $\lambda < 1/2$, the Prometheus–Pandora system is slowly chaotic, i.e., it resides in the adiabatic regime of chaos. The perturbation relative amplitudes are given by

$$\varepsilon_1 = \frac{C_2}{C_3}, \qquad \varepsilon_2 = \frac{C_4}{C_3}. \tag{8.15}$$

The amplitudes are similar in magnitude; this means that the resonance triplet is nearly symmetric. The mean perturbation amplitude is

$$\varepsilon = \frac{1}{2}(\varepsilon_1 + \varepsilon_2).$$

For any adiabatic resonance triplet, the maximum Lyapunov exponent is given by formula (3.22); therefore, the Lyapunov time can be calculated as

$$T_{\mathrm{L}} \approx \frac{T_{\mathrm{pert}}}{2\pi} \ln \left| \frac{16}{\lambda\varepsilon} \sin \left(\frac{\lambda}{2} \ln \frac{4}{\lambda|\varepsilon|} \right) \right|. \tag{8.16}$$

With the data on Prometheus and Pandora presented in Table 8.3, the adiabaticity parameter λ, the relative perturbation strengths, and the perturbation period $T_{\mathrm{pert}} = 2\pi/\Omega$ can be easily calculated: $\lambda = 0.395$, $\varepsilon_1 = 0.517$, $\varepsilon_2 = 0.646$, $\varepsilon = 0.582$, $T_{\mathrm{pert}} = 6.237$ yr.

Then, formula (8.16) gives the Lyapunov time $T_{\mathrm{L}} = 3.59$ yr. The uncertainty of this value can be estimated in the following way (Shevchenko 2008b): by varying the perturbation relative amplitude in the range from ε_1 to ε_2, one obtains the estimate of the Lyapunov time by formula (8.16) in the range of 3.5–3.7 yr; therefore, $T_{\mathrm{L}} = 3.6 \pm 0.1$ yr.

In direct numerical integrations, Goldreich and Rappaport (2003a) and Goldreich and Rappaport (2003b) obtained estimates for the Lyapunov time of the

Prometheus–Pandora system in a model setting restricted to mutual gravitational interactions between Prometheus and Pandora. According to their results, the Lyapunov time $T_L \approx 3.3$ yr. Later on, the Lyapunov time was estimated by Cooper and Murray (2004) also in direct numerical integrations, but in the full problem taking into account all important perturbations from other Saturnian satellites. Farmer and Goldreich (2006) made estimates of the Lyapunov time by means of numerical integration of Eq. (8.12). The known numerical-experimental estimates of T_L turn out to be similar:

- $\approx$3.0–4.0 yr (Goldreich and Rappaport 2003a, Figure 7, Goldreich and Rappaport 2003b, Figure 4),
- $\approx$3.3 yr (Cooper and Murray 2004, Figure 10),
- $\approx$3.3–4.2 yr (Farmer and Goldreich 2006, Figure 9).

The presented above analytical estimate, $T_L = 3.6$ yr, is within the range of the given numerical-experimental results. One should underline that it is just the small, $\sim$3 yr, value of the Lyapunov time that makes the chaos in this system directly observable.

The separatrix map parameter c, given by Eq. (1.32), turns out to be $\approx$1.154. The winding number of the slowly chaotic motion is given by Eq. (3.20); one has $Q \approx 0.233$. This means that the Prometheus–Pandora system resides between two low-order resonances with $Q = 1/5$ and $1/4$, and the system is rather far from them both. With respect to the 121/118 primary mean motion resonance, these two resonances are secondary, i.e., they represent resonances between the perturbation frequency and the frequency of phase librations on the primary resonance. The remoteness of the Prometheus–Pandora system from the major low-order secondary resonances means that its motion inside the chaotic layer is almost ergodic. Therefore, the ergodicity condition is fulfilled, and the analytical formulas (3.22) and (8.16) should be practically precise; thus, it is no wander that the analytical T_L estimate is practically the same as the numerical-experimental T_L values.

Now let us see how the chaotic domain width in the Prometheus–Pandora system can be analytically estimated, in observable variables, and the width can be related to available observational data.

When the model pendulum, in the course of its near-separatrix motion, passes through its lower position, the momentum variation near the separatrix is connected to the variation in the relative energy by the formula

$$\Delta p = \frac{\omega_0}{2} \Delta w, \tag{8.17}$$

which follows from the w definition and from the pendulum equation, given by the unperturbed part of Hamiltonian (1.6). Formula (8.17) is analogous to that derived by Vecheslavov and Chirikov (1998) in a study of splitting of separatrices of the standard map integer resonances.

Let us set $p \equiv \dot{\varphi}$ by definition, then

$$dw/dp = p_{sx}(\varphi = 0)/\omega_0^2 = 2/\omega_0.$$

The full width $\Delta p = \Delta\dot{\varphi}$ of the chaotic domain in the momentum is equal to the sum of the width of the unperturbed resonance

$$\Delta\dot{\varphi}_{res} = 4\omega_0$$

and twice the half-width

$$p_b = \frac{\omega_0}{2} w_b(\lambda, \varepsilon)$$

of the chaotic layer in the momentum, given by Eq. (8.17). One has

$$\Delta\dot{\varphi} = \Delta\dot{\varphi}_{res} + \omega_0 w_b(\lambda, \varepsilon) = (4 + w_b(\lambda, \varepsilon))\, \omega_0. \tag{8.18}$$

For any adiabatic chaotic resonance triad, the chaotic layer width is given by formula (3.17), where, at $\lambda \sim 0$, $W \approx 8\lambda\varepsilon$ asymptotically, as follows from Eq. (1.28). Hence

$$w_b(\lambda, \varepsilon) = 8\lambda \left| \varepsilon \operatorname{cosec}\left(\frac{\lambda}{2} \ln \frac{4}{\lambda|\varepsilon|}\right) \right|. \tag{8.19}$$

As estimated above, for the Prometheus–Pandora system, one has $\omega_0 = 2.55$ rad yr^{-1}, $\lambda = 0.395$, $\varepsilon = 0.582$; therefore

$$\Delta\dot{\varphi}_{res} = 10.19 \text{ rad yr}^{-1}, \quad w_b = 3.17, \tag{8.20}$$

and the full sum is

$$\Delta\dot{\varphi} = 18.3 \text{ rad yr}^{-1}. \tag{8.21}$$

The rates of change of the mean motions of Prometheus and Pandora are connected to that of the ψ resonant phase by the equations

$$\dot{n} = -\frac{(am'/a'm)}{118\,[1 + (am'/a'm)]}\dot{\psi}, \quad \dot{n}' = \frac{1}{121\,[1 + (am'/a'm)]}\dot{\psi} \tag{8.22}$$

(Goldreich and Rappaport 2003b). If, when calculating the maximum deviation range in the mean motions, one takes into account solely the unperturbed resonance width $\Delta\dot{\psi} = \Delta\dot{\psi}_{res} = 10.2$ rad yr^{-1}, then

$$\Delta n \approx 1.8 \text{ deg yr}^{-1}, \quad \Delta n' \approx 3.1 \text{ deg yr}^{-1}, \tag{8.23}$$

as found by Goldreich and Rappaport (2003b).

Conversely, in Eq. (8.21), the contribution of the near-separatrix chaotic layer width is taken into account. This contribution increases the expected full extent of the chaotic domain substantially:

$$\Delta\dot{\psi} = \Delta\dot{\varphi} = 18.3 \text{ rad yr}^{-1},$$

in accord (18 versus 16 rad yr^{-1}) with the chaotic scatter in the phase space section constructed in (Goldreich and Rappaport 2003b, Figure 8).[4]

Finally, as given by Eqs. (8.22), the expected full ranges of chaotic variations in the mean motions of Prometheus and Pandora turn out to be

$$\Delta n \approx 3.3 \text{ deg yr}^{-1}, \quad \Delta n' \approx 5.5 \text{ deg yr}^{-1}. \tag{8.24}$$

These ranges are about two times broader than those calculated in Goldreich and Rappaport (2003b) by considering solely the width of the unperturbed resonance; see Eqs. (8.23). The chaotic mean motion variation observed in the course of 15 yr is $\approx 20°$ (Bosh and Rivkin 1996; Nicholson et al. 1996; McGhee et al. 2001). The observed range is naturally smaller than the theoretical range, because the latter provides the maximum possible bounds, perhaps achievable on much longer timescales of observations.

The orbital dynamics of the Prometheus–Pandora system represents the first ever known example of directly observable chaos in the orbital dynamics of natural bodies in our Solar system. If one considers also rotational dynamics, then the Prometheus–Pandora system should be regarded as the second known example of directly observable dynamical chaos in the Solar system.

Indeed, the first example is chaotically rotating Hyperion, the 7th satellite of Saturn; see Sect. 7.1.1. If the orbital eccentricity of Hyperion remained unchanged, then its chaotic rotational regime were the final stage of the dissipative tidal dynamical spin-orbit evolution. The regular final state would not exist, because all relevant spin-orbit modes are attitude unstable, at the current orbital eccentricity (Wisdom et al. 1984; Wisdom 1987a). In other words, if one considers long-term dynamics and, therefore, takes into account tidal dissipation, then this chaotic rotational regime plays the role of a chaotic attractor. Whether the chaotic orbital motion of the Prometheus–Pandora system is a chaotic attractor in the same sense or it represents a transient chaotic state in the long-term dynamical history of the system, remains an open problem.

It is remarkable that, in addition to the orbital chaos, the rotational dynamics of both Prometheus and Pandora might also be chaotic, with the Lyapunov time of only ~1 d, see Sect. 7.1.1. Contrary to the case of Hyperion, the chaos in the rotational dynamics of these two moons, if exists, would be due to fine tuning of their dynamical and physical parameters rather than to the large extent of the chaotic domain in the rotation phase space. If, in addition to the orbital chaos,

[4] In Goldreich and Rappaport (2003b), the φ angle is designated as Φ_3.

the rotational chaos were observationally confirmed, then Prometheus and Pandora would be real "champions" of dynamical chaos in our Solar system, and not only in what concerns the emergence of double (orbital and rotational) chaos itself, but also in the degree of unpredictability of the both kinds of motion, orbital and rotational. The Lyapunov times of the chaotic orbital dynamics and the hypothetically chaotic rotational dynamics of these satellites are the smallest amongst the known Lyapunov times of natural bodies in the Solar system; therefore, their dynamics are most unpredictable. Of course, future observations and theoretical studies may reveal new minor bodies with even smaller Lyapunov times of orbital or rotational chaos; however, most probably, such bodies would be physically smaller.

For comparison, let us assess, following the work of Melnikov and Shevchenko (2005), how small might be the Lyapunov times for the considered above Miranda–Umbriel and Mimas–Tethys systems. For analytical estimates of the Lyapunov times in these systems, methods described above in Sect. 3.3 are used. Relevant numerical-experimental estimates of the Lyapunov times (Melnikov and Shevchenko 2005) are based on direct numerical integrations of the equations of orbital motion, accomplished in parallel with calculations of the Lyapunov exponents by the HQR method (Von Bremen et al. 1997) in software realization by Shevchenko and Kouprianov (2002), Kouprianov and Shevchenko (2003). The HQR method allows one to calculate the full spectrum of the Lyapunov exponents.

For the Miranda–Umbriel system, the Hamiltonian (8.1) parameters are given in Table 8.1; and in the Hamiltonian (1.6) it is set $k = 1/2$, $a = -\varepsilon_1$, and $b = \varepsilon_2$. The corresponding parameters of the separatrix algorithmic map (8.5) are given in Table 8.4. Here the perturbation frequencies $\widetilde{\Omega}$ are expressed in yr^{-1}. Accordingly, the Lyapunov exponents are also expressed in yr^{-1}, and the Lyapunov times are in years.

Based on these parameters, Melnikov and Shevchenko (2005) calculated theoretical L_{theor} values of the maximum Lyapunov exponent. Along with numerical-experimental L_{num} values, computed in direct integrations of the equations of motion, these estimates are presented in Melnikov and Shevchenko (2005). As follows from the obtained analytical and numerical results (which are in accord with each other), the typical Lyapunov times in the Miranda–Umbriel system in all four models range from $\sim$50 to $\sim$100 yr.

Concerning the Mimas–Tethys system, its Hamiltonian is given by Eqs. (8.9), and the parameters are presented in Table 8.2. There are three perturbing terms, instead of two perturbing terms in the model pendulum Hamiltonian (1.6) and in the Hamiltonian (8.1) for the Miranda–Umbriel system. To estimate the Lyapunov

Table 8.4 The separatrix algorithmic map parameters for the Miranda–Umbriel system (Melnikov and Shevchenko 2005)

Model	λ	W^+	W^-
1	3.717	0.0248	0.0425
2	2.985	0.0493	0.0848
3	2.000	0.1060	0.1820
4	1.372	0.1030	0.1790

times, first, the sum of all three perturbing terms (the "$i_2^2 e_2 + i_1 i_2 e_2 + i_1^2 e_2$" model) is taken; then, the sum of the first two perturbing terms (the "$i_2^2 e_2 + i_1 i_2 e_2$" model) is taken.

The first model corresponds to reality; and the second model is considered solely to estimate the role of the third perturbing term. In the both cases, the $i_1 i_2$ resonance is the guiding one. The ε_1 and ε_2 resonances (in our designations) correspond to the $i_2^2 e_2$ and $i_1 i_2 e_2$ resonances, whereas the ε_3 resonance corresponds to the $i_1^2 e_2$ resonance. In the second model, the Hamiltonian is completely similar to the Hamiltonian (1.6); therefore, one may estimate the maximum Lyapunov exponent in actually the same way as accomplished above for the Miranda–Umbriel system. One has $k = 1/2$, $a = \varepsilon_1$, and $b = \varepsilon_2$.

The parameters of the separatrix algorithmic map are given in Table 8.5. The resulting theoretical L_{theor} values of the maximum Lyapunov exponent, along with their numerical-experimental counterparts L_{num}, are considered and discussed in Melnikov and Shevchenko (2005). The obtained L_{num} and L_{theor} values turn out to be generally in accord. It follows that the typical Lyapunov times in the model Mimas–Tethys system range from ~300 to ~600 yr; see Melnikov and Shevchenko (2005) for details.

The Lyapunov times for the Miranda–Umbriel system are negligible compared to the time spent by the system in the i_1^2 resonance; the latter time, according to Dermott et al. (1988), Tittemore and Wisdom (1990), is 10^8–10^9 yr. The Lyapunov times for the Mimas–Tethys system are negligible compared to the time elapsed from the epoch of capture of the system in the $i_1 i_2$ resonance; the latter time, according to Champenois and Vienne (1999a), is $\simeq 2 \cdot 10^8$ yr.

Table 8.5 The separatrix algorithmic map parameters for the Mimas–Tethys system (Melnikov and Shevchenko 2005)

Model	λ	Perturbing resonances	W^+	W^-
1	0.293	"$i_2^2 e_2 + i_1 i_2 e_2 + i_1^2 e_2$"	−0.000454	0.001266
	0.293	"$i_2^2 e_2 + i_1 i_2 e_2$"	−0.001461	0.003794
2	0.293	"$i_2^2 e_2 + i_1 i_2 e_2 + i_1^2 e_2$"	−0.001934	0.005389
	0.293	"$i_2^2 e_2 + i_1 i_2 e_2$"	−0.006215	0.016146
3	0.571	"$i_2^2 e_2 + i_1 i_2 e_2 + i_1^2 e_2$"	−0.000904	0.008014
	0.571	"$i_2^2 e_2 + i_1 i_2 e_2$"	−0.001031	0.008775
4	0.571	"$i_2^2 e_2 + i_1 i_2 e_2 + i_1^2 e_2$"	−0.003617	0.032055
	0.571	"$i_2^2 e_2 + i_1 i_2 e_2$"	−0.004124	0.035101
5	0.571	"$i_2^2 e_2 + i_1 i_2 e_2 + i_1^2 e_2$"	−0.005064	0.044878
	0.571	"$i_2^2 e_2 + i_1 i_2 e_2$"	−0.005774	0.049142
6	0.571	"$i_2^2 e_2 + i_1 i_2 e_2 + i_1^2 e_2$"	−0.016278	0.144250
	0.571	"$i_2^2 e_2 + i_1 i_2 e_2$"	−0.018560	0.157956

8.2 Chaotic Dynamics of Asteroids and Kuiper Belt Objects

Tu t'es fatiguée à force de consulter : Qu'ils se lèvent donc et qu'ils te sauvent, ceux qui connaissent le ciel, qui observent les astres, qui annoncent, d'après les nouvelles lunes, ce qui doit t'arriver !

Esaïe 47:13

Historically, studies of dynamics of planet-crossing minor bodies provided first ever theoretical hints for identifiable chaotic behaviour in the motion of celestial bodies. Giovanni Valsecchi convincingly argued (Valsecchi 2007) that the history of studies of chaotic dynamics of minor bodies actually started with works by Anders Johan Lexell, Leonhard Euler's pupil, in the seventies of the eighteenth century and by Urbain Le Verrier in the forties and fifties of the nineteenth century. They explored dynamics of the famous Lexell comet (Lexell 1777a,b, 1778a,b; Le Verrier 1844, 1848, 1857). Giovanni Valsecchi wrote: "It can be said that the work of Lexell started the modern understanding of the dynamics of small solar system bodies" (Valsecchi 2007). This modern understanding consists in taking into account, when explaining the observed dynamics of such kind of objects, the essential role of resonances and close encounters with planets. In celestial-mechanical studies by Le Verrier (1844, 1848, 1857), a novel concept of the *sensitive dependence on initial conditions* appeared for the first ever time: tiny model variations (about several metres per second) of the Lexell comet velocity in its orbit perihelion resulted in *qualitative* changes of the comet's orbit. Thus, scientific grounds for exploration of dynamical chaos, whatever the phenomenon could have been called at that time, emerged in the science of celestial mechanics already in the middle of the nineteenth century.

In 1770, the Lexell comet passed at an unprecedentedly close distance from the Earth. Nine years later, it was most probably thrown away from the Solar system as a result of its close encounter with Jupiter. In the history of astronomical observations it became the first ever outstanding example of a celestial body closely encountering the Earth. This example graphically shows that the problem of estimating the degree of predictability of the motion of planet-crossing bodies is not solely theoretical, but is practically important.

Estimating the degree of predictability of the orbital motion of near-Earth asteroids (NEAs), as well as of other objects potentially hazardous to the Earth, is one of complicated aspects of the whole problem of the asteroid-cometary hazard. Some objects move in quite predictable orbits, others do not. Whipple (1995) wrote: "The existence of a significant population of extremely chaotic Earth-crossing asteroids must be factored into the thinking about the potential hazard posed by these objects. An asteroid with a Lyapunov time of 20 years may be considered as an example. If the initial error in its position is 100 km (a very optimistic assumption) then that error will grow to one Earth radius in 83 years and to an Earth–Moon distance in 165 years. Assessments of the threat from specific objects like this can be made for only short spans of time."

As follows from data contained in the AstDyS database (AstDyS 2020), where estimates of Lyapunov times (obtained in direct integrations on time intervals of 2 mln years) are provided, any asteroids with Lyapunov times less than 400 yr are absent in the main belt.

Conversely, asteroids and comets crossing the orbits of planets, such as near-Earth asteroids (NEAs), represent major classes of chaotic objects in the Solar system; they usually have Lyapunov times much smaller than those of the typical main belt asteroids, and their motion is practically unpredictable on timescales as small as tens of years.

8.2.1 Resonant Structure of the Asteroid Belt

In the dynamics of asteroids, the essential role of resonances became evident since the time of discovery of gaps (sparsely populated areas) in the belt of asteroids in 1867 by Daniel Kirkwood. More than a hundred and 50 years ago, he sagaciously wrote "... the tendency of Jupiter's influence would be to form gaps or chasms in the primitive ring" (Kirkwood 1867, p. 106). The distribution (histogram) of the main belt asteroids in the semimajor axis shows prominent minima, called *Kirkwood gaps*, which correspond to the 3/1, 4/1, 5/2, etc., mean motion resonances with Jupiter (Kirkwood 1867); see Fig. 8.5.

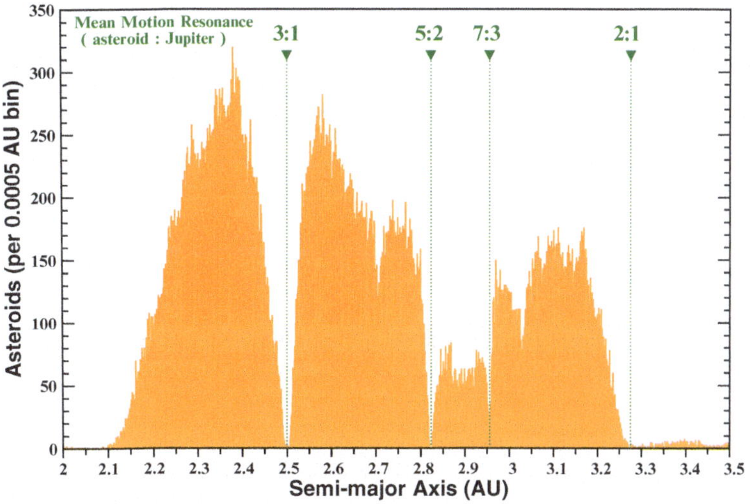

Fig. 8.5 Distribution of the main-belt asteroids in semimajor axis (number of objects per 0.0005 AU). Several major Jovian resonances are indicated by arrows (Credits: NASA/JPL/Caltech/Alan Chamberlain 2007)

Although obviously related to Jupiter, the actual dynamical origin of the Kirkwood gaps remained enigmatic since 1867 for more than a century, until Wisdom (1982) discovered that the motion of asteroids in a vicinity of the 3/1 resonance with Jupiter is typically chaotic. In massive numerical simulations, he observed that the asteroids exhibited sporadic excursions to high eccentricities, thus becoming Mars-crossers and consequently being potentially removed by encounters with this planet. Although nowadays the mechanism of opening the gaps, including the 3/1 one, is perceived to be more complicated (Morbidelli 2002), it is still based on the concept of chaotic diffusion in the space of asteroidal orbital elements.

The domain of chaos corresponding to the 3/1 mean motion resonance with Jupiter is shown in Fig. 5.2 on a representative set of initial values of the semimajor axis a and eccentricity e of asteroidal orbits. The plot is obtained in direct numerical integrations of the asteroidal equations of motion in the framework of the planar elliptic restricted three-body problem, with Jupiter's eccentricity set to its modern value, $e = 0.048$. The initial conditions resulting in chaotic orbits with and without jumps of eccentricity are shown in black and grey, respectively. The graph clearly reveals significant extents and intricate structure of the near-resonant chaotic zone.

The long-term behaviour of eccentricity of chaotic asteroids in the 3/1 mean motion resonance is illustrated in Fig. 8.6. The Hamiltonian intermittency pattern is readily recognizable in the plot. Note a periodic-like sequence of eccentricity jumps at the end of the integration time interval; it is due to temporary sticking of the trajectory to a particular chaos border in the divided phase space.

However, when perturbations from all planets of our Solar system are taken into account, sporadic jumps of eccentricity transform into a more complicated evolution. The clearing of the gap becomes more rapid and radical. A typical evolution of the eccentricity, inclination, and semimajor axis of an asteroid, starting in the 3/1 resonance with Jupiter, is shown in Fig. 8.7. It is obtained in a direct numerical integration of the equations of motion in the full many-body problem, including all planets of the Solar system. The asteroid's orbit eventually achieves small pericentric distance and, finally, the asteroid falls onto the Sun.

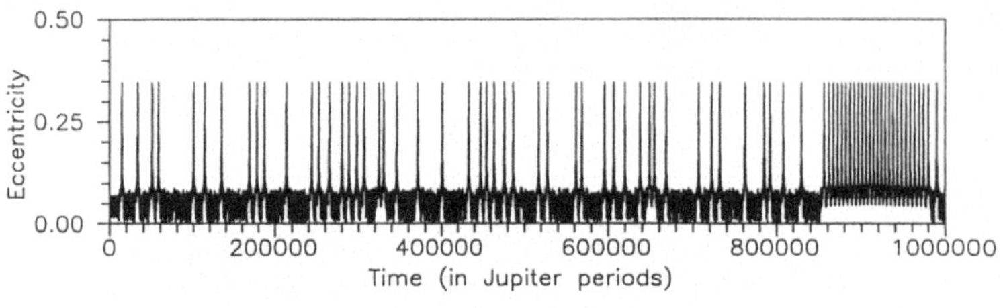

Fig. 8.6 Eccentricity versus time for an asteroidal trajectory in the 3/1 mean motion resonance with Jupiter; the plot is obtained by a direct numerical integration in the planar elliptic restricted three-body problem setting. Note a periodic-like sequence of eccentricity jumps at the end of the integration time interval (Figure 1 from Shevchenko and Scholl 1997, by permission from Springer Nature, © 1997)

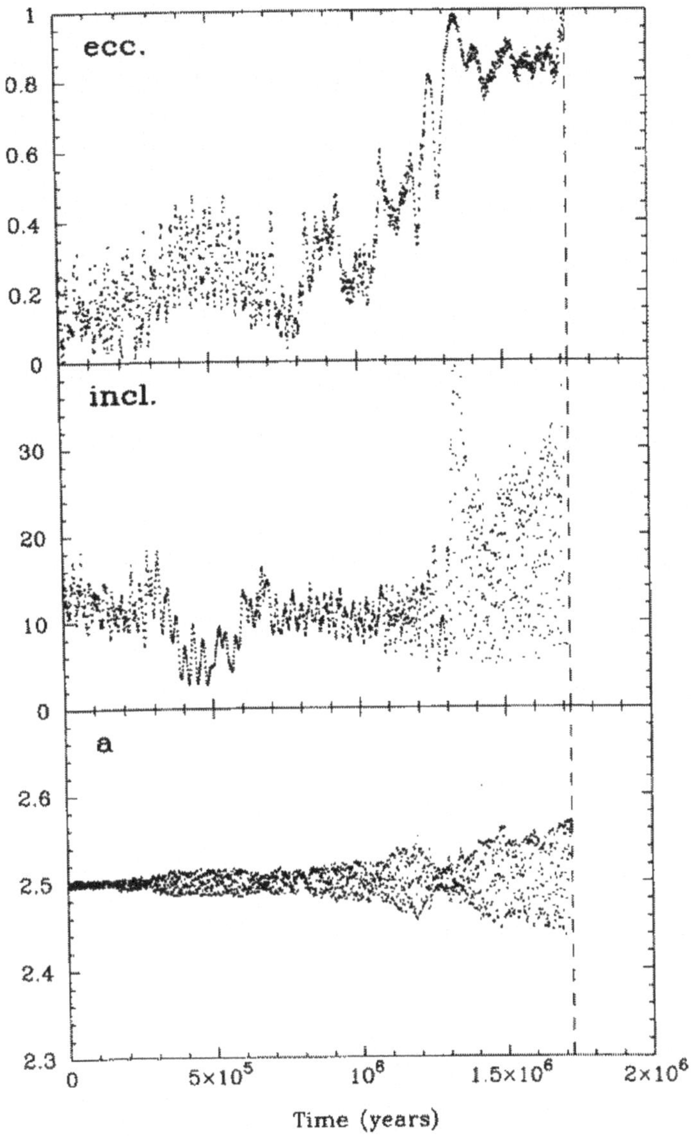

Fig. 8.7 A typical evolution of the eccentricity, inclination, and semimajor axis of an asteroidal orbit, starting in the 3/1 resonance with Jupiter; the perturbations from all planets are taken into account (Figure 9 from Farinella et al. 1994, by permission from Springer Nature, © 1994)

The orbital resonances in the motion of asteroids subdivide into the mean motion resonances and secular resonances. The mean motion resonances correspond to commensurabilities between the mean frequencies of the orbital motions of asteroids and a planet, and the secular resonances correspond to commensurabilities between the rates of the apsidal or nodal precession of an asteroid and a planet.

Apart from the well-known two-body mean motion resonances, an essential role in the dynamics of asteroids is played by the three-body mean motion resonances. In this case, the resonant phase represents a combination of the angular coordinates of an asteroid and two planets, e.g., an asteroid, Jupiter and Saturn. Atlases of two-body and three-body mean motion resonances in the Solar System are presented in Gallardo (2006, 2014).

Both for the two-body and three-body resonances, the equations of motion in typical cases are approximately reducible to the equation of the non-linear pendulum with periodic perturbations. Therefore, it becomes possible to estimate analytically the Lyapunov times of the motion (Shevchenko 2007a). It should be noted that the analytical estimating of the Lyapunov exponents represents a perspective tool for identification of asteroids in resonances: by means of comparison of analytical and numerical-experimental values of the Lyapunov exponents one may judge on plausibility of any such identification.

In Fig. 8.8, the distribution of osculating elements of all main belt asteroids (known in 1983) is shown, with separatrices (boundaries of the libration areas) of low-order mean motion Jovian resonances superimposed, as calculated in Dermott and Murray (1983). Inside the gross chaotic zone of the overlapping resonances, the separatrices of the 2/1 and 3/2 resonances are dashed. It is evident that the resonances are mostly cleared from asteroids. This is due to the subresonances

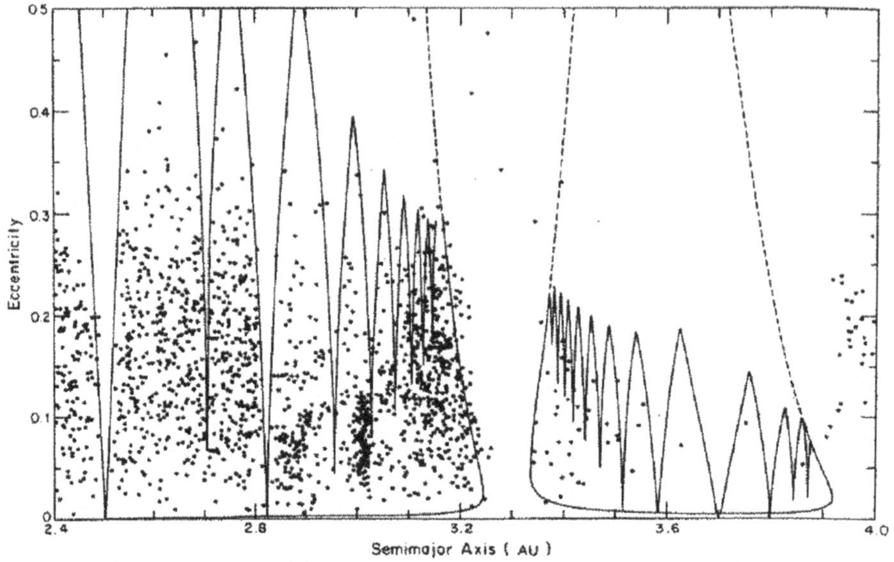

Fig. 8.8 Distribution of osculating elements of asteroids, with the separatrices of mean motion resonances superimposed (Figure 5 from Dermott and Murray 1983, by permission from Springer Nature, © 1983)

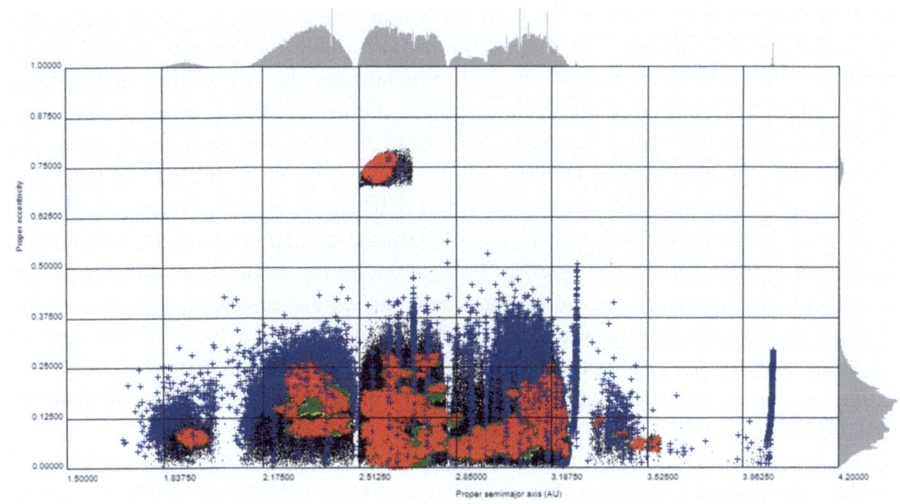

Fig. 8.9 Two-dimensional distribution of asteroids in the "semimajor axis—eccentricity" proper elements. Blue: resonant asteroids. Black: non-resonant asteroids. Red, yellow, and green: members of asteroid families in various definitions (Credits: AstDyS 2020)

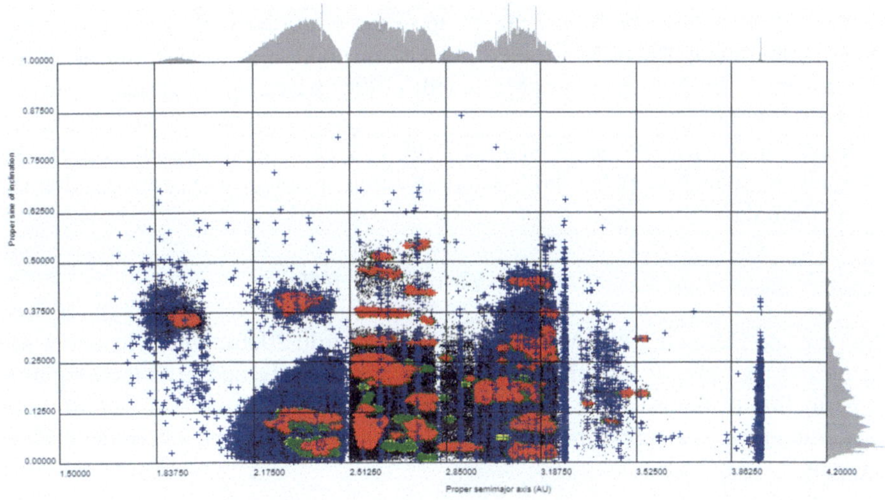

Fig. 8.10 Two-dimensional distribution of asteroids in the "semimajor axis—sine of inclination" proper elements. The colour designations are the same as in Fig. 8.9 (Credits: AstDyS 2020)

overlap inside multiplets of subresonances, corresponding to the given mean motion resonances; the overlap phenomenon will be discussed further on.

In Figs. 8.9 and 8.10, high-order two-body resonances and three-body resonances graphically manifest themselves as depopulated narrow vertical areas in the distributions of asteroids in the "semimajor axis—eccentricity" and "semimajor axis—inclination" proper elements. The *proper orbital elements* represent the elements' values after removal of any contributions due to perturbations from other bodies. The proper elements characterize the inherent properties of the orbit; rigourous definitions and procedures for calculations of the proper elements are given in Murray and Dermott (1999).

The asteroid distributions in Figs. 8.9 and 8.10 show that, due to two-body and three-body mean motion resonances, and also due to secular resonances (considered further on in Sect. 8.2.5), the asteroid belt is essentially structured. Moreover, the dynamical interaction of asteroid families with three-body and high-order two-body resonances represents an important dynamical factor in the asteroidal transport in the Solar system. A vivid example is provided by a chaotic diffusion process, due to the 5-2-2 three-body resonance, in the (490) Veritas asteroidal family (Cachucho et al. 2010).

It is usually not known which resonant multiplet causes chaotic behaviour of a particular asteroid, because three-body resonances (which are most numerous among the asteroidal resonances) are difficult to identify. A massive identification of the main belt asteroids in two-body and three-body resonances with planets was carried out in Smirnov and Shevchenko (2013). In direct numerical integrations, the long-term behaviour of resonant arguments was analysed for all asteroids from the AstDyS database (AstDyS 2020), containing hundreds of thousands of objects. The asteroids in pure three-body resonances up to the 6th order inclusive were shown to constitute about 1% of all studied objects, in good agreement with an earlier estimate made by Nesvorný and Morbidelli (1998) based on a sample of several hundreds asteroids.

Taking into account the D'Alembert rules (see, e.g., Morbidelli 2002), the resonant argument for any two-body asteroid–Jupiter resonance of order q is defined by the formula

$$\sigma = (p+q)\lambda_{\mathrm{J}} - p\lambda - q\varpi \tag{8.25}$$

(Murray and Dermott 1999; Morbidelli 2002; Gallardo 2006), where λ_{J} and λ are the mean longitudes of Jupiter and an asteroid, respectively, and ϖ is the longitude of perihelion of the asteroid; $q \geq 0$ is the resonant order, $p > 0$ is an arbitrary integer number. The approximate resonant semimajor axis of the asteroidal orbit is given by

$$a_{\mathrm{res}} \approx a_{\mathrm{J}}(1+\mu)^{-1/3}\left(\frac{p}{p+q}\right)^{2/3}, \tag{8.26}$$

where a_{J} is the semimajor axis of Jupiter's orbit, and μ is the mass of Jupiter in Solar units.

The resonant libration is defined as *transient*, if circulation of the resonant argument appears at any time during the time interval of integration. In Smirnov and Shevchenko (2013), to distinguish between the pure-resonant, transient-resonant and non-resonant behaviours, a technical parameter was introduced: the resonance minimum time, which was set to 20,000 yr. If the total time of libration was equal to the full time of computation (10^5 yr), then the asteroid was regarded to be in pure resonance. If not, but the total time of libration exceeded the resonance minimum time, then the asteroid was regarded to reside in transient resonance. If the total time of libration was less than the resonance minimum time, then the asteroid was regarded to be non-resonant.

Figures 8.11 and 8.12 show the time behaviour of the resonant arguments and orbital elements of 190 Ismene (pure resonance) and 1915 Quetzálcoatl (transient resonance), as two typical examples of asteroids in two-body resonances with Jupiter. In fact, a half ($\approx$53%) of all asteroids identified by Smirnov and Shevchenko (2013) to be in pure two-body resonances with Jupiter are Trojans. Pure Trojans plus pure Hildas (asteroids in the 3/2 resonance with Jupiter) constitute $\approx$85% of all asteroids residing in pure two-body resonances.

The three-body resonant argument in the planar asteroid–Jupiter–Saturn problem is given by

$$\sigma_{p_J, p_S, p} = m_J \lambda_J + m_S \lambda_S + m\lambda + p_J \varpi_J + p_S \varpi_S + p\varpi, \tag{8.27}$$

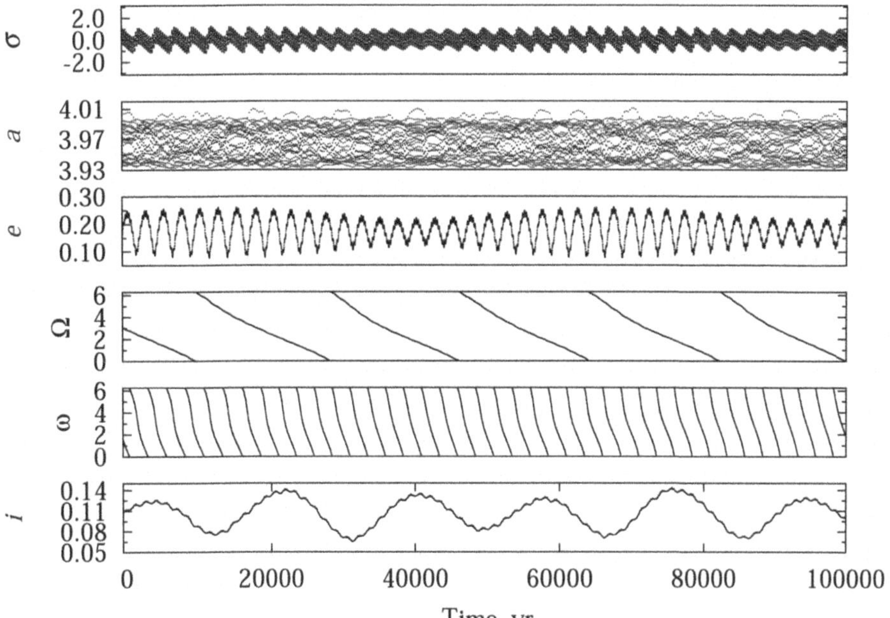

Fig. 8.11 The orbital elements and resonant argument of 190 Ismene, as a function of time. Ismene is in the 3/2 resonance with Jupiter; the resonance is pure (Figure 8 from Smirnov and Shevchenko 2013, with permission from Elsevier)

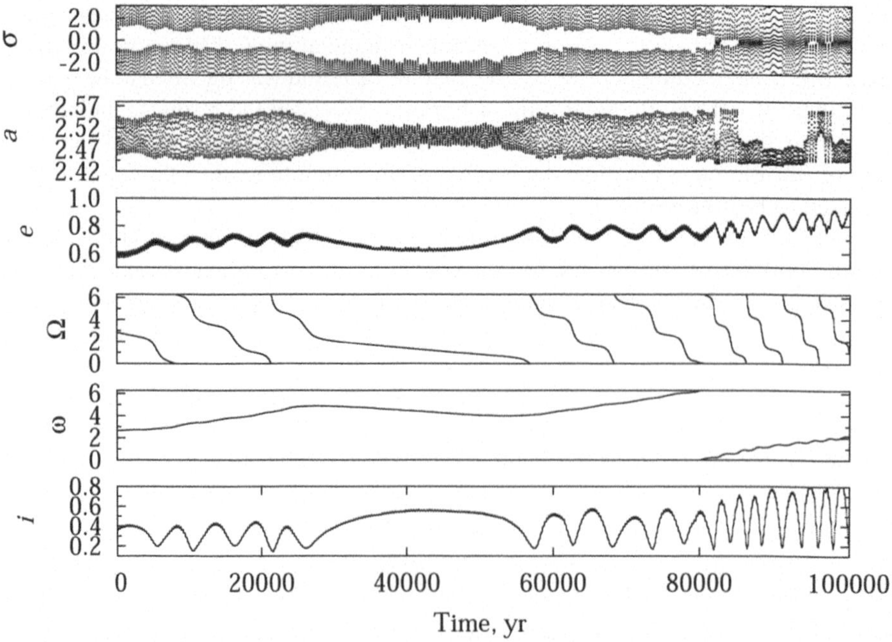

Fig. 8.12 The orbital elements and resonant argument of 1915 Quetzálcoatl, as a function of time. Quetzálcoatl is in the 3/1 resonance with Jupiter; the resonance is transient (Figure 9 from Smirnov and Shevchenko 2013, with permission from Elsevier)

where λ_J, λ_S, λ, ϖ_J, ϖ_S, ϖ are the mean longitudes and longitudes of perihelia of Jupiter, Saturn, and an asteroid, respectively; m_J, m_S, m, p_J, p_S, p are integers satisfying the D'Alembert rule

$$m_J + m_S + m + p_J + p_S + p = 0. \tag{8.28}$$

On the D'Alembert rules see Morbidelli (2002), and on the D'Alembert functions in general see Kholshevnikov (1997, 2001).

The *order* q of a three-body mean motion resonance is defined as the absolute value of the algebraic sum of the coefficients at the mean longitudes in the resonant argument:

$$q = |m_J + m_S + m|. \tag{8.29}$$

The order q is important, because the coefficient of the leading resonant term in the perturbing function contains the eccentricity e in the power of q (Nesvorný and Morbidelli 1998). The corresponding subresonance width is proportional to the square root of this coefficient. Therefore, the value of q determines the strength of the guiding subresonance. Note that, in case of two-body resonances, the role of the two-body resonance order q is analogous: the coefficient of the leading resonant term is proportional to e^q (Nesvorný and Morbidelli 1998).

Three-body mean motion resonances correspond to commensurabilities between the mean frequencies of the orbital motions of an asteroid and two planets, here Jupiter and Saturn:

$$m_J \dot{\lambda}_J + m_S \dot{\lambda}_S + m \dot{\lambda} \approx 0, \tag{8.30}$$

where $\dot{\lambda}_J$, $\dot{\lambda}_S$, $\dot{\lambda}$ are the time derivatives of the mean longitudes of Jupiter, Saturn, and asteroid; m_J, m_S, m are integers. The set of possible combinations of the m_J, m_S, and m integers is limited by the inequalities

$$q \leq q_{\rm max}, \tag{8.31}$$

and

$$|m_J|, \ |m_S|, \ |m| \leq M_{\rm max}, \tag{8.32}$$

where $q_{\rm max}$ and $M_{\rm max}$ were set in Smirnov and Shevchenko (2013) to 6 and 8, respectively.

In Smirnov and Shevchenko (2013), solely the resonances with Jupiter and Saturn were studied, and the resonances were considered in the planar problem, i.e., the longitudes of nodes in the expression for the resonant argument were ignored. The maximum considered order $q_{\rm max}$ of the three-body resonances was set equal to 6.

An example of a pure three-body resonance is given in Fig. 8.13, where the long-term orbital behaviour of the asteroid 463 Lola, residing in the 4-2-1 resonance with Jupiter and Saturn, is graphically presented. An example of a transient three-body resonance is given in Fig. 8.14, where the time behaviour of 490 Veritas, residing in the 5-2-2 resonance, is illustrated.

According to Smirnov and Shevchenko (2013), the fraction of asteroids in three-body resonances (transient plus pure) up to the 6th order inclusive turned out to be ≈4.4% of the total studied set of ≈250,000 asteroids. The fraction of asteroids in pure three-body resonances of the same orders is ≈0.94% of the total studied set. The top three most populated three-body resonances (pure plus transient ones) turned out to be the 5-2-2, 4-2-1, and 3-2-1 ones. The top three most populated pure resonances are the 4-2-1, 3-1-1, and 5-2-2 ones. By using a high-order extrapolation (in the form of a power law) of the q dependence of the number of identified resonant objects, the actual total fraction of asteroids residing in the pure three-body resonances of all orders can be estimated as ≈1.1% of the whole set.

In the transient plus pure resonances, the identified three-body resonant asteroids are ≈2.5 times more abundant than the two-body resonant ones. In the pure resonances, the abundances are comparable, but, if one excludes Trojans and Hildas, the abundance of three-body-resonant asteroids becomes overwhelming. If one extrapolates abundances by taking into account higher-order resonances, this

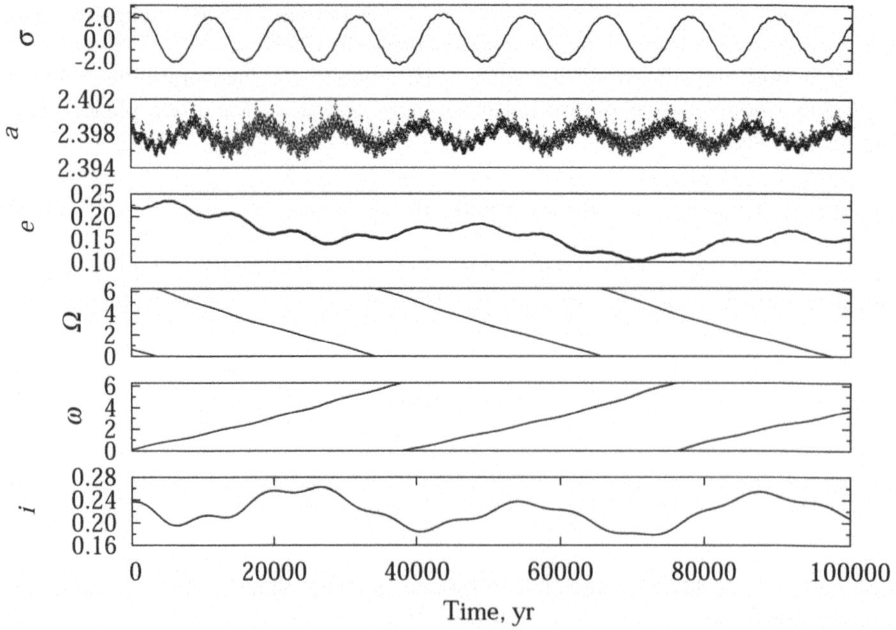

Fig. 8.13 The orbital elements and resonant argument of 463 Lola, as a function of time. Lola is in the 4-2-1 resonance with Jupiter and Saturn; the resonance is pure (Figure 1 from Smirnov and Shevchenko 2013, with permission from Elsevier)

overwhelming domination would even increase. Therefore, the analysis of Smirnov and Shevchenko (2013) quantitatively confirms the assertion by Nesvorný and Morbidelli (1998) that "the three-body mean motion resonances seem to be the main actors structuring the dynamics in the main asteroid belt."

The omnipresence of chaotic two-body and three-body mean motion resonances in the asteroid belt is conveniently illustrated by the asteroidal maximum Lyapunov exponent "scan," across the whole belt, in the initial asteroidal semimajor axis with all other initial orbital elements fixed to some values. Such a scan, constructed by Morbidelli and Nesvorný (1999) in direct numerical integrations of the asteroidal motion, is presented in Fig. 8.15. The initial asteroidal eccentricity is fixed to 0.1, and the inclination to zero. The dynamical model comprises perturbations from the four giant planets. The maximum Lyapunov exponent is measured in yr^{-1}. In the scan, the chaotic resonances reveal themselves as peaks, whose height characterizes the chaoticity degree and whose width characterizes the resonance strength (the radial extent of the corresponding multiplet of overlapping subresonances). The flat ground level of the scan is due to the finite integration time, which was set to 2.3 Myr. The peaks' designations mostly indicate two-body Jovian resonances and Jupiter–Saturn-asteroid three-body resonances; the exceptions are: S6/1 is the 6/1 resonance with Saturn, and 4J-2U-1 is a Jupiter–Uranus-asteroid three-body resonance. The realm of the ν_6 secular resonance is designated by "ν6." The

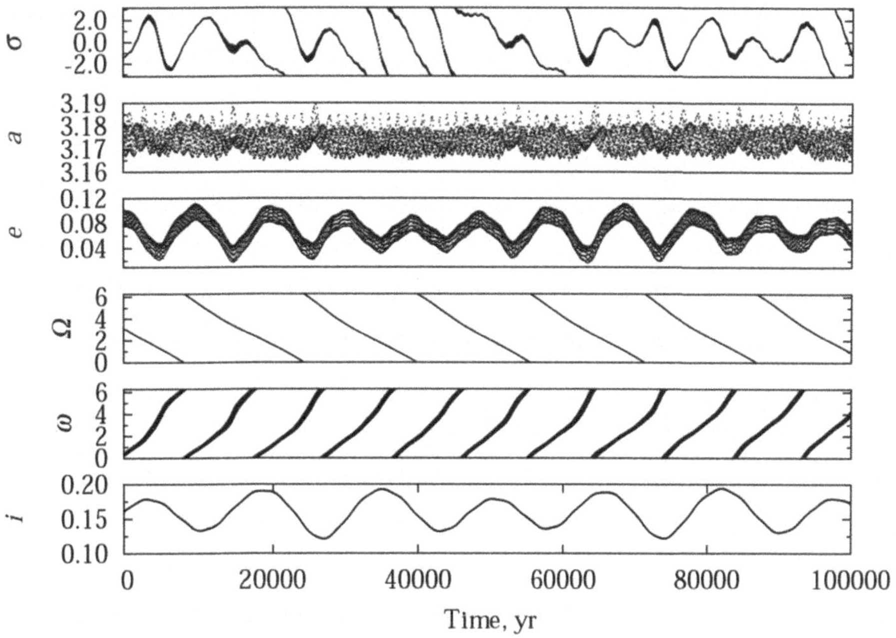

Fig. 8.14 The orbital elements and resonant argument of 490 Veritas. Veritas is in the 5-2-2 resonance with Jupiter and Saturn; the resonance is transient (Figure 2 from Smirnov and Shevchenko 2013, with permission from Elsevier)

resonant structure of the asteroid belt clearly manifests itself in Fig. 8.15. The abundance of chaotic resonances obviously rises on approaching the orbit of the main perturber (Jupiter).

8.2.2 The Kuiper Belt Objects

The *Kuiper belt* represents a broad ring of primordial planetesimals (trans-Neptunian objects, TNOs), situated outer to Neptune's orbit; see Fig. 8.16. The Kuiper belt has a significant resonant component. It also contains a lot of highly-eccentric bodies, whose orbits are essentially chaotic.

In Fig. 8.16, the distribution of the observed trans-Neptunian objects (TNOs) is presented in the "semimajor axis—eccentricity" plane. The plot graphically illustrates the complex structure of the Kuiper belt, featuring, first of all, its resonant populations. The resonant structure of the belt clearly reveals itself. The vertical lines indicate mean motion resonances with Neptune, including the 1/1 resonance corresponding to the Neptune Trojans (at 30 AU). The objects of the *scattered disc* are located in the region between two curves (dotted) of constant perihelion

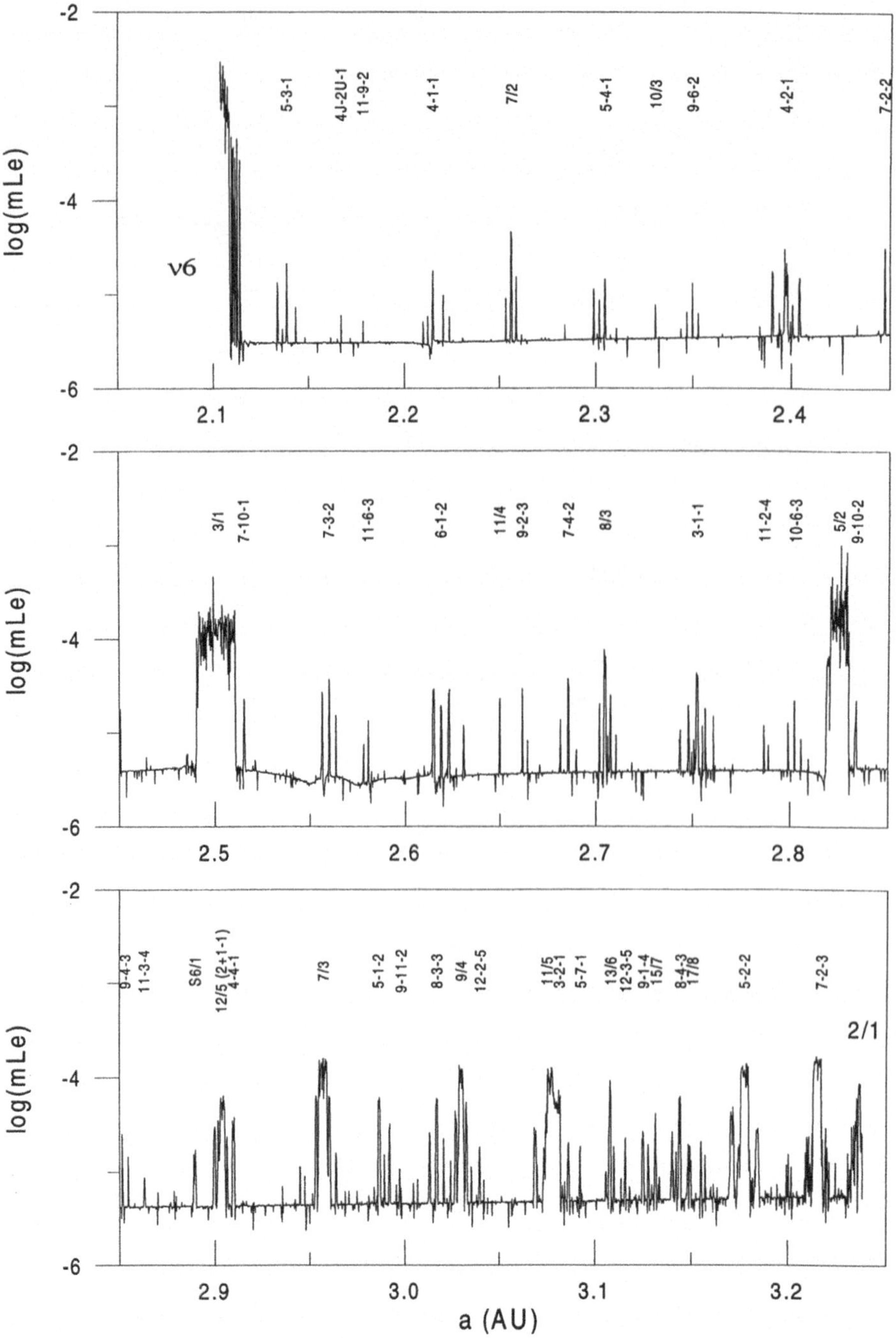

Fig. 8.15 The asteroidal maximum Lyapunov exponent as a function of the initial asteroidal semimajor axis a, across the asteroid belt; see text for details (Figure 1 from Morbidelli and Nesvorný 1999, with permission from Elsevier)

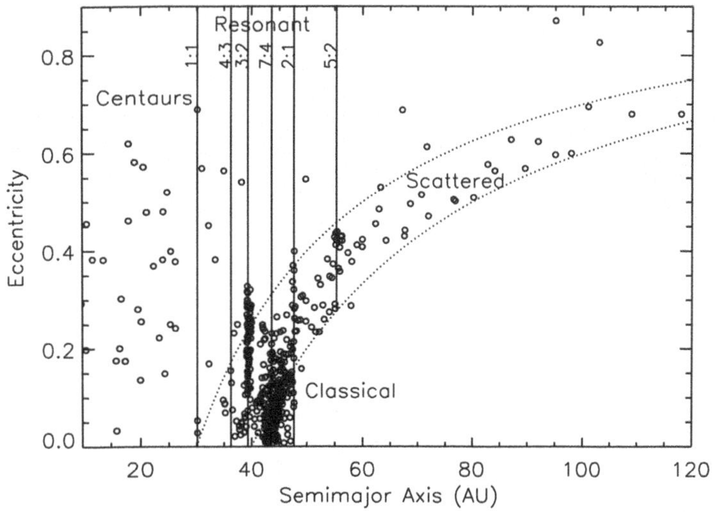

Fig. 8.16 The Kuiper belt dynamical structure. Resonant and highly-eccentric populations are obviously present (Figure 1 from Sheppard 2006, by permission from ASP)

distances, equal to 30 and 40 AU. The "extended scattered disc" objects have pericentric distances greater than 45 AU. Classical TNOs are distributed at smaller eccentricities, their orbital semimajor axes range up to 50 AU.

By debiasing the detection statistics, the 5/2 resonance with Neptune was found to host a population comparable with that in the 3/2 resonance (Gladman et al. 2012); see Fig. 8.17. According to Gladman et al. (2012), the population of TNOs in the 5/2 resonance with Neptune is estimated to be as large as in the 3/2 resonance, whereas other (non-half-integer) resonant populations are radically smaller. One of the most distant known resonant TNOs resides in the 27/4 resonance with Neptune (Gladman et al. 2012).

The abundance of potential two-body and three-body mean motion resonances in the Kuiper belt is illustrated by a model TNO's maximum Lyapunov exponent scan, across the whole Kuiper belt, in the initial TNO's semimajor axis with all other initial orbital elements fixed to some values. Such a scan, constructed by Nesvorný and Roig (2001), is presented in Fig. 8.18. Its construction is analogous to that of the scan for the asteroid belt, presented in Fig. 8.15. The peaks' designations indicate two-body resonances with Neptune and Uranus (marked, accordingly, with "N" and "U") and Neptune–Uranus–TNO three-body resonances. In Fig. 8.18, the resonant structure of the Kuiper belt graphically manifests itself; note the greater chaoticity of the "inner" ($a \lesssim 43$ AU) Kuiper belt.

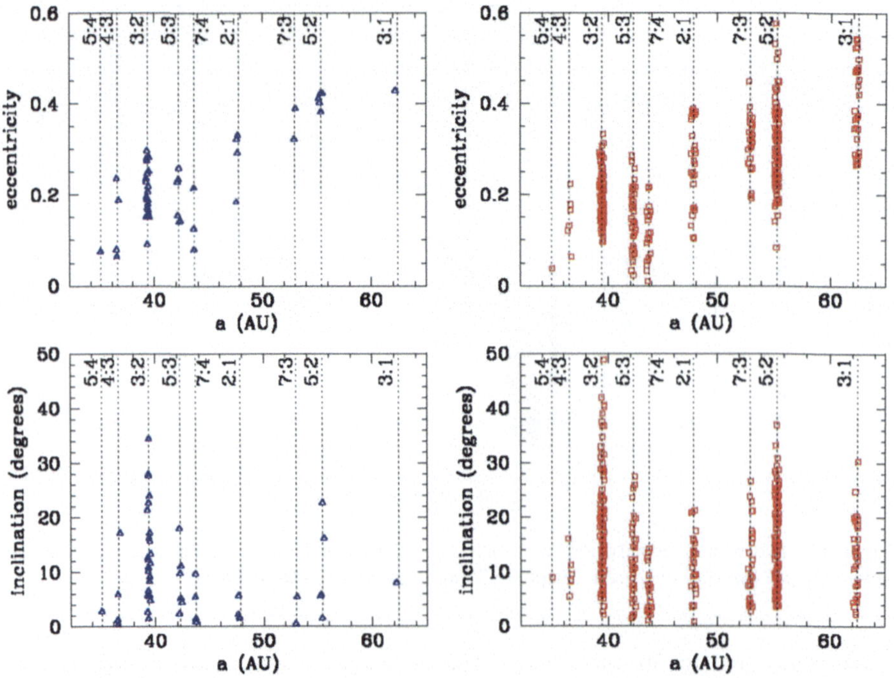

Fig. 8.17 Resonant populations of the Kuiper belt. Left panels: before debiasing the statistics; right panels: after the debiasing (Figure 8 from Gladman et al. 2012, by permission of IOP Publishing/AAS)

8.2.3 *Two-Body Resonances*

To estimate Lyapunov timescales of asteroidal orbits in mean motion resonances with Jupiter, a resonance model derived in Holman and Murray (1996); Murray and Holman (1997) can be straightforwardly used (Shevchenko 2007a). This model is described further on, in Sect. 13.2.1, in a generalized form suitable for any passively gravitating particle moving around a primary and perturbed by a secondary in an outer mildly eccentric orbit. Thus, we work in the framework of the planar restricted three-body "Sun–Jupiter-asteroid" problem.

Equations (13.9)–(13.13) are used to estimate parameters of any particular resonance. The units are chosen in such a way that the system total mass (Sun plus Jupiter), the gravitational constant $\mathcal{G}$, Jupiter's semimajor axis a_J are all equal to one; $\mu = 1/1047.355$, $\mu_1 = 1 - \mu$. Jupiter's mean longitude $l_J = n_J t$, and eccentricity $e_J = 0.048$. Jupiter's mean motion $n_J = 1$, i.e., the time unit equals $\frac{1}{2\pi}$th part of Jupiter's orbital period. The initial eccentricity of the asteroidal orbit is set to 0.01.

For any mean motion resonance, the guiding subresonance in the multiplet is identified (as that with maximum $|\varphi_{k+q,k+p,k}|$), and its two closest neighbours are considered as the perturbing resonances. The ratio of the obtained Ω and

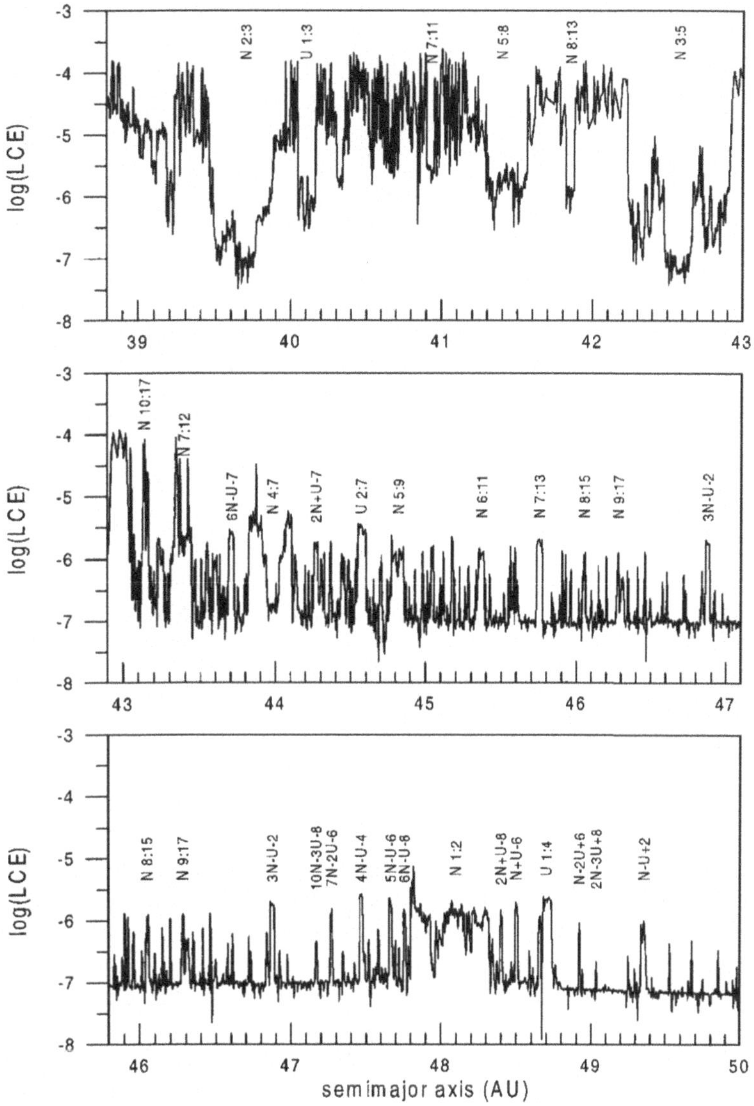

Fig. 8.18 Same as Fig. 8.15, but for the Kuiper belt; see text for details (Figure 2 from Nesvorný and Roig 2001, with permission from Elsevier)

ω_0 frequencies gives the adiabaticity parameter λ. Then, the resonance type is determined: non-adiabatic chaotic triad, non-adiabatic chaotic duad, adiabatic chaotic triad, or adiabatic chaotic duad. If the amplitudes of the neighbours differ from each other less than by a factor of two, the model resonance is considered to be a triad, otherwise a duad.

Table 8.6 Lyapunov timescales T_{L} (yr) in mean motion resonances (Shevchenko 2007a)

$\frac{k+q}{k}$	λ	$\log_{10} T_{\mathrm{L}}^{\mathrm{num}}$ [a]	$\log_{10} T_{\mathrm{L}}^{\mathrm{theor}}$	Res. type
3/1	0.093	3.8–4.3	4.3	ACD
5/2	0.192	3.5–3.8	4.1	ACT
7/3	0.415	3.8–4.2	4.0	ACT
9/4	0.932	3.9–4.3	4.2	NACD
11/5	1.970	3.9–4.3	4.3	NACT
9/5	0.323	3.6–3.8	3.7	ACT
7/4	0.166	3.2–3.3	3.7	ACT
12/7	0.594	3.6–4.0	3.9	NACD
5/3	0.101	2.5–3.3	3.7	ACD
8/5	0.156	2.5–3.3	3.6	ACT
11/7	0.264	3.3–3.6	3.5	ACT

[a]Morbidelli and Nesvorný (1999), Holman and Murray (1996)

The resulting estimates are given in Table 8.6. In the table, the resonance types are designated as "NACT" (non-adiabatic chaotic triad), "NACD" (non-adiabatic chaotic duad), 'ACT" (adiabatic chaotic triad), and "ACD" (adiabatic chaotic duad). The analytical estimates of Lyapunov times are made by means of formulas (3.26), (3.28), (3.30), and (3.31), respectively.

For comparison, we take the data on the numerical T_{L} values from (Morbidelli and Nesvorný 1999, Figure 1)[5] and (Holman and Murray 1996, Figure 6), where the motion close to the mean motion resonances was investigated in direct numerical integrations.

From Table 8.6 it follows that the analytical Lyapunov timescale estimates are generally in accord with the numerical-experimental ones. However, some differences can be clearly seen, especially in the domain of adiabatic chaos. This should be attributed to the imperfectness of the model (13.9) in application to the given problem; mostly due to the $\varphi_{k+q,k+p,k}$ coefficients being treated as constants.

Following Shevchenko (2007a), let us consider Lyapunov timescale of the motion of a real asteroid, namely, 522 Helga. This asteroid is famous to be the first ever actual example of *stable chaos* in the dynamics of asteroids (Milani and Nobili 1993): Helga's Lyapunov time is relatively small ($\approx$6900 yr), but its orbit does not exhibit any gross changes on timescales of millions of years, as follows from direct numerical integrations.

522 Helga is known to reside in the 12/7 mean motion resonance. We take necessary data on the semimajor axis a, eccentricity e, and perihelion frequency $g = \dot{\varpi}$ from the AstDyS database (AstDyS 2020). T_{pert} is defined by the perihelion frequency g. One finds that the guiding subresonance in the resonance sextet is the

[5]Reproduced in Fig. 8.15 in this book.

third one ($p = 2$); therefore, the perturbing neighbours in the model (13.9) have numbers $p = 1$ and 3. Since $\epsilon q = 0.624 < 1$, the model is valid. The separatrix map parameters $\lambda = 2.325$ and $\eta = 0.812$. Therefore, the resonance type is the non-adiabatic chaotic triplet. Applying formula (3.26), one obtains $T_L = 9700$ yr.

On the other hand, the standard map theory, given by Eq. (3.38), provides a lower bound for the Lyapunov time. According to Eq. (3.39), the stochasticity parameter $K = (2\pi/\lambda)^2$, and for 522 Helga one has $K \approx 7.3$. Equation (3.38) gives $T_L \approx 5100$ yr (Shevchenko 2014). We see that the theoretical estimates of Helga's Lyapunov time range from 5000 to 10,000 yr; this generally agrees with results of direct numerical integrations in the full dynamical problem (taking into account the perturbations from all planets), which give 6900 yr (Milani and Nobili 1993) and 6860 yr (AstDyS 2020).

The standard map theory seems to provide T_L estimates closer to the actual numerical-experimental values. The cause is that the number of subresonances in the multiplet is large and the relative perturbation strength ε is not far from unity, i.e., to the value characteristic for the standard map Hamiltonian (1.24).

8.2.4 Three-Body Resonances

An essential role in orbital dynamics of Solar system bodies, and in particular asteroids, is played by the three-body resonances, defined above in Sect. 8.2.1. The three-body resonances can be analytically described in the perturbed pendulum model (Murray et al. 1998; Nesvorný and Morbidelli 1998, 1999). The Hamiltonian of the motion of a zero-mass test particle near a three-body resonance $\{m_J m_S m\}$ with Jupiter and Saturn in the planar-elliptic problem can be approximately expressed as

$$H = \alpha S^2 + \sum_{p_J, p_S, p} \beta_{p_J p_S p} \cos \sigma_{p_J p_S p} \tag{8.33}$$

(Nesvorný and Morbidelli 1999), where the resonant argument

$$\sigma_{p_J p_S p} = m_J l_J + m_S l_S + ml + p_J \varpi_J + p_S \varpi_S + p\varpi$$

is conjugated to the momentum-like S variable; $\alpha = -(3/2)n^2 a_{\mathrm{res}}^{-2}$, n is the asteroid's mean motion. It is assumed that the time derivatives of l_J, l_S, ϖ_J, and ϖ_S are constants. Analytical expressions for $\beta_{p_J p_S p}(e)$ for some important three-body resonances are given in (Nesvorný and Morbidelli 1999, Tables 3–6).

From the Hamiltonian (8.33) it is clear that any $\{m_J m_S m\}$ three-body resonance is split in a cluster of subresonances with various $\{p_J p_S p\}$ combinations. The frequency of small-amplitude phase oscillations on subresonance $\{p_J p_S p\}$ is given by

$$\omega_0 = 2\pi n (3\beta_{p_J p_S p})^{1/2} a_{\mathrm{res}}^{-1} \tag{8.34}$$

(Nesvorný and Morbidelli 1999). The perturbation frequency Ω generally represents an algebraic combination of the perihelion frequencies of Jupiter, Saturn, and the asteroid. The combination is defined by the guiding subresonance choice, as specified below. The ratio of Ω and ω_0 gives the value of the adiabaticity parameter λ.

Consider asteroids that were identified by Nesvorný and Morbidelli (1999) to reside in three-body resonances. To assess the coefficients of resonant terms, we use analytical data given in Tables 3–6 in Nesvorný and Morbidelli (1999). The *guiding subresonance* in the multiplet is identified as the subresonance that has the maximum value of $|\beta_{p_J p_S p}|$. Its two closest neighbours are considered as the perturbing resonances. The theoretical estimates of Lyapunov times are accomplished by means of formulas (3.26), (3.28), (3.30), and (3.31), as in case of two-body resonances in the previous section. The formula is chosen according to the resonance multiplet type.

The identification of the guiding subresonances in the multiplets shows that the three-body resonances can be subdivided into two distinct types: (1) in which the perturbation frequency Ω in model (3.23) is equal to $\dot{\varpi} - \dot{\varpi}_J$, and (2) in which Ω is equal to $\dot{\varpi}_S - \dot{\varpi}_J$ (Shevchenko 2007a). The 5-2-2 and 3-1-1 resonances belong to the first type, while the 2+2-1 and 6+1-3 resonances to the second. We set $\dot{\varpi}_J = 4.257''$/yr and $\dot{\varpi}_S = 28.243''$/yr (Bretagnon 1990). The data on $\dot{\varpi}$, a, and e are taken from the AstDyS database (AstDyS 2020).

The theoretical estimates of Lyapunov times are presented in Table 8.7. Some of them agree with the numerical ones (in particular, that of Veritas), others do not. The latter case is presented by Genua and Paine-Gaposchkin; one may judge that these objects most probably do not reside in the chaotic domains of the prescribed resonances. In this way, analytical estimating the Lyapunov times may allow one to discern between possible models of chaos in the motion of actual asteroids.

Table 8.7 Lyapunov timescales T_L (yr) of asteroids in three-body mean motion resonances (Shevchenko 2007a)

Asteroid	Resonance $\{m_J m_S m\}$	λ	T_L^{num}, yr [a]	T_L^{num}, yr [b]	T_L^{theor}, yr	Res. type
258 Tyche	$2+2-1$	0.536	35,900	–	43,100	NACT
485 Genua	$3-1-1$	0.376	6550	6500	35,700	ACD
1642 Hill	$3-1-1$	0.643	36,100	–	43,300	NACD
936 Kunigunde	$6+1-3$	0.624	22,200	–	54,600	NACD
490 Veritas	$5-2-2$	0.546	10,200	8500	9100	NACD
2039 Paine-Gaposchkin	$5-2-2$	0.449	22,000	–	6020	ACD
3460 Ashkova	$5-2-2$	0.433	65,100	8300	5940	ACD

[a] AstDyS (2020)
[b] Nesvorný and Morbidelli (1998, 1999), Milani et al. (1997)

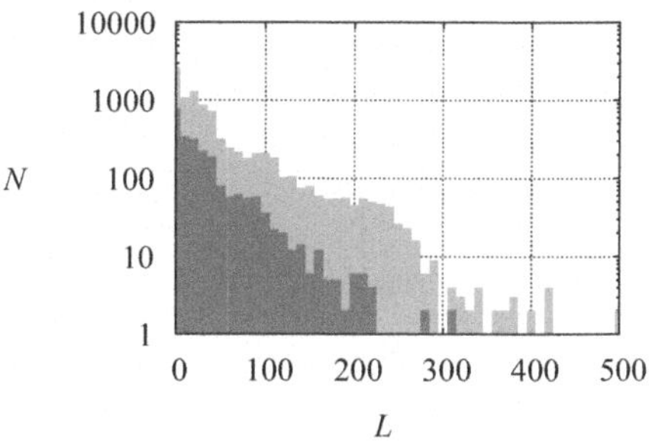

Fig. 8.19 Histograms of resonant asteroids in the maximum Lyapunov exponent L. N is the number of asteroids per bin. Grey: asteroids in transient resonances; black: asteroids in pure resonances (Figure 6 from Smirnov and Shevchenko 2013, with permission from Elsevier)

The AstDyS database (AstDyS 2020) provides information on the maximum Lyapunov exponents for almost all asteroids contained in it. It is instructive to check how the AstDyS data on Lyapunov exponents correlate with the pure/transient division of resonances to which resonant asteroids, identified by Smirnov and Shevchenko (2013), belong. In pure resonances, the motion is expected to be mostly regular, whereas in transient resonances it is expected to be chaotic. Indeed, sporadic transitions from libration to circulation and vice versa are characteristic for dynamics in near-separatrix chaotic layers.

The resulting differential distributions of resonant asteroids in the maximum Lyapunov exponent L are presented in Fig. 8.19. N is the number of asteroids in the $(L,\ L + \Delta L)$ bin, where $\Delta L = 10$, and L are measured in units of $(\text{mln yr})^{-1}$. (Note that in the AstDyS database they are given in units of yr^{-1}.) The histogram for asteroids in transient resonances is depicted in grey, and that for asteroids in pure resonances in black. At $L > 500\ (\text{mln yr})^{-1}$ (the range not shown in the plot), there are only few objects; such highly chaotic objects are in transient resonances.

As one would expect, the distribution of the transiently resonant objects in Fig. 8.19 is much more extended to greater L values, in comparison with the pure-resonant objects. This distinction is uniform for all resonant groups: in Fig. 8.20, where the distribution of resonant asteroids on the "semimajor axis a—maximum Lyapunov exponent L" plane is shown, the open circles (representing the transiently resonant objects) are distributed to much greater heights, in comparison with the black circles (representing the pure-resonant objects), in all resonant groups present in the plot. From the plot it is evident that solely transiently resonant objects possess extremely large Lyapunov exponents, corresponding to the Lyapunov time down to $\approx$570 yr.

Averaged over all identified pure-resonant asteroids, the maximum Lyapunov exponent is $L \approx 34\ (\text{mln yr})^{-1}$, and, averaged over all identified transient-resonant asteroids, it is $L \approx 50\ (\text{mln yr})^{-1}$. The Lyapunov times $T_{\text{L}} \approx 3 \cdot 10^4$ and $\approx 2 \cdot 10^4$ yr, respectively. In the AstDyS database, asteroids with numerically measured Lyapunov exponents have, on average, the maximum Lyapunov exponent $L \approx 21\ (\text{mln yr})^{-1}$; this corresponds to the Lyapunov time $\approx 5 \cdot 10^4$ yr. Therefore,

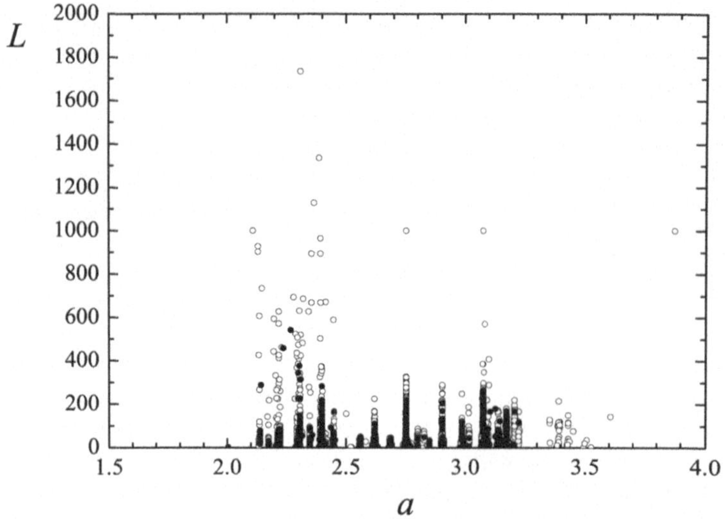

Fig. 8.20 Distribution of resonant asteroids on the "semimajor axis a—maximum Lyapunov exponent L" plane. Open circles: asteroids in transient resonances; black circles: asteroids in pure resonances (Figure 7 from Smirnov and Shevchenko 2013, with permission from Elsevier)

the pure-resonant and transient-resonant asteroids both turn out to be on average more chaotic than a randomly chosen asteroid (Smirnov and Shevchenko 2013).

8.2.5 Secular Resonances

Along with two-body and three-body mean motion resonances, *secular resonances* play an essential role in forming the dynamical structure of the asteroid belt (Murray and Dermott 1999; Morbidelli 2002). Recall that secular resonances correspond to commensurabilities between the rates of apsidal or nodal precession of orbits of an asteroid and a planet.

The Lagrange planetary equations of the lowest order in the eccentricities and inclinations are in the basis of the classical secular planetary theory. Their solution splits in two independent sub-solutions, one in the (e, ϖ) elements (eccentricity and longitude of pericentre) and one in the (i, Ω) elements (inclination and longitude of ascending node); see, e.g., Murray and Dermott (1999). Using these solutions, the locations of the main secular resonances in the proper elements space can be identified. If one takes into account higher-order terms in the eccentricity and inclination and/or the second order terms in the mass parameter, the two sub-solutions are no more decoupled and the theory becomes more complicated; see Froeschlé and Morbidelli (1994) and Knežević and Milani (1994).

A simplest secular resonance takes place if the frequency of the proper longitude of pericenter $\dot{\varpi}_{\mathrm{proper}}$ or the frequency of the proper longitude of ascending node $\dot{\Omega}_{\mathrm{proper}}$ of a particle are in a resonant combination with one of the eugenfrequencies of the system of perturbers. For an asteroid, a secular resonance takes place, for example, if any of its proper frequencies of the apsidal or nodal precession equals any of the proper frequencies of the planetary system. The secular-resonant relations specify two-dimensional surfaces in the three-dimensional $(a,\ e,\ i)$ proper element space. The locations of principal *linear secular resonances* in the asteroid belt, called the ν_5, ν_6, and ν_{16} resonances, were calculated by Williams (1969) and Williams and Faulkner (1993). In these resonances, $\dot{\varpi}_{\mathrm{proper}} - g_5 = 0$, $\dot{\varpi}_{\mathrm{proper}} - g_6 = 0$, and $\dot{\Omega}_{\mathrm{proper}} - s_6 = 0$, respectively. Here g_k and s_k are the planetary frequencies, $k = 1, 2, \ldots, 8$. The subscript in the resonance designations gives the ordinal number of the corresponding planetary frequency: $\nu_1 = g_1, \ \ldots, \ \nu_{10} = g_{10}, \ \nu_{11} = s_1, \ \ldots, \ \nu_{18} = s_8$.

In Fig. 8.21, the distribution of the main-belt asteroids in the "semimajor axis—inclination" plane of proper elements is shown along with the curves corresponding to several major secular resonances; the curves data were calculated for $e_{\mathrm{proper}} = 0.1$ in Milani and Knežević (1990). From the plot it is clear that the main belt's inner edge is formed by the ν_6 resonance, and the outer edge by the 2/1 mean motion resonance; see also Fig. 8.15. The distributions of the main-belt asteroids

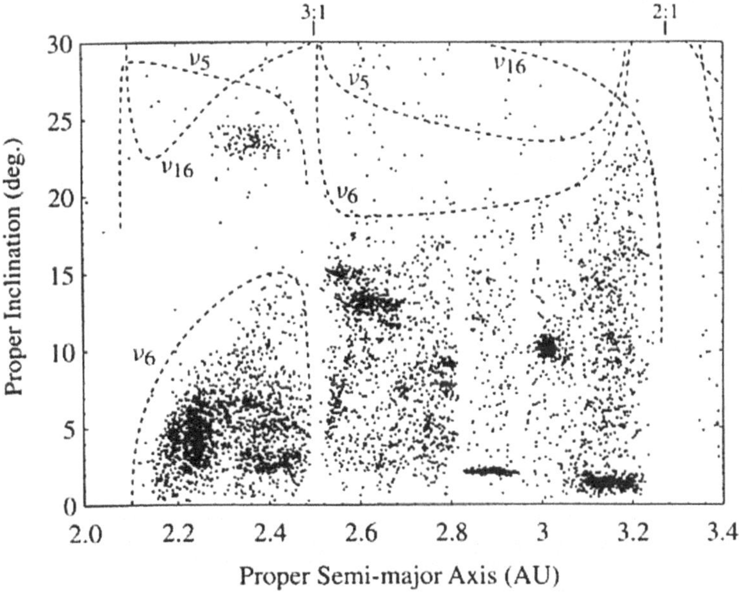

Fig. 8.21 Dashed curves: the ν_5, ν_6, and ν_{16} secular resonances in the "proper semimajor axis—proper inclination" plane, at $e_{\mathrm{proper}} = 0.1$. Dots: distribution of asteroids (Figure 7.19 from Murray and Dermott 1999, by permission from © Cambridge University Press)

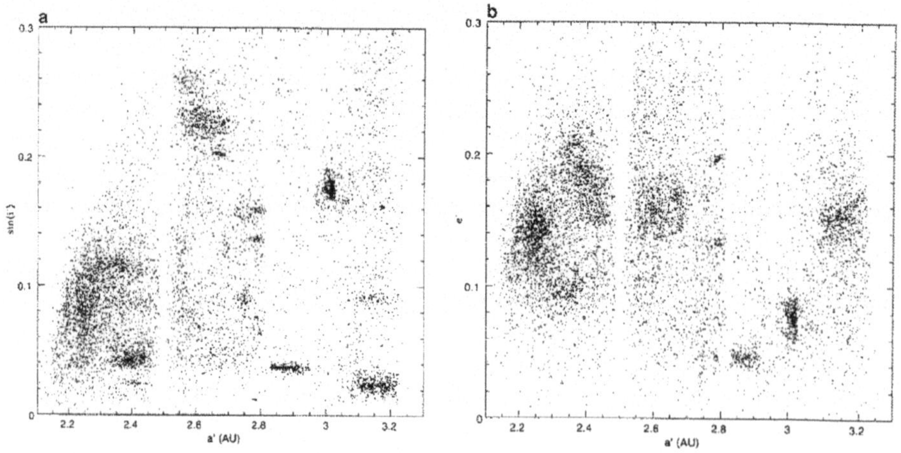

Fig. 8.22 Distribution of 12487 asteroids in two planes of proper elements (Figure 1 from Zappalà et al. 1995, with permission from Elsevier)

in inclination at $i < 30°$ are controlled by the ν_5, ν_6, and ν_{16} resonances. The depletion of the ν_5 resonance is mostly due to the Lidov–Kozai effect, described further on in Sect. 15.

The so-called *non-linear secular resonances*[6] correspond to terms of higher order in the eccentricities and inclinations in the expansion of the perturbing function. According to Knežević and Milani (1994), the following non-linear secular resonances are important in the asteroid belt: $\dot{\varpi}_{\rm proper} + \dot{\Omega}_{\rm proper} - g_5 - s_6$, $\dot{\varpi}_{\rm proper} + \dot{\Omega}_{\rm proper} - g_6 - s_6$, $\dot{\varpi}_{\rm proper} + \dot{\Omega}_{\rm proper} - g_5 - s_7$, $\dot{\varpi}_{\rm proper} - 2g_6 + g_5$, $\dot{\varpi}_{\rm proper} - 2g_6 + g_7$, $\dot{\varpi}_{\rm proper} - 3g_6 + 2g_5$, $\dot{\Omega}_{\rm proper} - s_6 - g_5 + g_6$, $2\dot{\varpi}_{\rm proper} + \dot{\Omega}_{\rm proper} - 2g_6 - s_6$, and $3\dot{\varpi}_{\rm proper} + \dot{\Omega}_{\rm proper} - 3g_6 - s_6$.

In Fig. 8.22, distribution of some twelve thousand asteroids on two planes of proper semimajor axes, eccentricities, and inclinations is presented; and, in Fig. 8.23, the locations and widths of major secular resonances are shown, for a model with the proper eccentricity set equal to 0.1. Comparing Figs. 8.22 (left panel) and 8.23, one may see that, in areas where major secular resonances overlap, the asteroidal population is strongly depleted. The population is also depleted at the resonances themselves, due to overlap of their subresonances. In this way, the secular resonances play a major role in formation of the asteroid belt's dynamical structure.

[6]Not to confuse with general non-linear resonances, introduced in Sect. 1.1.

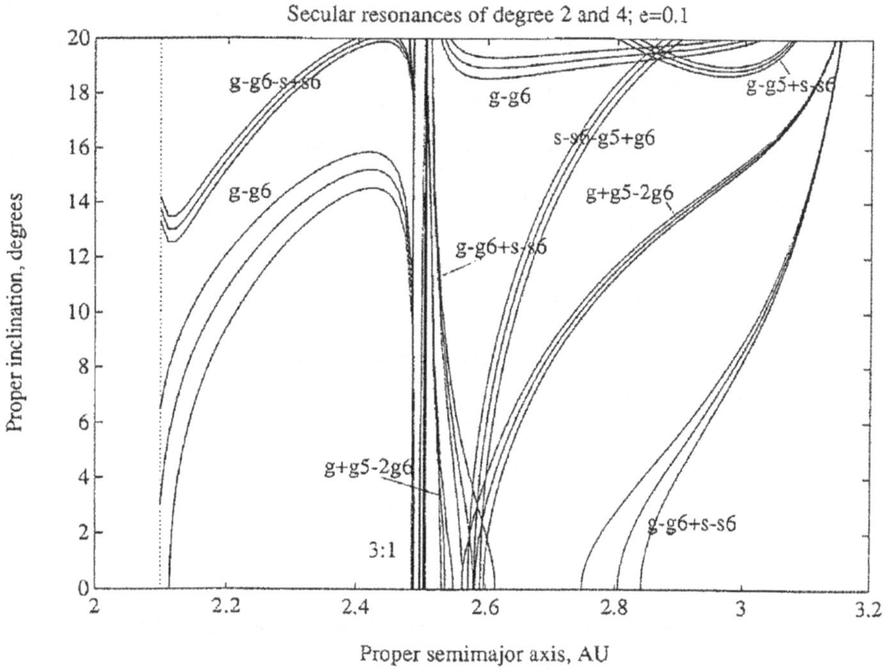

Fig. 8.23 The locations and widths of major secular resonances in the "proper semimajor axis—proper inclination" plane, at $e_{\rm proper} = 0.1$; compare with Fig. 8.22a (Figure 5 from Milani and Knežević 1990, by permission from Springer Nature, © 1990)

8.3 Cometary Dynamics

The cometary motion and, generally speaking, any perturbed highly-eccentric motion is generally chaotic, even when close encounters with perturbers are absent. The fundamental cause for the cometary chaos lies in the accumulation of the $p/1$ mean motion resonances between the "planet" (inner perturber) and the comet (test particle) in the vicinity of the parabolic trajectory, which plays the role of a separatrix between the bounded (elliptic) and unbounded (hyperbolic) types of motion. This was revealed in the works by Petrosky (1986) and Chirikov and Vecheslavov (1986); in the both works, an especial area-preserving two-dimensional map, now called the *Kepler map*, was independently derived to provide analytical means for description of the chaotic motion of comets in perturbed highly-eccentric orbits.

The goal of the work by Chirikov and Vecheslavov (1986) was to characterize the qualitative dynamics of the Halley comet, while awaiting its apparition in 1987. As well as Hyperion's rotation served as the first-ever example of observed chaotic rotational motion in the Solar system, the motion of comet Halley served the first-ever example of revealed (although not directly observed) chaotic orbital motion. The chaos in the Halley comet long-term orbital dynamics was revealed by Chirikov and Vecheslavov (1986, 1989) in an analysis of the historical chronicles of the comet's apparitions. Since then, the modern Kepler map theory allows one to characterize the chaotic diffusion phenomenon in the cometary dynamics and dynamics of meteor streams (Emelyanenko 1992; Liu and Sun 1994; Malyshkin and Tremaine 1999; Zhou and Sun 2001; Zhou et al. 2000, 2002).

8.3.1 Comets: Highly Chaotic

The Kepler map is derived in the assumption that the planetary perturbation is concentrated at the time moment when the comet passes its orbital pericentre; and away from the pericentre it moves in a non-perturbed Keplerian orbit. The map is given by Eqs. (1.45) or (1.46) with $\gamma = 3/2$:

$$\begin{aligned} y_{i+1} &= y_i + \sin x_i, \\ x_{i+1} &= x_i + \lambda |y_{i+1}|^{-3/2}, \end{aligned} \tag{8.35}$$

where y is the comet's normalized orbital energy, x is the normalized time, and λ is the map's sole constant parameter, analogous to the adiabaticity parameter of the ordinary separatrix map. Each iteration of the map corresponds to one consecutive orbital revolution of the comet. This means that the map time unit, corresponding to one iteration, is not constant: the increment of actual time per iteration is $\Delta x_i = x_i - x_{i-1}$.

The chaos described by the Kepler map is generically non-adiabatic, with the adiabaticity parameter $\lambda \gg 1$ (Shevchenko 2011b). This allows one to approximate the Kepler map locally in the y variable by the standard map with good accuracy. Then, using the standard map theory, the local and global Lyapunov and diffusion timescales of the motion can be estimated straightforwardly.

According to Fig. 8.24, the Kepler map's chaotic trajectories have the Lyapunov time $T_{\mathrm{L}} \sim 1$ in the units of the map iterations; therefore, any two chaotic trajectories with slightly different initial conditions would substantially diverge already on the timescale of several map iterations.

First estimates of the Lyapunov and diffusion timescales were made in the pioneering works of Chirikov and Vecheslavov (1986, 1989) in application to the Halley comet. As given by formula (11.99), the Kepler-map-based lower bound for the Halley comet's Lyapunov time is ~30–40 yr (Shevchenko 2007a). This lower bound characterizes the Lyapunov timescale averaged over all possible dynamical states of the comet inside the Kepler map's chaotic layer. Using data of Chirikov and Vecheslavov (1986); Vecheslavov and Chirikov (1988); Chirikov and Vecheslavov (1989) on the amplitude of perturbation of the comet's energy, one finds $\lambda \approx 1.2 \cdot 10^4 \gg 1$; therefore, the local approximation of the Kepler map by the standard map, used in deriving the lower bound, is valid.

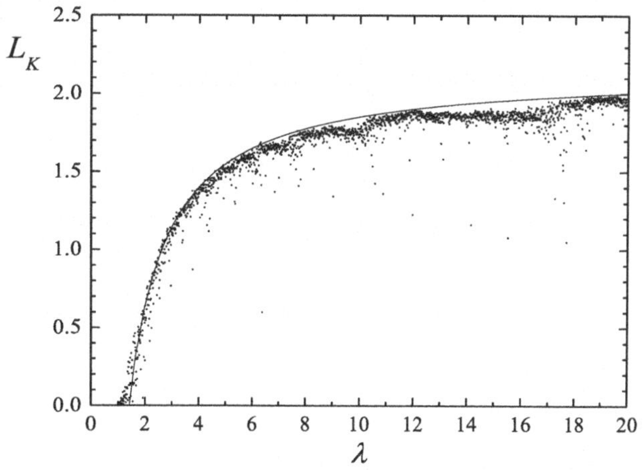

Fig. 8.24 Dots: the λ dependence of the maximum Lyapunov exponent L_K of the Kepler map. Solid curve: a fitting rational function (Figure 3 from Shevchenko 2007a, by permission from © Cambridge University Press)

This lower bound is in accord with results of numerical integrations of the comet's dynamics performed in the full problem settings (Muñoz-Gutiérrez et al. 2015; Boekholt et al. 2016; Pérez-Hernández and Benet 2019). The integrations usually give the Lyapunov time of the order of hundreds of years. Note that the numerical-experimental estimates typically has a local character, as they are naturally obtained on restricted time intervals, limited from above at least by the time of comet's ejection from the Solar system.

It is expectable that relation (11.99) applies for the motion of any long-periodic comet, or any Halley-type comet, or any asteroid in a highly-eccentric orbit with a similar kind of perturbation; i.e., the Lyapunov time of such an object can be in practice approximately determined by its orbital period solely.

As a whole, the planet-crossing comets and planet-crossing asteroids, near-Earth asteroids (NEAs) among them, form the most chaotic classes of orbiting bodies in the Solar system. The Lyapunov times of these objects in comparison with the Lyapunov times of the ordinary main-belt asteroids can be rather small, down to several years. Whipple (1995) explored the chaotic orbital motion of 175 actual asteroids with the pericentric distance $q < 1.6$ AU (corresponding to the Solar system inner part), and found that the Lyapunov times could be as small as 10 yr, due to encounters with the terrestrial planets. The T_L values range from $\sim$10 to $\sim$20,000 yr. 34 of these 175 objects are so chaotic, that the errors in determination of their orbits double in less than 70 yr.

To date, no known asteroid or comet seems to violate the bound (11.99). In this respect, a "Lyapunov time—semimajor axis" diagram, constructed in (Whipple 1995, Figure 2) is of particular interest; in this diagram, one may see that all the highly-chaotic objects presented in the diagram have $T_L \geq 10$ yr, and, since their semimajor axes $a < 3.5$ AU, the bound (11.99) is in no way violated.

Tancredi (1995, 1999) considered the orbital evolution of two groups of planet-crossing bodies, namely, (1) 145 Jupiter family comets and (2) 307 NEAs. The NEAs were defined as inactive objects with aphelia $Q > 1$ AU and perihelia $q < 1.5$ AU. In direct integrations, the T_L values were found to be in the range 30–200 yr for the first group, and 10–300 yr, mostly 50–150 yr, for the second one. The minimum observed value was $\approx$10 yr.

Therefore, the planet-encountering asteroids and comets definitely belong to the most chaotic classes of objects in the Solar system; their Lyapunov times can be as low as several years. This is in accord with the straightforward estimate (11.99).

8.3.2 Lévy Flights at the Edge of Escape

Chaotic behaviour of planet-crossing, and not only planet-crossing, small bodies in celestial-mechanical problems can be naturally considered in the framework of the abstract theory of Lévy flights. The Lévy flights, i.e., the increments (in a kind of a random walk) that have a heavy-tailed distribution, is a well-studied subject with various applications; see discussion in Shevchenko (2010). They were thoroughly studied in connection with abstract problems of Hamiltonian dynamics; see Denisov et al. (2002) and references therein. In celestial mechanics, Lévy random walks arising due to close encounters with perturbing bodies were analysed in studies of the long-term cometary orbital evolution (Zhou et al. 2002).

Lévy flights are typically considered in random walks with steps possible in at least two (forward and back) directions. Conversely, the Lévy distributions analysed below are one-sided: the increments are positive. In a general statistical setting, one-sided Lévy flights are considered in Koren et al. (2007a,b), where exact results for the first passage time and leapover statistics are given.

Both kinds of Lévy flights studied below in the framework of the three-body problem are not directly due to encounters of bodies. One of these two kinds is due to encounters of a trajectory with the separatrix in the phase space of motion; hereafter we call such flights the *Lévy flights of the first kind*, or LF1s, since this phenomenon corresponds to the Hamiltonian intermittency of the first kind. Conversely, the *Lévy flights of the second kind*, or LF2s, are due to sticking of a trajectory to chaos borders in the divided phase space of motion; these flights arise due to the second kind Hamiltonian intermittency. They effect the duration of long Poincaré recurrences, but not (practically) the orbital periods (Shevchenko 2010).

In the orbital behaviour of a tertiary in the planar restricted three-body problem, which we consider in what follows, LF1s emerge in the form of sudden jumps in the orbital semimajor axis and period, while LF2s emerge in the form of long sequences of orbital revolutions with almost constant orbital semimajor axis and period.

The Lévy flights emerge independently of whether any encounters between bodies take place or not. For example, in the cometary dynamical model described by the Kepler map (Sect. 8.3), the perihelion distance of the comet can be by any amount greater than the semimajor axis of the perturber's orbit, but encounters with

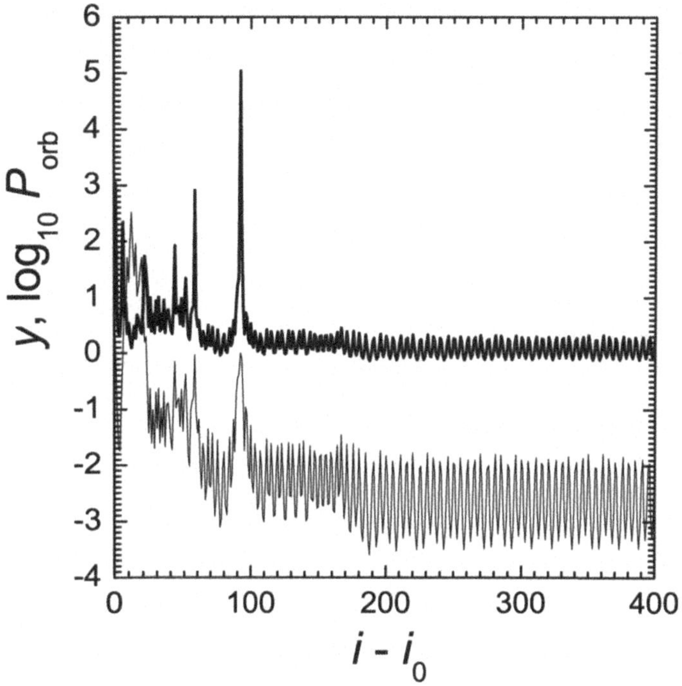

Fig. 8.25 A chaotic trajectory of the Kepler map (8.35) with $\lambda = 5$. LF1s are prominent as the peaks of P_{orb} at $i - i_0 \lesssim 100$, and an LF2 is obvious as the oscillatory "plateau" of y and P_{orb} at $i - i_0 \gtrsim 200$. Thin curve: y; bold curve: $\log_{10} P_{orb}$. The i_0 value is a large iteration number chosen so that to exhibit the given part of the trajectory. (Reprinted Figure 1 with permission from Shevchenko 2010. © 2010 by the American Physical Society)

the separatrix, with or without the separatrix crossings, may still take place, if the cometary orbit is chaotic.

In the dynamics of the generalized separatrix map (1.45), LF1s and LF2s coexist, if λ is large enough, as we shall see further on. In Fig. 8.25, a fragment of a chaotic trajectory of the map (1.45) with $\gamma = 3/2$ is shown. At $i - i_0 \lesssim 100$, LF1s are prominent; they are presented by narrow peaks in the variation of the orbital period P_{orb} of the tertiary. At $i - i_0 \gtrsim 200$, an LF2 is obvious; is betrays itself by the oscillatory low "plateau" in the variation of y and P_{orb}. When y comes close to $y = 0$, a jump in the orbital period is observed. When the trajectory sticks to a chaos border in the divided phase space, both the energy and the orbital period oscillate near some low constant value.

The stickiness effect, immanent to Hamiltonian dynamics in conditions of divided phase space (Chirikov 1990), determines the distribution character for Poincaré recurrences on large timescales: it is algebraic (Chirikov and Shepelyansky 1981, 1984). The algebraic decay in recurrence statistics in Hamiltonian systems with divided phase space was considered, in particular, in Chirikov and Shepelyansky (1981, 1984), Chirikov (1990), Chirikov (2000), Cristadoro and Ketzmerick (2008), starting with the pioneering work by Chirikov and Shepelyansky (1981). Chirikov (1990), using his resonant theory of critical phenomena in Hamiltonian dynamics, justified the value of $3/2$ for the critical exponent α in the recurrence length integral distribution

$$F \propto T_{\mathrm{r}}^{-\alpha}. \tag{8.36}$$

The $F(T_{\mathrm{r}})$ integral distribution is defined here as the relative fraction of the recurrences with the duration greater than T_{r}, in the whole sample. An illustration to the law (8.36) is given in Fig. 4.4.

In Cristadoro and Ketzmerick (2008), the algebraic decay of Poincaré recurrences was explored statistically on the basis of massive numerical data on behaviour of various Hamiltonian systems. These numerical experiments showed system-dependent power-law exponents, but the mean universal exponent α turned out to be well-defined and equal to 1.57 ± 0.03, somewhat different from the standard $3/2$ value. Venegeroles (2009) reports $\alpha = 1.54 \pm 0.07$, resulting from averaging independent massive numerical results; see Table 1 and references in Venegeroles (2009).

In celestial mechanics, the algebraic decay was observed in numerical experiments by Shevchenko and Scholl (1996, 1997) on asteroidal dynamics in the 3/1 mean motion resonance with Jupiter. The experiments were performed in the framework of the restricted three-body Sun–Jupiter–asteroid problem; it was found that the distribution tail for intervals T_{r} between asteroidal eccentricity jumps is algebraic:

$$F \propto T_{\mathrm{r}}^{-\alpha} \tag{8.37}$$

with $\alpha \approx 1.5$–1.7. This was interpreted as an effect of the chaotic orbits sticking to chaos borders in the divided phase space of motion (Shevchenko and Scholl 1997). In other words, this is the second kind Hamiltonian intermittency effect.

On the other hand, Dones et al. (1996) reported on algebraic tails of integral distributions with the power-law index $\alpha = 0.8 \pm 0.2$, in studies of escape time statistics of the highly-eccentric chaotic cometary dynamics in the Solar system, with perturbations due to the four giant planets all taken into account. This behaviour was interpreted in Shevchenko (2010) as an effect of the first kind Hamiltonian intermittency. The reason is that the integral distribution decay with $\alpha \approx 1.5$ is expected when LF2s dominate over LF1s.

The decay law with $\alpha \approx 0.7$ was observed in the behaviour of the Kepler map by Borgonovi et al. (1988). They analysed long-time decay properties of a Kepler map describing a one-dimensional model of hydrogen atom in a microwave field, and by means of rigorous analytical deduction found the $T_{\mathrm{d}}^{-2/3}$ law for the first approximation for the time decay of the survival probability in case of the escape times measured in real (constant) time units. This law was confirmed in Borgonovi et al. (1988) in numerical experiments; see Figure 1 in Borgonovi et al. (1988). On the opposite, when the escape times were measured in the map (fictitious) time units, the usual $T_{\mathrm{d}}^{-3/2}$ law was observed; see Figure 2 in Borgonovi et al. (1988).

Along with the rigorous treatment by Borgonovi et al. (1988), several heuristic deductions of the $T_{\mathrm{d}}^{-2/3}$ law were given in related problems (Schlagheck and Buchleitner 2001; Hut 1993; Malyshkin and Tremaine 1999). Hut (1993) derived an heuristic $T_{\mathrm{d}}^{-2/3}$ law as a lower bound for the time decay of the survival probability in a general *hierarchical resonant scattering* setting for the three-body interaction (Heggie 1975; Hut 1993). Malyshkin and Tremaine (1999) derived the $T_{\mathrm{d}}^{-2/3}$ law for the time decay of the survival probability in cometary ensembles in the Solar system. Schlagheck and Buchleitner (2001) obtained the $T_{\mathrm{d}}^{-2/3}$ law for the time decay of the survival probability in an autoionizing configuration of chaotic helium.

In these approaches, two basic assumptions were made explicitly or implicitly:

(i) the distribution of ejection energies is flat or smooth in the neighbourhood of the zero energy threshold,
(ii) the asymptotic decay of the survival probability is the same as the tail of the distribution of the orbital periods of the escaping body.

Due the complete ergodicity of the motion near the threshold, the first assumption is reasonable, but the second one is solely hypothetical. However, given that all results by Schlagheck and Buchleitner (2001), Hut (1993), Malyshkin and Tremaine (1999) coincide with the result of the rigorous treatment in Borgonovi et al. (1988), one may regard the second assumption to be also plausible.

Using the same two assumptions, an asymptotic distribution of the phase increments per iteration (which are the orbital periods of the escaping body in case of $\gamma = 3/2$) can be found in the general case of arbitrary γ in the map (1.45) (Shevchenko 2010). The phase increment is

$$P = \Delta x_i = x_i - x_{i-1} = \lambda |y_i|^{-\gamma} \tag{8.38}$$

in the original time units. In case of $\gamma = 3/2$, the original time unit is equal to the central binary's orbital period divided by 2π.

Therefore, $|y_i| = \lambda^{1/\gamma} P^{-1/\gamma}$. The near-separatrix motion is locally ergodic, i.e., any significant regular islands are absent. Indeed, in the local (in y) approximation of the map (1.45) by the standard map, the standard map stochasticity parameter K tends to infinity when y tends to zero. The ergodicity implies that, if $y \ll 1$, then the distribution function of y is flat: $f(y) = \text{const}$. Therefore,

$$\text{const}\, \mathrm{d}|y| \propto P^{-\frac{1}{\gamma}-1} \mathrm{d}P, \tag{8.39}$$

and the differential distribution function of P has the form

$$f(P) \propto P^{-\frac{1}{\gamma}-1}. \tag{8.40}$$

Then, the integral distribution

$$F(P) \propto P^{-\frac{1}{\gamma}}.$$

In case of the Kepler map, $\gamma = 3/2$ and $P = P_{\rm orb}$ (the tertiary's orbital period). Then, the differential distribution

$$f(P_{\rm orb}) \propto P_{\rm orb}^{-5/3},$$

and the integral distribution

$$F(P_{\rm orb}) \propto P_{\rm orb}^{-2/3}.$$

Kepler's third law gives

$$f(a) \propto a^{-2}$$

for the differential distribution of the orbital semimajor axis. Thus, the distribution of the orbital size is also heavy-tailed. The process of orbital disruption is characterized by the Lévy flights in both tertiary's orbital period and size.

By assumption (ii), the law (8.40) coincides with the asymptotics of the survival probability time decay. Alternatively, the same law follows from an analytical treatment, analogous to that given in Borgonovi et al. (1988), for arbitrary γ values instead of 3/2. If one adapts the rigorous deduction, accomplished by Borgonovi et al. 1988 in case of $\gamma = 3/2$, to the case of arbitrary γ, one arrives at the following recurrence time distribution:

$$F(T_{\rm r}) \propto T_{\rm r}^{-\frac{1}{\gamma}}.$$

Borgonovi et al. (1988) explained the dominance of LF1s over LF2s (in our notations) in actual time statistics in the dynamics of the Kepler map, using an argumentation based on the infinite measure of the extended near-separatrix phase space. If one uses this argumentation in the general case of (1.45) with arbitrary $\gamma > 0$, then it follows that LF1s should dominate for all $\gamma \geq 1$ at least. However, while at $\gamma \geq 1$ LF1s indeed dominate, the transition to this domination occurs at $\gamma < 1$. Let us estimate this critical γ value. At $\gamma < \gamma_{\rm crit}$, when LF2s dominate, the slope index is critical (as it is conditioned by the *critical structure* at the chaos border):

$$\alpha = \alpha_{\rm crit} \approx 3/2.$$

At $\gamma > \gamma_{\text{crit}}$, when LF1s dominate, the slope index $\alpha = 1/\gamma$. These two $\alpha(\gamma)$ curves intersect at

$$\gamma = \gamma_{\text{crit}} = 1/\alpha_{\text{crit}} \approx 2/3.$$

What is the physical cause for the behaviour change at this point? For the recurrences forming LF2s with durations greater than T, the *total sojourn time*

$$\sim TF(T)$$

(Chirikov 2000), i.e., $\sim T^{-\alpha_{\text{crit}}+1}$. Analogously, in case of LF1s, the total sojourn time is

$$\sim TF(T) \propto T^{-\frac{1}{\gamma}+1}.$$

LF1s asymptotically dominate, if the second sojourn time is greater than the first one, i.e., $T^{-\frac{1}{\gamma}+1} > T^{-\alpha_{\text{crit}}+1}$. Hence, the condition for the LF1 domination is

$$\gamma > \alpha_{\text{crit}}^{-1}.$$

This is just what has been derived above for the point of intersection of the two $\alpha(\gamma)$ curves.

Therefore, if the adiabaticity parameter λ is large enough (as discussed below), the critical non-trivial value γ_{crit} of the γ parameter exists: the maps with $\gamma > \gamma_{\text{crit}}$ have LF1s dominating over LF2s in actual time statistics; whereas at $\gamma < \gamma_{\text{crit}}$ LF2s dominate in both the actual time and map-time statistics.

One has $\gamma_{\text{crit}} = 2/3$ if $\alpha = 3/2$, and $\gamma_{\text{crit}} \approx 0.637$, if $\alpha = 1.57$ (as numerically estimated in Cristadoro and Ketzmerick 2008). For example, LF1s dominate over LF2s in the actual time statistics of Poincaré recurrences in the Fermi map dynamics ($\gamma = 1$, Lichtenberg and Lieberman 1992) and the Kepler map dynamics ($\gamma = 3/2$); whereas for the Markeev maps ($\gamma = 1/4$ and $1/3$, Markeev 1995, 1994), the $\hat{L}$-map ($\gamma = 1/2$, Zaslavsky et al. 1991), and the separatrix map (1.31), the tails of the recurrence distributions are LF2-dominated (if λ is large enough) and their slopes do not depend on the choice of units (map units or actual time units) in which the recurrence lengths are measured.

What is the boundary value of the adiabaticity parameter λ, separating the cases with and without global fractal chaos border? It can be estimated from the λ dependence of the maximum Lyapunov exponent L of the map (1.45). The L value increases with λ, while λ is small, but then saturates at some constant value; see Fig. 8.24 for the $\gamma = 3/2$ case. The saturation takes place when the role of the chaos border becomes important in the dynamics. As follows from Fig. 8.24, the boundary value of λ for this map is $\approx$2–3. In case of the ordinary separatrix map (1.31), the transition value of λ is $\approx$0.5–1; see Fig. 5.6.

By means of constructing the dependences $L(\lambda)$ at arbitrary $0 < \gamma < 2$, it can be shown that the boundary value of λ does not change much with γ and, by the order of magnitude, is ~ 1 (Shevchenko 2010). Therefore, if $\lambda \gg 1$, then LF1s and LF2s coexist in the map (1.45) dynamics. If $\lambda \ll 1$, then only LF1s are possible.

8.4 Dynamical Environments of Small Bodies

Although hundreds of binary and multiple asteroids have been discovered up to now, one should not expect that chaos in such systems can be often observed. Indeed, the binary asteroids whose type of inner dynamics is already identified all have regular behaviour. Moreover, as a rule, they possess double synchronization: the periods of rotation of both components about their mass centres are equal to the period of the components' orbital revolution about the system barycentre. Thus, both components are always facing each other by the same sides, i.e., one deals with the systems that have reached their final stage of tidal evolution.

8.4.1 Chaotic Zones Around Contact Binaries

In the nineties of the twentieth century, a new potential class of chaotic celestial bodies was revealed: it was found out that the orbital dynamics of a satellite of a rotating irregularly-shaped asteroid can be highly chaotic (Chauvineau et al. 1993; Scheeres 1994). Since then, dynamics of satellites orbiting around rotating irregularly-shaped bodies was thoroughly and extensively studied, because several asteroids with satellites were discovered and a number of space missions were planned and under way to asteroids and cometary nuclei. For the sake of space missions, understanding of dynamical environments of target bodies was required, both in view of possible existence of moons, fragments, and debris orbiting around the bodies, and, no less important, in view of stability/instability of planned passive orbits (around the target bodies) of the space probes themselves; see Scheeres (1994, 2012) and references therein.

A number of models to describe gravity fields of such objects were proposed and used: a model of a uniform-density triaxial ellipsoid (Chauvineau et al. 1993; Mysen et al. 2006; Olsen 2006; Mysen and Aksnes 2007), a polyhedron model (Werner and Scheeres 1996), a "molecule of gravitating points" (Petit et al. 1997), a rod model (Bartczak and Breiter 2003), a dumbbell (bilobed) model (Marchis et al. 2014; Feng et al. 2016; Lages et al. 2017, 2018b; Lages and Shevchenko 2020). Resonant phenomena, along with determination of stability/instability zones, were studied (Scheeres 1994; Hu and Scheeres 2004; Mysen et al. 2006; Olsen 2006; Mysen and Aksnes 2007; Scheeres 2012; Feng et al. 2016), featuring the role of resonances between the orbital particle's motion and the rotational motion of the irregularly-shaped, elongated or bilobed host body (Mysen et al. 2006; Olsen 2006; Mysen and Aksnes 2007; Lages et al. 2017, 2018b; Lages and Shevchenko 2020).

Dynamical environments of asteroids Castalia, Eros, Hektor, Toutatis and a number of other minor irregularly-shaped minor bodies were explored in model numerical simulations (Scheeres et al. 1996, 1998, 2000; Marchis et al. 2014; Yu and Baoyin 2012). Many asteroids and cometary nuclei are bilobed, i.e., they resemble contact binaries or dumbbells. A well-known example is the nucleus of the 67P/Churyumov–Gerasimenko comet, which was the target of the *Rosetta* space mission (Jorda et al. 2016). Another example is asteroid 25143 Itokawa, which was the target of the *Hayabusa* space mission (Jorda et al. 2016).

Chaotic dynamical environments of asteroid Ida, and, more generally, small dumbbell-shaped objects, can be conveniently described by the Kepler map (8.35) in a generalized version, which allows for the arbitrary rate of rotation of the central object (Lages et al. 2017). Indeed, the Solar system small bodies, like asteroids, trans-Neptunian objects, cometary nuclei, and planetary satellites, with physical radii smaller than 300–500 km, typically have irregular shapes; see Fig. 7.10.

Many of them resemble dumbbells, and are therefore called *contact binaries*. Rotation of a gravitating dumbbell creates around it a zone of chaotic orbits. In Lages et al. (2017), the extent of this zone was determined analytically and numerically.

Any gravitating binary, such as a binary star or a binary asteroid, that has the mass parameter $\mu \gtrsim 0.05$, is surrounded by a circumbinary chaotic zone, where all circumbinary orbits of the orbiting particles with any initial eccentricity are chaotic (Shevchenko 2015); see Sect. 11.1.5. In Lages et al. (2017), the Kepler map theory was generalized to describe the motion of a particle in the gravitational field of a spinning body modelled by a dumbbell with masses m_1, m_2 separated by constant distance (dumbbell size) d. Thus, an irregularly-shaped body is modelled by two contact uniform-density spheres (equivalent to two point masses). The dumbbell rotates around its mass centre with angular frequency ω, which can be different from the Keplerian frequency ω_0 for the contact binary's masses m_1 and m_2.

The Keplerian rate of rotation of a contact binary, consisting of two tangent equal-sized equal-mass spheres, is given by

$$\omega_0 = (\pi \mathcal{G} \rho / 3)^{1/2}, \tag{8.41}$$

where $\mathcal{G}$ is the gravitational constant, ρ is the body's density (Scheeres 2007). This Keplerian rate is critical: it corresponds to centrifugal disintegration of the contact binary; at greater rotation rates the contact binary cannot exist. For the typical density $\rho = 1$ g/cm^3, one has $\omega_0 = 2.5 \cdot 10^{-4}$ s^{-1}. This corresponds to the rotation period $\approx$7 hr.

From the generalized Kepler map theory it follows that rotating gravitating dumbbells create circumbinary zones of chaos. The chaotic zone is formed by the overlap of accumulating integer and half-integer orbit-spin resonances with the rotating contact binary (Lages et al. 2017, 2018b). The chaotic zone swells significantly if the dumbbell rotation rate is decreased. As found by Lages et al.

(2017), the chaos zone swells more than twice if the rotation rate is decreased ten times with respect to the ω_0, the centrifugal breakup threshold, given by formula (8.41).

The generalized Kepler map, derived in Lages et al. (2017), can be used to describe chaotic dynamics around bilobed cometary nuclei and contact-binary TNOs. For cometary nuclei, the circumbinary chaos can be described based on accessible observational data for five comets whose nuclei are well-documented to resemble dumbbells (Lages et al. 2018b). Using the map, the sizes of chaotic zones around the cometary nuclei, as well as the Lyapunov timescales inside the chaotic zones, were estimated in Lages et al. (2018b). In case of Comet 1P/Halley, it turned out that its circumnuclear chaotic zone seems to engulf an essential part of the Hill sphere of the comet, at least for orbits of moderate to high eccentricities.

All five cometary nuclei with well-documented shape parameters resemble dumbbells. Four of them were visited and closely observed by space probes: 1P/Halley by *Vega*-1, *Vega*-2, *Giotto* in 1986; 19P/Borrelly by *Deep Space*-1 in 2001; 67P/Churyumov–Gerasimenko by *Rosetta*-1 in 2014–2016; and 103P/Hartley by *Deep Impact EPOXI* space mission in 2010. No circumnuclear material (moons, fragments, large particles) was observed, apart from obviously replenishable clouds of "grains"; see, e.g., Bertini et al. (2015) and references therein. The observed absence of circumnuclear material is in accord with the theoretical results of Lages et al. (2018b) on large extents of the circumnuclear chaotic zones.

There exist physical mechanisms of material introduction in orbits around cometary nuclei; see Fulle (1997), Scheeres and Marzari (2000), A'Hearn et al. (2011), Keller et al. (2017). However, Lages et al. (2018b) find that, even in the absence of any forces other than gravitational, no material may sustain in long-lived stable orbits inside the circumnuclear zones of chaos. The clearing process is two-fold: if any particle chaotically diffuses in pericentric distance down to almost zero, it eventually hit the host body, is absorbed and thus removed; and if the particle diffuses in eccentricity up to unity, it is consequently ejected to outer space and thus removed.

8.4.2 Ida and Dactyl

Properties of the chaotic environments of small bodies are illustrated in Lages et al. (2017), in examples of the dynamical environments of asteroids 243 Ida and 25143 Itokawa. Among small irregularly-shaped asteroids, 243 Ida is famous to have a small natural satellite, discovered by the *Galileo* space probe. This tiny moon, named Dactyl, moves in an orbit prograde with Ida's rotation. The orbit has very small inclination, $i < 8°$, with respect to Ida's equatorial plane (Petit et al. 1997).

Petit et al. (1997) made a unique estimate of the Lyapunov time of the chaotic dynamics of the asteroid's moon: in direct numerical integrations they found that the orbital motion of Dactyl can be chaotic with the Lyapunov time ranging from

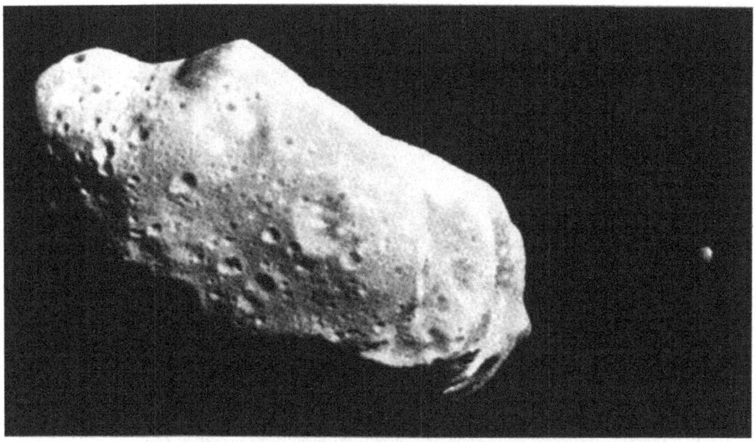

Fig. 8.26 Ida and Dactyl. An image taken by the *Galileo* space probe (A fragment of Figure 1 from Belton et al. 1995, by permission from Springer Nature, © 1995)

9 d to 4 yr. In conditions of the strong dynamical chaos, Dactyl should eventually either fall on Ida's surface, or escape to outer space.

Figure 8.26 shows 243 Ida and its satellite Dactyl. The image was taken by the *Galileo* space probe on 28th August 1993, when it was on its way to Jupiter. Ida's diameters, in the triaxial ellipsoid model, are $\approx 56 \times 24 \times 21$ km, whereas the diameter of Dactyl ≈ 1.4 km.

Ida resembles an aggregate of two merged bodies with the ratio of masses $m_2/m_1 \simeq 1$ (Petit et al. 1997). Therefore, it can be approximately described as a symmetric dumbbell with the mass parameter $\mu = m_2/(m_1 + m_2) \approx 1/2$ (Lages et al. 2017). Ida's density ρ and rotation period $P_{\mathrm{rot}} = 2\pi/\omega$ can be estimated as equal to $2.24\,\mathrm{g \cdot cm^{-3}}$ and 4.63 h, respectively (Petit et al. 1997; Vokrouhlický et al. 2003). Ida's mass in the same model is $3.6 \cdot 10^{19}$ g (Petit et al. 1997).

In units of ω_0, Ida's rotation rate is $\omega \simeq 0.953\omega_0$; i.e., it is almost at the brink of centrifugal decay. Any twin binary, consisting of two tangent spherical masses m, has $\rho\pi d^3/3 = 2m = M$, where M and d are the full mass and size of the dumbbell, respectively (the size is the distance between the mass centres of the dumbbell components). For Ida, therefore, $d \simeq 24.9$ km (Lages et al. 2017).

For the Ida–Dactyl system, it was analytically demonstrated in Lages et al. (2017) that Dactyl orbits chaotically, not far from the outer border of the chaotic zone created by rotating Ida. The stability diagram, in the "pericentric distance—eccentricity" coordinates, for dynamics around Ida is shown in Fig. 8.27. Chaos is determined by the maximum Lyapunov exponent Λ values for the particle's orbits with initial values of pericentric distance q and eccentricity e. The critical eccentricity e_{cr} at which integer resonances between orbiting Dactyl and rotating Ida overlap is shown by solid black line (e_{cr} as a function of q), and the critical eccentricity at which half-integer resonances bifurcate is shown by dashed black

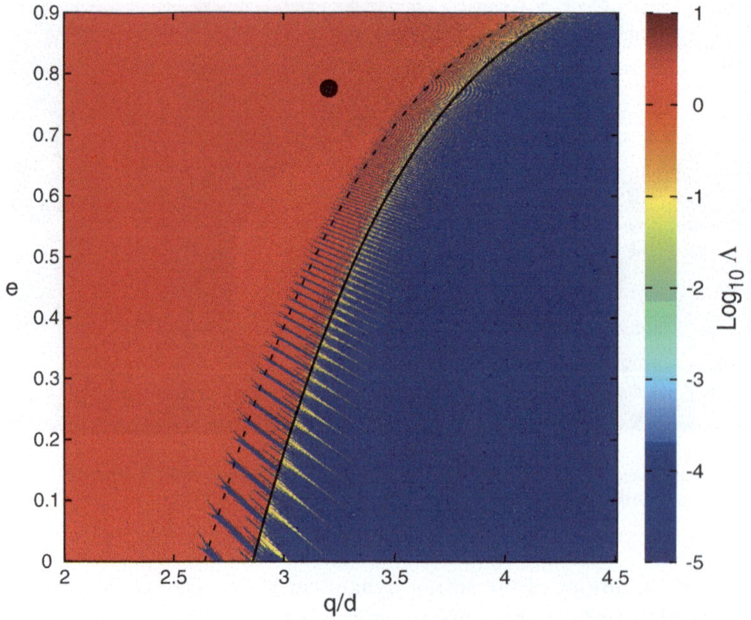

Fig. 8.27 A stability diagram for dynamics around Ida. Λ is the maximum Lyapunov exponent. Red colour means chaos. Solid black curve: the critical curve for overlap of integer resonances. Dashed black curve: the critical curve for bifurcation of half-integer resonances. Black dot: the most probable location of Dactyl (Figure 7 from Lages et al. 2017, by permission of IOP Publishing/AAS)

line. The most probable Dactyl's position, as identified in Petit et al. (1997), is depicted by a black dot.

In Fig. 8.27, the 51/1 and 52/1 "resonant teeth" (corresponding to the 51/1 and 52/1 orbit-spin resonances between orbiting Dactyl and rotating Ida) delimit the resonant cell where Dactyl might be located. Higher-order orbit-spin resonances densely accumulate at higher eccentricities in the diagram. From Fig. 8.27, it is clear that the motion of Dactyl is chaotic (Lages et al. 2017), in accord with results of direct numerical integrations performed in Petit et al. (1997). By calculating the $e_{\rm cr}(\omega, q)$ dependence, the size of the chaotic zone around the asteroid can be found; it is given by the root of the equation $e_{\rm cr}(q) = 0$ at $\omega \simeq 0.953\omega_0$. In case of Ida, the root is $q \simeq 2.85d \approx 70$ km (Lages et al. 2017). This estimate for the chaotic zone extent is in good qualitative agreement with the numerical-experimental findings on the stability limit for Dactyl's orbit size found in Petit et al. (1997).

The "ragged" structure of the chaos border in Fig. 8.27 is conditioned by the orbit-spin resonances with the rotating central body. This structure is approximately described by the separatrices of these resonances; they are explicitly presented in Fig. 8.28.

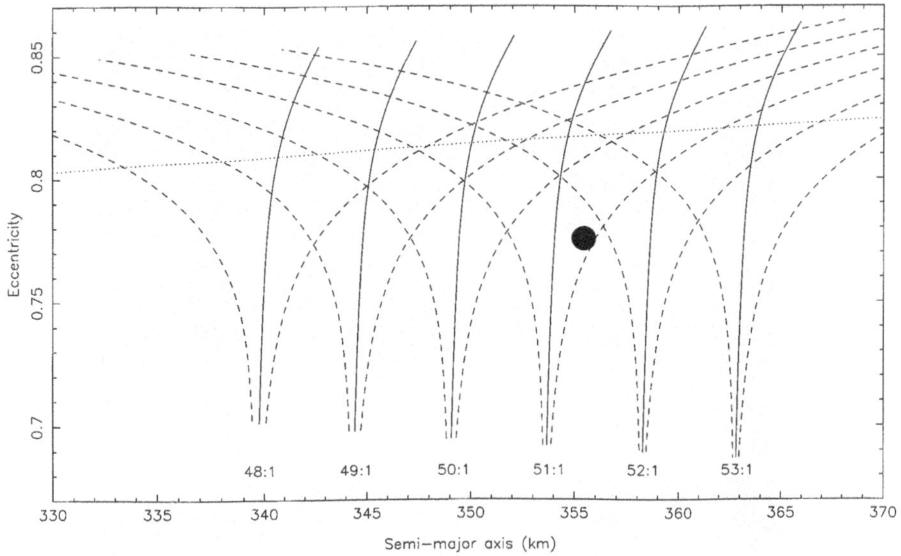

Fig. 8.28 A stability diagram in the "semimajor axis—eccentricity" coordinates for the initial conditions of a particle orbiting Ida. Solid curves: locations of integer orbit-spin resonances with rotating Ida. Dashed: separatrices of the resonances. Dotted: the approximate border of global chaos. Black dot: the most probable location of Dactyl (Figure 19 from Petit et al. 1997, with permission from Elsevier)

Time-limited observations from the *Galileo* space probe did not provide certain data on the stability of Dactyl's orbit. Dactyl can be either indeed chaotic and therefore short-lived, or, conversely, it could appear in the chaotic part of the stability diagram due to observational uncertainties in its orbital elements.

8.4.3 Trans-Neptunian Objects, 2014 MU69 Among Them

Apart from cometary nuclei, typical Kuiper belt objects (KBOs) have distinctive abilities for clearing their immediate dynamical environments (Lages et al. 2018b). In Fig. 8.29, the chaotic zone, where the clearing takes place, is represented graphically in the "dumbbell rotation rate—particle's pericentric distance" coordinates. The rotation rate ω is measured in units of its critical value ω_0, given by formula (8.41). The pericentric distance q is in units of the binary's size d, i.e., the distance between the mass centres of its components. The range of typical rotation rates of KBOs is from $\sim$1 to $\sim$6 rev/d, i.e., typical rotation periods range from $\sim$4 to $\sim$30 h (Thirouin et al. 2014, Figures 7 and 8). This corresponds to $\omega/\omega_0 \in (\sim 0.2, \sim 1.0)$.

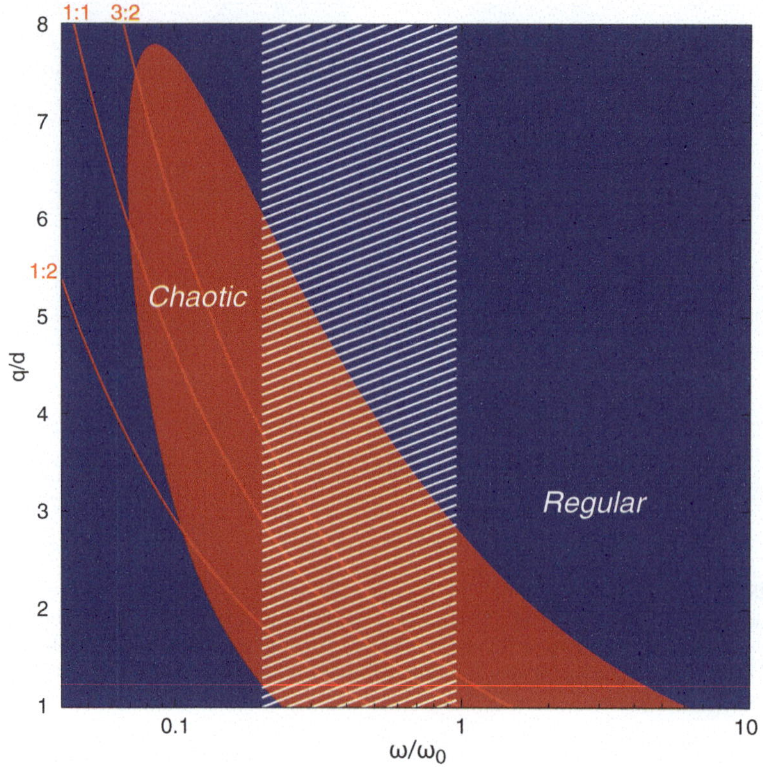

Fig. 8.29 Extents of the chaotic zone (shown in red) around a contact binary as a function of the binary's rotation rate ω, in ratio to the critical rate ω_0. The pericentric distance q is in units of the contact binary size d (the distance between the components' mass centres). The chaos borders are drawn according to Lages et al. (2018b). White shaded area delimits the range of KBOs' typical rotation rates, according to data given in Thirouin et al. (2014) (Figure 1 from Lages and Shevchenko 2020, by permission from © Cambridge University Press)

The area bounded by these limits in Fig. 8.29 is white-shaded. Locations of the 1/2, 1/1, and 3/2 main orbit-spin resonances between particle orbits and the rotation of the central body are shown as red curves. The stability diagram demonstrates that typical KBOs may have rather extended circumnuclear chaotic zones: the pericentric distance q of the chaotic orbits inside the zone may range up to $\sim 6d$.

A distant TNO, 2014 MU69, was chosen to represent the second target of the *New Horizons* space mission. MU69's orbit around the Sun has semimajor axis $a = 44$ AU, eccentricity $e = 0.05$, and inclination is negligible. Orbital period is 296 yr.

The rendezvous of the *New Horizons* space probe with MU69 took place on 1 January 2019. As images got by *New Horizons* demonstrate (see Fig. 8.30), MU69 turns out to be a classical contact-binary object. The figure of MU69 perfectly fits the binary model, presented, in particular, in (Scheeres 2007, Figure 5) and in (Lages

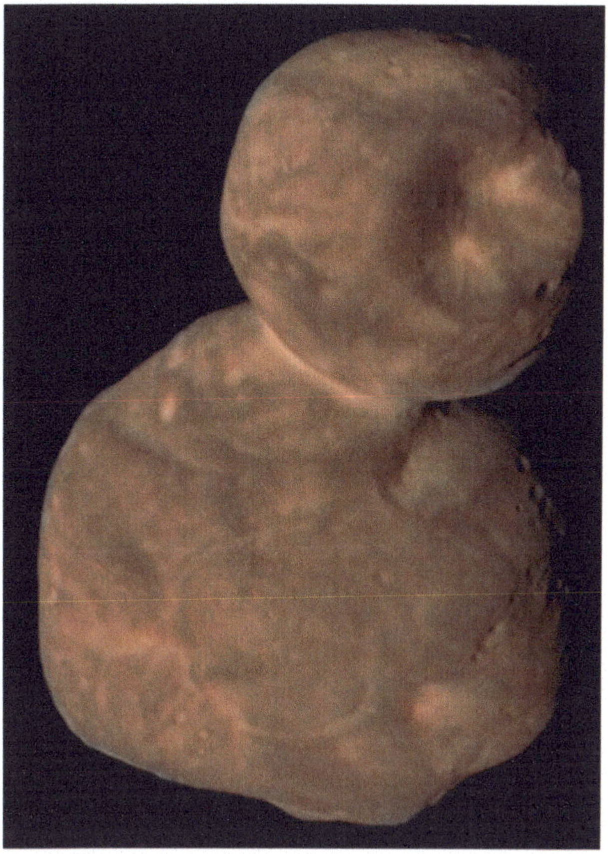

Fig. 8.30 The Kuiper belt object 2014 MU69. An image compiled from data transmitted by the *New Horizons* space probe on January 1, 2019 (Credits: NASA/Johns Hopkins University Applied Physics Laboratory/Southwest Research Institute/Roman Tkachenko)

et al. 2017, Figure 1). Note that, according to a further analysis of the transmitted sequence of the TNO's images, the companions are somewhat flattened (Stern et al. 2019; Spencer et al. 2020).

The components of MU69 have sizes (diameters) ~20 and ~18 km. Adopting their density to be that of water ice (1 g/cm^3), one finds that the Hill sphere radius of MU69 is $R_H \sim (m/(3M))^{1/3}a \sim (R/R_{\rm Sun})a \sim 3 \cdot 10^5$ km. Here $m \sim \rho d^3$, $M = M_{\rm Sun} = 2 \cdot 10^{33}$ g, $R_{\rm Sun} = 7 \cdot 10^5$ km. Therefore, the cleared zone around MU69 may range from zero up to ~300 km in radial distance; the latter corresponds to ten times maximum the full object size. This is thousand times less than the object's Hill radius and ten times less than the minimum distance between MU69 and the *New Horizons* probe during the flyby.

Tantalizingly, chaotic clearing processes affect both targets of the *New Horizons* mission, but in different ways: Pluto is not able to clear any radial neighbourhood of its orbit, and on this reason it was deprived of the planetary status (IAU General Assembly 2006); conversely, MU69 is able to create a clearing, but of another (circumnuclear) kind (Lages and Shevchenko 2020).

Chapter 9
Orbital Dynamics of Planets

The planets seem to interfere in their curves,
But nothing ever happens, no harm is done.

Robert Frost,

On looking up by chance at the constellations

And at the highest, planetary, level of the structural hierarchy of the Solar system dynamical chaos also manifests itself. At the end of eighties—beginning of nineties of the twentieth century, first ever estimates of the Lyapunov time of the Solar planetary system were obtained in massive and complicated numerical experiments (Sussman and Wisdom 1988, 1992; Laskar 1989, 1990, 1994). It turned out that it is not at all infinite, and, therefore, the dynamics of the Solar planetary system is chaotic. Moreover, the Solar system's Lyapunov time is, in fact, small: $T_{\rm L} \sim 10^6$–10^7 yr, i.e., it is by three orders of magnitude less than the Solar system age. According to Sussman and Wisdom (1988, 1992), the Lyapunov time of the Solar system outer part (ranging from Jupiter to Pluto) is $\sim 10^7$ yr, and for the full system, including its inner part, $T_{\rm L} \sim 5 \cdot 10^6$ yr, either with or without Pluto.

Note that the boom in the Kuiper belt observational studies resulted in the decrease of the official number of planets in the Solar System from nine to eight: Pluto, by a resolution of the 26th IAU General Assembly in 2006 in Prague (IAU General Assembly 2006), was attributed to a new class of celestial bodies, that of dwarf planets, to which also large asteroids such as Ceres were attributed. The decision was triggered by discoveries, in the Kuiper belt, of several large TNOs, including one larger than Pluto, namely Eris, whose physical size exceeds that of Pluto by some 10%.

Notwithstanding its small Lyapunov time, our Solar system is known to be extremely stable in the sense that it does not disintegrate on timescales of billions of years; this was established and confirmed in Sussman and Wisdom (1988, 1992) and Laskar (1989, 1990, 1994). Due to the small Lyapunov time, the angle variables of the system cannot be predicted on timescales greater than $\sim 10^6$–10^7 yr, but the planetary semimajor axes, eccentricities, and inclinations stay in restricted bounds.

I. I. Shevchenko, *Dynamical Chaos in Planetary Systems*, Astrophysics and Space Science Library 463, https://doi.org/10.1007/978-3-030-52144-8_9

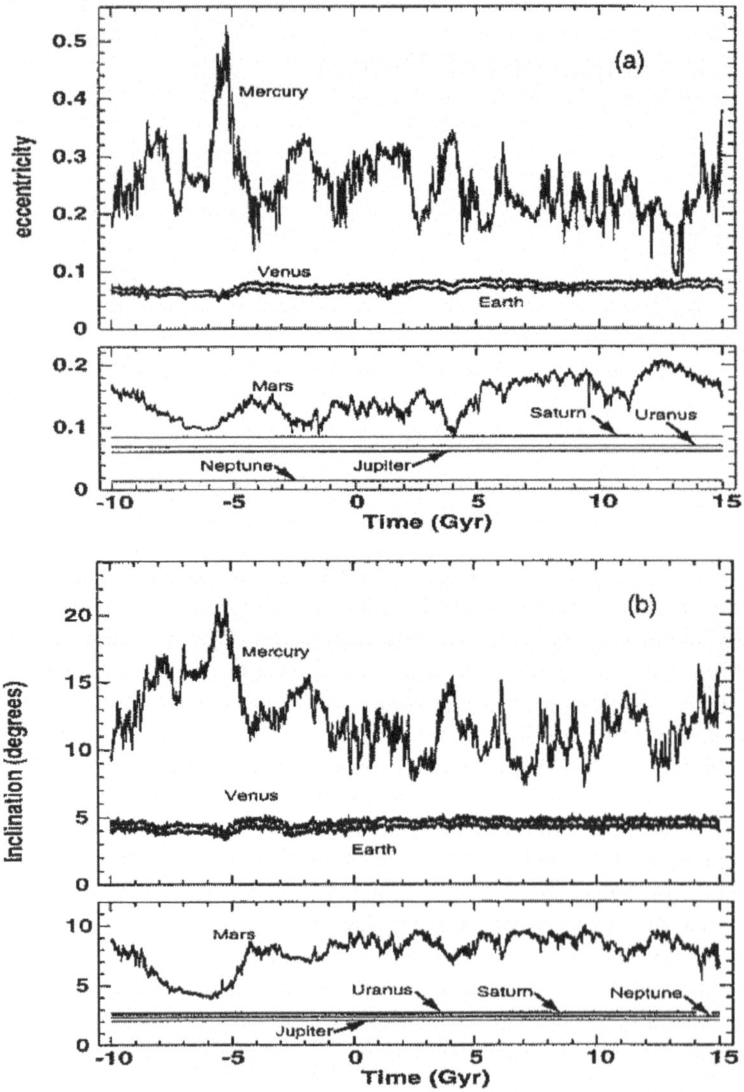

Fig. 9.1 Long-term evolution of the maximum eccentricities (panel **a**) and inclinations (panel **b**) of the Solar system planets (Figure 1 from Laskar (1994), reproduced with permission © ESO)

However, as revealed in Laskar (1994), Mercury may escape from the system (and become a *rogue planet*, in modern terms), on a timescale of several billion years.

Laskar (1994) considered the Solar system planetary dynamics in long-term numerical integrations of averaged equations of motion. In Fig. 9.1, maximum eccentricities and inclinations, exhibited by each of the planets at consecutive intervals of 10 million years, are plotted as a function of time on the whole explored time span of 25 billion years. In case of giant planets, these quantities are practically constant.

At a first glance, it might seem that the basic contribution to chaos must be brought by the planets with relatively small masses, i.e., the terrestrial group planets, as well as Pluto (ranked as planet until 2006). However, if the dynamical model were limited to solely four giant planets, then, as it was revealed by Sussman and Wisdom (1992) and confirmed by Murray and Holman (1999), chaos remains and, moreover, the Lyapunov time practically does not change, staying at $T_L \approx 5$–7 million years.

9.1 Chaotic Giant Planets

Either with inner rocky planets and small Pluto or without them, the computed models turned out to possess practically (by an order of magnitude) the same Lyapunov times. This means that the giant-planet system is itself chaotic. What is the dynamical cause for this chaotic behaviour? Murray and Holman (1999) attributed the origin of chaos in the giant-planet system to interaction of subresonances in a multiplet corresponding to a particular three-body Jupiter–Saturn–Uranus resonance.

It is well-known that our Solar system is quite close to the 5/2 Jupiter–Saturn two-body mean motion resonance, and it is also not far from the 7/1 Jupiter–Uranus two-body mean motion resonance; but neither of the corresponding resonant arguments librate, i.e., the system is out of the both resonances. Murray and Holman (1999) argued that a three-body linear combination of the two resonant arguments may actually librate, i.e., the 3J-5S-U7 three-body resonance

$$3\dot{\lambda}_J - 5\dot{\lambda}_S - 7\dot{\lambda}_U \sim 0 \tag{9.1}$$

may actually be present in the Solar system. Here λ_J, λ_S, and λ_U are the mean longitudes of Jupiter, Saturn, and Uranus, respectively; $\dot{\lambda}_J$, $\dot{\lambda}_S$, and $\dot{\lambda}_U$ are the corresponding mean motions; $\dot{x}$ means the time derivative of x.

According to the D'Alembert rules, the resonance (9.1) possesses a lot of eccentricity-type, inclination-type, and eccentricity-inclination-type subresonances. This is just their interaction and/or overlap may cause the giant planets' chaotic behaviour. The possible two-body and three-body mean motion resonances in the giant-planet system are straightforwardly revealed in the Lyapunov timescale "scan," in some range of the initial asteroidal semimajor axis of Uranus, with all other initial conditions in the system fixed to some values. Such a scan for a_U, constructed by Murray and Holman (1999) in direct numerical integrations of the planetary system, is presented in Fig. 9.2; its close-up is given in Fig. 9.3. The scan was performed on a fine grid of the a_U initial values; for each set of initial conditions, the planetary system was integrated on the time interval of 200 Myr, and the Lyapunov time was found. The scanned interval in a_U ranges from 18.98 to 19.40 AU.

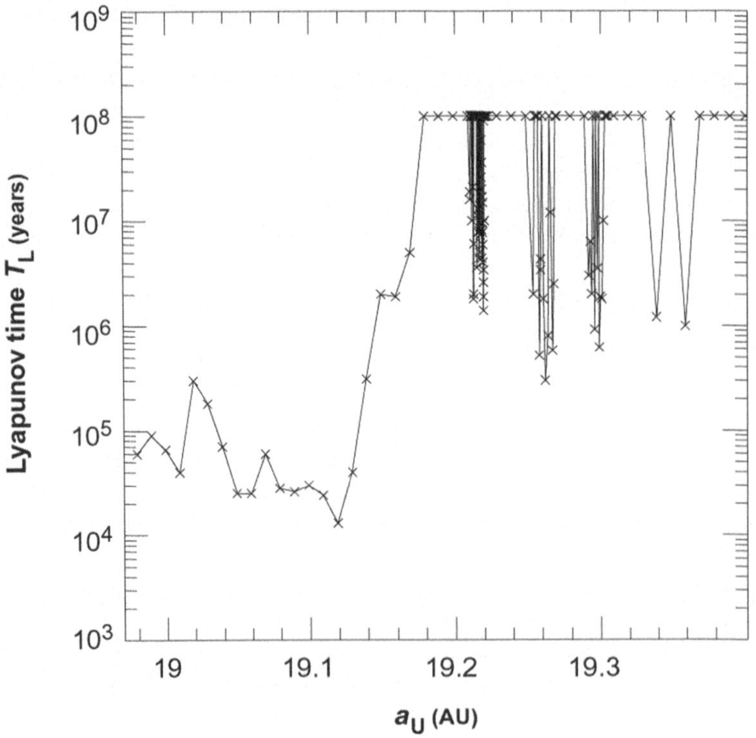

Fig. 9.2 The Lyapunov time T_L as a function of the initial semimajor axis a_U of Uranus, with all other initial conditions fixed. (Figure 1 from Murray and Holman (1999). Reprinted with permission from AAAS)

All strong-enough chaotic resonances in this range were identified. In Fig. 9.2, the strong chaos at $a_U \lesssim 19.13$ AU is due to the potential 2/1 Uranus–Neptune two-body resonance; and, at a_U from 19.13 to 19.17 AU, it is due to the potential 7/1 Jupiter–Uranus two-body resonance. Chaotic zones associated with three-body mean motion resonances (involving either Jupiter, Saturn, and Uranus; or Saturn, Uranus, and Neptune) are present at $a_U \approx 19.22$, 19.26, 19.29, and 19.34 AU.

At a finer resolution of the scan (Fig. 9.3), at a_U from 19.216 to 19.221 AU, a cluster of interacting subresonances of the three-body resonance (9.1) is readily identifiable. The subresonances do not quite overlap; that is why they are intermeshed with thin regular-looking zones. The actual location of Uranus is shown by the solid vertical line.

From Fig. 9.3 it is clear that the actual degree of the Solar system chaoticity (the maximum Lyapunov exponent) has, in some sense, arbitrary character: if the orbital semimajor axis of Uranus differed from its present value by only several physical sizes of this planet, then the chaoticity would sharply decrease. Indeed, the physical diameter of Uranus is $\sim 10^{-4}$ AU, whereas the accuracy with which the actual orbital

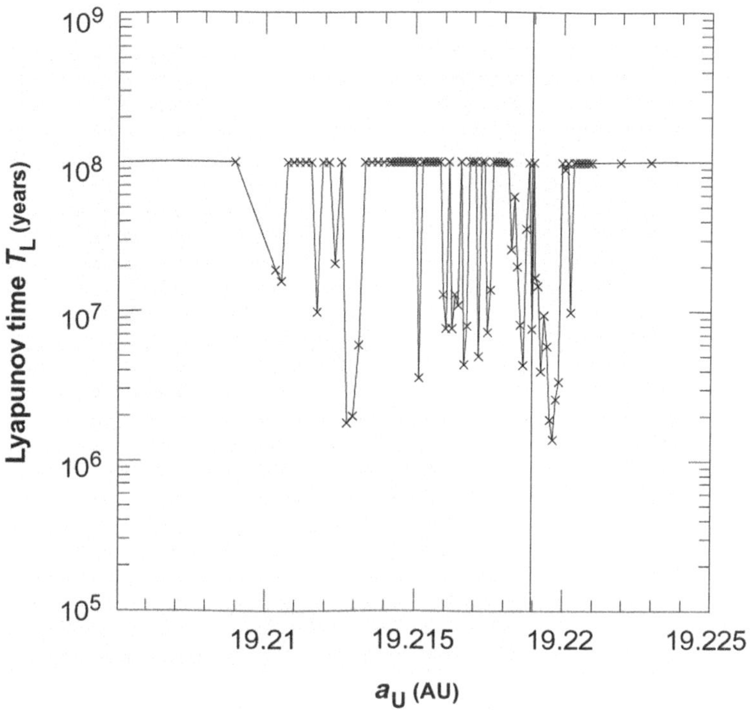

Fig. 9.3 Same as Fig. 9.2, in a higher resolution in a_U. Solid vertical line: the actual location of Uranus (Figure 2 from Murray and Holman (1999). Reprinted with permission from AAAS)

semimajor axis of Uranus is known is $\sim 10^{-5}$ AU. This circumstance explains why the Solar system Lyapunov timescales obtained in numerical integrations were observed to sharply depend on variations of initial conditions, whatever tiny these variations were in the pioneering numerical-experimental works in this field (Sussman and Wisdom 1988, 1992).

Massive direct numerical integrations performed by Guzzo (2006), Hayes (2007), and Hayes et al. (2010) showed that the chaos-order interplay could be present even on much finer scales. The location of the actual Solar system with respect to the neighbouring chaotic resonances is conveniently illustrated in Fig. 9.4. The FLI dynamical chart (on the FLI techniques, see Sect. 2.2) was constructed by Guzzo (2006) by means of varying the initial Jovian and Saturnian semimajor axes a_5 and a_6 on a fine two-dimensional grid of their values, whereas other initial orbital elements of these two planets and the initial conditions for other planets in the adopted Sun–Jupiter–Saturn–Uranus–Neptune model were fixed. The integration time was set to 1 Myr for all orbits.

In Fig. 9.4, yellow colour corresponds to dynamical chaos, signalled by higher values of the FLI, and dark orange (background) colour corresponds to regular behaviour, signalled by low values of the FLI. The yellow colour evidently traces

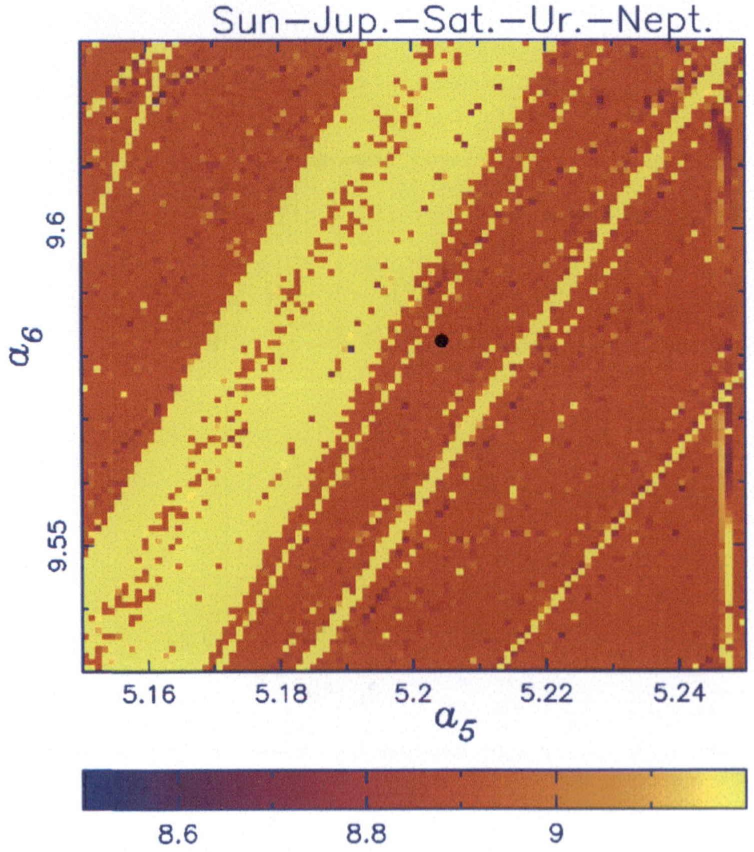

Fig. 9.4 A dynamical chart for the giant-planet system, obtained by varying the initial semimajor axes a_5 and a_6 of Jupiter and Saturn, whereas other initial conditions are fixed. Yellow colour corresponds to chaos, signalled by higher values of the FLI; dark orange (background) colour corresponds to regular dynamics, signalled by low values of the FLI. Black dot: the actual location of the Solar system (Figure 1f from Guzzo (2006), with permission from Elsevier)

major chaotic resonances, and the most pronounced among them is the Jupiter–Saturn 5/2 resonance. The Jupiter–Saturn–Uranus three-body resonance, revealed by Murray and Holman (1999), forms the thin yellow band closest to the actual Solar system position, which is marked by the black dot. The diagram graphically demonstrates how our Solar system is lucky to be located in a stable zone, although a mesh of chaotic resonances is around.

In the eighteenth century, Pierre Simon Laplace found out that the 5/2 Jupiter–Saturn near-resonance was responsible for observed anomalies in the orbital motion of these two planets; he revealed that these anomalies had a periodic character and, therefore, did not lead to any long-term monotonous changes of the orbits. This discovery led Laplace to the concept of determinism. In 1999, by considering the same 5/2 Jupiter–Saturn near-resonance, in an ensemble with other relevant resonances, Murray and Holman (1999) succeeded to put forward the described above explanation of the giant-planets' chaos. Murray and Holman (1999) wrote: "We find it ironic that the 2:5 resonance plays such a strong role in producing chaos among the outer planets, thereby placing a limit on our ability to state the positions of the jovian planets in the distant future."

Part III
Dynamics of Exoplanets

Up to the beginning of nineties of the last century, only nine planets (including Pluto, regarded as a planet at that time) were observed to exist. Since then, the number of known planets has risen immensely, by two orders of magnitude, due to discoveries of planets in systems other than ours. The discoveries of exoplanets grow in number like an avalanche. New effective observational tools and techniques (mostly space-based) are used more and more broadly. Up to the beginning of 2020, more than 4000 exoplanets have been discovered (and much more await confirmation),[1] belonging to ~3000 exoplanetary systems. The distances to known exoplanet systems range from 4 light years to ~30,000 light years, the maximum and minimum differing by four orders of magnitude. The star closest to our Sun, Proxima Centauri, which is only 1.32 pc far from us, possesses at least one planet (Anglada-Escudé et al. 2016). The farthest known systems with confirmed planets are SWEEPS-04 and SWEEPS-11 (Sahu et al. 2006, 2008).

Resonances, instabilities, and the chaotic behaviour caused by interaction of resonances play an essential role in the dynamics of planetary systems at various stages of their evolution. In concert with cosmogonical factors and migration processes, they determine the observed system architectures. In this part of the book, the dynamics of exoplanetary systems is considered. We concentrate on resonances and manifestations of dynamical chaos; problems on resonant and chaotic orbital dynamics of exoplanetary systems of various types are explored. The analysed exosystems include multiplanet systems[2] of single stars and planetary systems of binary stars. Theoretical methods and criteria for revealing the stability or instability in various planetary configurations are described.

[1] This rapid growth has been mostly due to the success of the *Kepler* space observatory mission.

[2] A planetary system is called multiplanet if it contains more than one planet.

Chapter 10
Exoplanets: An Overview

Night, full of constellations.
What fate, what destinations,
You sparkle broadly, Book,
For liberty or yoke?

Velimir Khlebnikov,
Night, full of constellations (1912)
(Translated from Russian by I.I. Shevchenko.)

In this Chapter, we describe history and methods of exoplanet discoveries, the planet definition; typology and properties of exoplanets; types of exoplanets, physical types of planetary systems. Architectures and dynamical configurations of exoplanet systems are characterized, and actual examples of resonant and multi-resonant systems are provided.

10.1 History and Discoveries

In ancient times, the planets were distinguished by properties of their motion in the projection onto the celestial sphere. Indeed, in translation from ancient Greek, "planet" means "wandering." Unlike the ordinary stars, the Solar system planets wander over the celestial sphere, sometimes writing out loops; that is why they were named so. Nowadays, the planets are implied to constitute an astrophysical class of objects whose properties occupy an intermediate domain between stars (including brown dwarfs), on one hand, and dwarf planets and various "minor bodies," on the other hand.

Most known exoplanetary systems are not at all similar to our own Solar system. In contrast to the Solar system planets, many exoplanets have large orbital eccentricities; what is more, giant exoplanets are usually observed to have orbits very close to their parent stars; orbital periods of a few days and even less are observed. Such objects are called *hot Jupiters* or *hot Neptunes*, depending on planet

I. I. Shevchenko, *Dynamical Chaos in Planetary Systems*, Astrophysics and Space Science Library 463, https://doi.org/10.1007/978-3-030-52144-8_10

size. The origin of hot Jupiters is enigmatic, though they are abundant. Supermassive rocky planets (so-called *super-Earths*) are present in many exoplanet systems, but there are no such objects in the Solar system. However, multiplanet systems quite similar to our Solar system do exist, e.g., Gliese 581,[1] 47 UMa, μ Arae (equivalently, HD 160691).[2]

The first ever detected two exoplanets were discovered in the early 1990s by Wolszczan and Frail (1992). These planets orbit around PSR B1257-12, a neutron star with mass 1.4 in Solar units. They were discovered by a technique which is now called *pulsar timing*: in the observed radio pulses (with a proper period of only 6 milliseconds, corresponding to the star's rotation period), periodic modulations were detected, and they were interpreted as due to the presence of two planets with orbital periods 66.54 and 98.21 d and masses 3.4 and 2.8 in Earth units, respectively. (More rigorously, the mass estimate in this case represents the product $m_{\rm pl} \sin i$, where $m_{\rm pl}$ is the planet's mass, i is the inclination of the planet's orbital plane to the celestial plane.) It was also suggested that there exists a third orbiting body, with a Lunar mass and orbital period of 25 d. Later on, a planet orbiting a similar star, pulsar PSR B1620-26, was detected, much more massive ($\sim$2.5 in Jovian units) than these three ones. It is a wonder how planets of a neutron star could have survived the supernova explosion of the host star? Or, maybe, they have formed only after this catastrophic event? The astrophysicists have not come to a definite conclusion on this subject up to now.

The first ever detected exoplanet orbiting a main sequence (Solar-like) star was discovered in 1995 by Michel Mayor and Didier Queloz. For this discovery they were awarded with Nobel prize 2019. The planet is hosted by the star 51 Pegasi.

Discoveries of planets of many other stars were reported in the consequent years; see Fig. 2 in Preface. Bursts in the number of discoveries correspond to periods of operation of specialized space missions, such as *Kepler* space observatory. As of the end of 2019, $\sim$4000 exoplanets were discovered. They belong to $\sim$3000 exoplanetary systems, of which: $\sim$700 are multiplanet (have two or more planets), $\sim$140 are planetary systems of multiple stars; 23 are circumbinary (in which the planets orbit around the central binary star). In the whole set of discovered exoplanets, about one third are members of multiplanet systems (Rein 2012). However, this low percentage is undoubtedly determined by an effect of observational selection: it is not easy to detect low-mass planets; instead, giant planets are discovered first of all. What is more, the discoveries are statistically biased to planets orbiting close to their parent stars. The observed deficit of multiplanet systems is mostly conditioned by a lack of observable planets of relatively small size, in particular, Earth-sized planets, due to the relative insufficiency of power of the observational techniques used now.

As an outcome of these selection effects, many newly discovered exoplanets turned out to be hot Jupiters, i.e., planets of Jovian masses (or even much more

[1]The "Gliese" designation means that the star is from the catalogue of close-in stars compiled by Wilhelm Gliese.

[2]The "HD" designation means that the star is from the star catalogue compiled by Henry Draper.

massive), but orbiting close to their parent stars. The hot Jupiters can be defined as Jupiters whose orbital radii are less than 0.1 AU, or orbital periods less than $\approx$10 days (Gaudi et al. 2005). For hot Neptunes and hot super-Jupiters, the orbit size limitations are the same.

This is just due to the proximity of their orbits to the parent stars that the hot Jupiters could have been discovered using the so-called *radial velocity method* (which is described below). Later on, however, using the transit method (also described below), super-Earths and even Earth-sized planets were discovered. Note that planetary systems with planets in close orbits, e.g. systems containing hot Jupiters, often have quite exotic configurations, completely different from the configuration of the Solar system.

Methods of detection of exoplanets are naturally subdivided into direct and indirect ones. The direct methods are based on direct observations of the proper and reflected radiation from planets; indirect methods are all the rest. The direct methods, in particular, include: direct acquisition of planetary images, differential spectrophotometry during transits (the passages of a planet across the disc of the parent star), coronography, polarimetry. Direct methods are extremely difficult to apply due to the huge contrast of the detected radiation fluxes from the parent star and its planet. For example, for a distant observer, the contrast of visual light fluxes from the Sun and from the Earth is $\sim 10^{10}$. In infrared, the contrast is only somewhat smaller. Direct observations are possible when a planet belongs to a sub-star (brown dwarf). The first planet that was discovered directly was the giant 2M1207-39b (Chauvin et al. 2004, 2005).

The indirect methods include: astrometry of stars (measurements of small oscillations of a star in the celestial plane due to the presence of a planet or planets), Doppler spectrometry (measurements of periodic variations of the radial velocity of the star), pulsar timing—measurements of variations of times of radio pulses from pulsars, eclipse timing—measurements of variations of times of eclipses in eclipsing binary stars, observations of microlensing events, observations of transits (passages of a planet across the disc of the parent star; such passages cause observable short-term weak decreases in the total flux from the star). The latter method is adjoined by the TTV (transit timing variations) analysis. The TTV consists in an analysis of deviations from the strict periodicity of transits, due to perturbations in the planetary motion.

Let us consider the most effective modern methods, as well as some promising ones. The method of Doppler spectrometry (briefly, the RV-method, from "radial velocity") was the first one to have been successfully used to detect exoplanets. It consists in measuring periodic variations (due to the presence of a planet or planets) of the host star's radial velocity. The method can be used for stars of spectral classes from F to M. The basic formula of the RV-method is

$$m_{\mathrm{pl}} \sin i = 0.035 P^{1/3} \Delta v, \tag{10.1}$$

where the planet mass m_{pl} is in Jovian units, Δv is the observed radial velocity variation half-amplitude, in metres per second, P is the observed variation period, in years; i is the inclination of the planet's orbital plane to the celestial plane. At $i = 0$ the effect is absent.

A distant observer (residing far from our Solar system) may potentially reveal the presence of the Earth orbiting the Sun by detecting radial periodic oscillations, caused by the Earth, in the radial velocity of the Sun. The oscillation amplitude would be ∼10 cm/s. This is too small to allow detection, if the observer is equipped by similar instruments that we have on the Earth; at present, their best resolution is ∼80 cm/s. Therefore, RV-discoveries of Earth-like planets orbiting in habitable zones of Solar-like stars are not yet possible.

The first ever detected exoplanet orbiting a Solar-like star, 51 Peg b, was discovered just by the RV-method. Its host star is almost a twin of the Sun: its spectral class is G5V, and mass is 1.06 in Solar units. Planet 51 Peg b is a representative of the broad class of hot Jupiters.

The transit method of searching for exoplanets was first proposed by Otto Struve in fifties of the twentieth century. The method consists in identification and analysis of small periodic dips in a star's lightcurve, emerging due to a planet's passage across the star's disc. Of course, the transits can only be observed if the inclination of the planet's orbit relative to the line of sight is small enough. Astronomers have long observed transits in the Solar system—passages of planets (Venus and Mercury) across the Sun disc. A transit of Venus causes the total light flux from the Sun to decrease by 0.01%. In observations of transits of Earth-like exoplanets of main sequence stars, the "transit signal" is of the same order. In case of giant planets, the flux drops by ∼1%. The first ever detected exoplanet transit was identified in an analysis of the lightcurve of the star HD 209458 (Charbonneau et al. 2000), whose planet had been previously discovered by the RV-method.

If a planetary system contains more than one planet, or the host star is multiple, then the time intervals between the consecutive planetary transits, observed in the system, are non-constant: due to variations of the perturbed orbital elements, the transit time oscillates with respect to the strictly periodic nominal signal. The analysis of the variability of transit times gave rise to the transit timing variations (TTV) method. Theoretical studies (Agol et al. 2005; Holman and Murray 2005) showed that the TTV modelling provides a virtually complete information on the masses and orbital elements of the transiting planets.

TTVs were first ever discovered and modelled in systems with several transiting planets (Lissauer et al. 2011a). In further applications of the TTV technique, Nesvorný et al. (2012) managed to discover a non-transiting planet by analysing its TTV signal. Thus, according to a graphical note by Morbidelli (2014), "The TTV analysis brings Celestial Mechanics back to the glorious time when Le Verrier predicted the existence and position of Neptune from an analysis of anomalies in the motion of Uranus. The "miracle" of Le Verrier now repeats routinely."

For details of planet discovery methods, see Perryman (2018). Each of the described above methods has its advantages and disadvantages. Using different methods in parallel gives most effective results. For example, one may combine the transit method and the RV-method: since the position of the orbital plane is known in a transit system, there is no need to hypothesize on the inclination i and therefore one may directly determine the planet mass $m_{\rm pl}$, instead of the $m_{\rm pl} \sin i$ product. Then, the knowledge of the size (from the transit data) and mass makes it possible to estimate the planet density.

10.2 Definition of a Planet

In view of the current rapid observational and theoretical progress in exoplanetary studies, the term *planet* obviously requires a rigorous definition, distinguishing the planets from other classes of celestial bodies.

Actually, the planet definition problem independently emerged in the beginning of the century, in the framework of the Solar system studies. It was broadly and thoroughly discussed at the 26th General Assembly of the International Astronomical Union (Prague, 2006). The discussion was triggered by novel observations of the Kuiper belt, where several new Pluto-sized objects and even larger bodies (such as Eris) had been discovered several years before the Assembly took place. The question was whether to class such objects as planets, analogously to Pluto, or to introduce a new classification which would class Pluto and these new bodies as different-type objects. The second way was chosen.

A simple straightforward approach would be to distinguish planets from "smaller bodies" by ranging all them in size. However, this would not work out because there is a number of planetary satellites larger than Mercury (say, Titan) in the Solar system. Taking this and other subtleties into account, the Assembly adopted the following definition: the *planet* is a celestial body that (1) orbits around the Sun; (2) has mass large enough for the self-gravity to dominate over the rigid-body forces, so that the body acquires a hydrostatic-equilibrium (nearly spherical) figure; (3) has purged a neighbourhood of its orbit, so that no planetesimal material is left in this neighbourhood.[3]

This definition emerged as a result of a profound scientific debate (see Stern and Levison 2002; Soter 2006; and references therein). In particular, it implicitly appeals to the stability conditions of the close-to-coorbital motion. Indeed, point (c), distinguishing planets from dwarf planets, is implicitly based on a criterion for emerging the close-to-coorbital chaotic band (considered further on in Sect. 11.1.2);

[3] In the original formulation, Resolution 5A of the 26th IAU GA reads: "The IAU therefore resolves that planets and other bodies in our Solar System, except satellites, be defined into three distinct categories in the following way: (1) A "planet" is a celestial body that (a) is in orbit around the Sun, (b) has sufficient mass for its self-gravity to overcome rigid body forces so that it assumes a hydrostatic equilibrium (nearly round) shape, and (c) has cleared the neighbourhood around its orbit. (2) A "dwarf planet" is a celestial body that (a) is in orbit around the Sun, (b) has sufficient mass for its self-gravity to overcome rigid body forces so that it assumes a hydrostatic equilibrium (nearly round) shape, (c) has not cleared the neighbourhood around its orbit, and (d) is not a satellite. (3) All other objects, except satellites, orbiting the Sun shall be referred to collectively as "Small Solar-System Bodies"." (IAU General Assembly 2006).

this implies existence of a *clearing effect* in the radial neighbourhood of the planetary orbit.

The field of applicability of the stated definition is restricted to the Solar system. By substituting a "star" for "the Sun" in it, one may seemingly arrive at the needed general definition incorporating exoplanets. However, nowadays, there is no problem in distinguishing exoplanets from "dwarf exoplanets" or "minor exobodies," due to obvious observational biases in exoplanet catalogues. Rather, the problem is to distinguish them "from above," i.e., from larger bodies. That is why additional astrophysical criteria come into play.

Generally, a celestial body is regarded to be a planet, if it is not massive enough to maintain the reaction of fusion of deuterium nuclei in its interims, i.e., its mass $m \lesssim 0.013 m_{\rm Sun}$, or, equivalently, $m \lesssim 13 m_{\rm Jupiter}$. The objects intermediate in mass between planets and stars ($13 m_{\rm Jupiter} \lesssim m \lesssim 80 m_{\rm Jupiter}$) are called *brown dwarfs*. They are not massive enough to maintain the reaction of fusion of ordinary hydrogen, but can maintain the reaction of fusion of deuterium; the temperature at the centre of such a body is less than $6 \cdot 10^6$ K.

10.3 Typology and Properties of Exoplanets

10.3.1 Mass Function and Mass-Radius Relationship

The observed mass distribution of exoplanets peaks at Jovian masses, due to the prevalence of hot Jupiters. Of course, this is a selection effect, because the most-massive planets are discovered first of all. However, lists of discovered planets with Neptunian and smaller masses, due to the current progress in observations, grow permanently; the smallest detected exoplanet is a *sub-Mercury Kepler*-37b, which is only slightly larger than the Moon and has mass ~ 0.01 in Earth units (Barclay et al. 2013). The mass function, uncorrected for the observational biases, can be fitted in its tail by a hyperbolic law, i.e., the differential distribution $\propto 1/m$; see Fig. 10.1. The sharp decline at large masses is obvious. This effect is most likely real, because observations of planets of such large masses are not prone to any known observational biases.

Apart from the mass function, the mass-radius relationship is also of fundamental importance. In a mass-radius plot for 138 planets with known masses and sizes (Fig. 10.2), a clear break-up is evident, corresponding to the transition from terrestrial planets and ice giants ($m < 150 m_{\rm E}$) to gas giants ($m > 150 m_{\rm E}$), where $m_{\rm E}$ is the Earth mass. In the first case, the solid-state isodens would have a slope of $1/3$ in the logarithmic frames, but the observed slope is $\approx 1/2$. This deviation indicates an increasing contribution of volatiles as mass increases. In the second case, with mass increasing, the planetary radius keeps an almost constant value or even decreases; this indicates a contribution of degenerate electron gas (Weiss et al. 2013). Besides, in the second case, enlargement of planets at high insolation levels

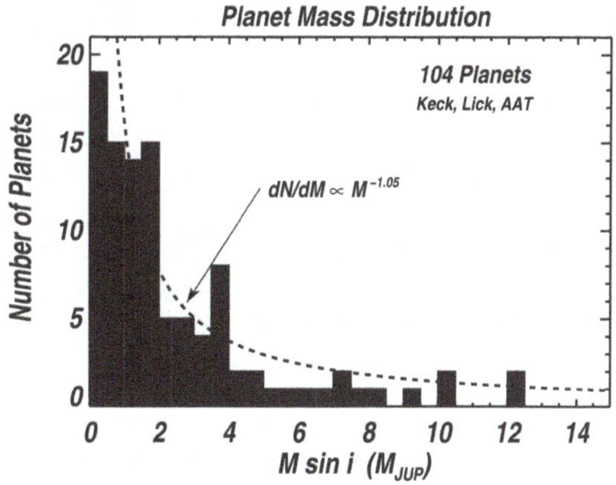

Fig. 10.1 The differential distribution of planets in mass. The maximum is at Jovian masses, due to the hot Jupiters domination. The sharp decline at large masses is obvious (Figure 1 from Marcy et al. (2005), by permission from © Oxford University Press)

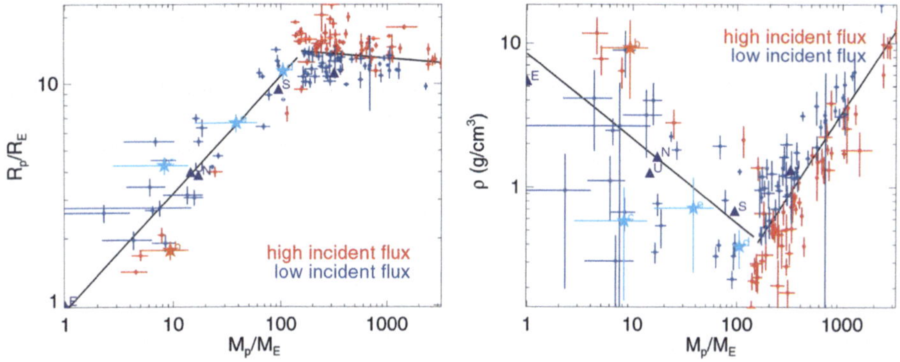

Fig. 10.2 Mass–radius and mass–density relationships (Figure 12 from Weiss et al. (2013), by permission of IOP Publishing/AAS)

is statistically significant. It is believed that, due to high temperatures, the gas giants swell.

Distributions of the observed binary star systems and star-planet systems (for Solar-like stars) in terms of minimum mass in the system, constructed jointly for these two classes, demonstrate that the planetary mass function is distinctly separate from the stellar one: the maxima at $m_{\mathrm{min}} \sim 0.001$ and $m_{\mathrm{min}} \sim 1$ (in Solar units) are separated by a profound dip with a minimum at $m_{\mathrm{min}} \sim 0.01$ (Udry et al. 2007). Of course, this feature emerges because the mechanisms of formation of planets and

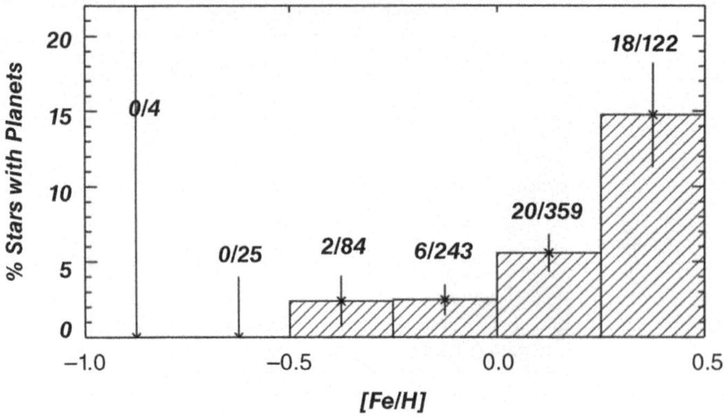

Fig. 10.3 Correlation between stellar metallicity and presence of planets (Figure 4 from Fischer and Valenti (2005), by permission of IOP Publishing/AAS)

stars are completely different. The so-called *brown dwarf desert*[4] is prominent in the distributions.

The spectrum of masses and sizes of known exoplanets is broad: it covers three orders of magnitude in masses and more than one order of magnitude in sizes; see Fig. 10.2. In the whole set of discovered exoplanets, the largest (in size) planet is TrES-4, which is 19.8 times larger than the Earth and 1.8 times larger than Jupiter. However, it is comparable to Jupiter in mass; therefore, it represents an extremely low-density gas giant with mean density $\sim$0.3 g/cm^3; compare with Jupiter's 1.33 g/cm^3. For the planets with known sizes and masses, this density value is the lowest one (Mandushev et al. 2007).

An important parameter of host stars is their metallicity. Usually it is defined as equal to the relative content of iron, which correlates with the content of other heavy elements. According to observational data, there can be a statistical correlation between the metallicity of a star and the presence of planets orbiting around it (Fig. 10.3): stars with low metallicity are observed to be relatively planetless, whereas for stars with metallicity higher than the Solar value, the probability for the planetary presence increases. It may indicate that planet-hosting stars are born in molecular clouds with relatively high metallicity (Fischer and Valenti 2005): with the increase in metallicity, the protoplanet disc's heavy component fraction increases. This fraction includes rocky material, forming large embryos, on which volatiles condense and the disc's gas accretes; in this way, gas-ice giant planets form.

[4]The statistical paucity of brown dwarfs in close (with radius $\lesssim$ 5 AU) orbits around Solar-like stars.

10.3.2 Types of Exoplanets

From the very beginning of exoplanetary studies, it became clear that there exist classes of exoplanets whose physical and dynamical properties are very different from those immanent to the planets in our Solar system. In our system, the four giant planets—Jupiter, Saturn, Uranus, and Neptune—have orbital radii ranging from 5 to 30 AU. In contrast, the observed exoplanetary systems frequently contain giant planets whose orbital radii are by two orders of magnitude smaller. As already mentioned above, such planets, orbiting within the radial vicinity < 0.1 AU of their host stars, are called hot Jupiters, on obvious reasons. The orbital periods of hot Jupiters can be as small as a few days.

The problem with hot Jupiters is that, according to modern theoretical views, they cannot form in situ, because the planet-forming accretion of matter cannot take place in such close vicinities of the parent star (Bodenheimer et al. 2000). Therefore, they should have been transported to their observed orbital locations, either due to a slow radial inward *migration* in a thick protoplanetary disc, or due to some other dynamical mechanism provided by perturbing bodies, such as other massive planets or/and a stellar companion of the host star. In the latter case, the Lidov-Kozai effect, considered further on in Chap. 15, may contribute decisively to any scenario of the hot Jupiters formation. Another problem concerns a mechanism for stalling the planets close to the host star (although their might be no permanent stalling, and hot Jupiters are eventually absorbed by the star). Tidal effects may contribute much to the dynamics in such close-to-star vicinities (Batygin et al. 2009; Lovis et al. 2011; Van Laerhoven and Greenberg 2012; Correia et al. 2013).

ESA's *CoRoT* mission and, in particular, NASA's *Kepler* mission became major achievements in the field of discoveries and explorations of exoplanets with sizes and masses of the Earth's kind. The data obtained from these space telescopes made it possible to distinguish five categories of relatively small planets: Mercuries (0.02–0.4 R_E), sub-Earths (0.4–0.8 R_E), Earths (0.8–1.25 R_E), super-Earths (1.25–2.6 R_E), Neptunes (2.6–6.0 R_E), where R_E is Earth's radius.

In a more general classification, major classes of the star-hosted planets comprise: sub-Earths, Earths, super-Earths, Neptunes, Jupiters, and super-Jupiters. Super-Jupiters are defined as giant planets with masses greater than 2–3, up to $\sim$10, in Jovian units. Super-Earths and super-Jupiters are not represented in our Solar system. At masses greater than 13 M_J, the super-Jupiters are classed as brown dwarfs. Irrespective of mass, the radii of super-Jupiters do not differ much from that of Jupiter, i.e., the density accordingly rises with mass. Any exoplanet, once its mass and/or size is measured, can be attributed to one of these classes.

According to Stevens and Gaudi (2013), the boundaries in mass between the planet classes, in Earth mass units, can be defined as:

- sub-Earths—Earths: 0.1,
- Earths—super-Earths: 2,
- super-Earths—Neptunes: 10,
- Neptunes—Jupiters: 100,

- Jupiters—super-Jupiters: 1000,
- super-Jupiters—brown dwarfs: 4000.

For example, sub-Earths have masses less than 0.1, and Earths have masses greater than 0.1 but less than 2, in the Earth units. These divisions are approximate and conditional, and may slightly vary from survey to survey. Main physical and orbital parameters of planetary class prototypes (Earth, Neptune, and Jupiter) present in our Solar system are listed in Appendix B for comparison.

In our Solar system, Mercury can be attributed to the class of sub-Earths; Mars, whose mass is 0.11 in Earth units, is just at the boundary between sub-Earths and Earths; Venus and Earth are Earths; Saturn, whose mass is 95 Earth units, is just at the boundary between Neptunes and Jupiters; Jupiter is Jupiter; Uranus and Neptune are Neptunes.

According to Borucki et al. (2011), the boundaries in physical radius between the planet classes, in Earth radius units, can be defined as:

- Earths—super-Earths: 1.25,
- super-Earths—Neptunes: 2,
- Neptunes—Jupiters: 6.

For example, super-Earths have physical radii greater than 1.25 but less than 2, in Earth radius units. These divisions are also approximate and conditional. In the given divisions, Saturn, whose radius is 9.1 Earth units, definitely belongs to the class of Jupiters.

An important planetary parameter is the I/R ratio of masses of water (ice) and solid matter (metals and rocks). According to Sotin et al. (2007) and Küchner (2003), three types of planets can be distinguished, depending on the I/R value:

- I/R $\sim 10^{-4}$: rocky planets with water content as on the Earth;
- I/R $\sim$ 0.3–0.5: planets with a liquid ocean under an ice cover, similar to Jovian moons Europe and Ganymede;
- I/R $\sim$ 1: completely oceanic planets.

Super-Earths, important in view of their enhanced potential habitability (due to the enlarged, compared to the Earth, surface area), can also be defined more strictly as the planets with masses from $\sim$1.5 to 13 Earth masses, with non-dominant atmospheres. The atmosphere is regarded as non-dominant if its height is much less than the planet's radius.

Super-Earths may form due to dispersal of gas shells of giant planets by the UV radiation of massive stars neighbouring to their host systems. Therefore, the presence of planetary systems containing super-Earths can be expected in regions of formation of massive stars; whereas the presence of systems with giant planets can be expected in regions of formation of low-mass stars (Boss 2006). Super-Earths may also form in the course of rapid radial inward migration of giant protoplanets, when the process of contraction of the planet "is not in pace" with the migration process: the tidal destruction of the protoplanet (which is a massive gas clot with a core) leaves finally the sole core (Nayakshin 2011).

There exist observable planets that do not belong to any stars. These solitary objects travelling in interstellar space are called *free-floating planets* (FFPs), or *rogue planets*, or *orphan planets*. First-ever detected FFPs were discovered in the σ Orionis star cluster (Zapatero Osorio et al. 2000). This class of objects may originate due to escape processes in planetary systems. A significant fraction of planetary populations formed in systems of binary stars can be ejected (Zinnecker 2001). Examples of such processes are familiar to researchers in celestial mechanics. For example, it is known that Mercury may escape from our Solar system, in the course of the long-term chaotic diffusion of its orbit in eccentricity; however, the escape timescale is large and may take billions of years (Laskar 1994).

10.3.3 Physical Types of Planetary Systems

When statistics of discovered planets are corrected for observational biases, it turns out that at least 25% of all Solar-like stars in our Galaxy should actually host planetary systems (Grether and Lineveawer 2004). Therefore, planetary systems of main-sequence stars should be ubiquitous. Greaves et al. (2007) distinguish four major observed types of stars with planets and discs:

1. Stars with debris discs (i.e., residual discs, consisting of small bodies and dust, presumably left after the planetary formation epoch), but without observed planets. An example is τ Ceti.
2. Stars with both planets and massive debris discs. An example is ε Eridani.
3. Stars with gas giants in orbits larger than 0.1 AU ("cold Jupiters") and without significant debris discs. An example is our Sun.
4. Stars with gas giants in orbits with radii smaller than 0.1 AU, and without debris discs. An example is 51 Pegasi.

10.4 Architecture and Dynamical Configurations

To date, thousands of exoplanetary systems have been discovered. Many of them are observed to be multiplanet systems, i.e., they host two or more planets. The orbital resonance and near-resonance phenomena are ubiquitous in them. In several multiplanet systems, planets were verified to be in mean motion resonances with each other (Wright et al. 2011; Fabrycky et al. 2012). The planetary resonances are believed to be a natural outcome of the primordial migration of planets due to their interaction with the protoplanetary disc; see, e.g., Wang et al. (2012).

Apart from the well-known two-body mean motion resonances, three-body mean motion resonances can be important. The three-body resonances in the dynamics of asteroids and planets of the Solar system represent, historically, a well-studied phenomenon; it was considered in the second part of this book. With the arrival of massive observational data on multiplanet systems, studies of three-body resonances in their dynamics were initiated (Quillen 2011).

Systems with planets in the 2/1 mean motion resonance are represented, for example, by Gliese 876 and HD 82943; in the 3/1 resonance—by 55 Cnc. In the Gliese 876 system, three planets are involved in two 2/1 resonances, thus forming the Laplace resonance 4:2:1 (Martí et al. 2013); recall that, in the Solar system, the Laplace resonance governs the dynamics of three Galilean satellites. Moreover, a closely packed multiplanet resonant system, *Kepler*-223, exhibiting the 8:6:4:3 mean motion resonance, was reported (Lissauer et al. 2011a,b).

The presence of mean motion resonances and their interaction implies possibility for chaos in the dynamics of exoplanets, as, for example, in case of the *Kepler*-36 system (Deck et al. 2012).

Planetary systems of binary stars might be no less ubiquitous that those of single stars. Indeed, more than a half of all observed main-sequence stars belong to multiple (mostly binary) star systems (Duquennoy and Mayor 1991; Mathieu et al. 2000). The dynamics and formation of circumbinary planets is a particular theoretical challenge, due to stringent stability conditions. One of the currently known circumbinary planets, *Kepler*-16b, orbits around a system of two main-sequence stars (Doyle et al. 2011). Although orbiting in a dangerous vicinity to the chaotic circumbinary zone, *Kepler*-16b survives because it is safe inside a cell bounded by the chaotic 5/1 and 6/1 resonances with the central binary (Popova and Shevchenko 2013). It is likely that *Kepler*-16b, as well as other discovered planets in circumbinary systems, suffered inward migration: its formation in situ is problematic due to hostile conditions for planetesimal accretion, in particular, high encounter velocities of planetesimals and low planetesimal spatial concentration in the planet's current orbital location (Meschiari 2012; Paardekooper et al. 2012).

10.5 Resonant and Multi-Resonant Systems

In Sect. 8.2 above, we considered, in particular, the resonant and chaotic dynamics of asteroids in our Solar system. The asteroidal resonances in our system have been explored for more than 150 years since the discovery of the gaps in the main belt by Kirkwood (1867). This discovery was based on statistics of $\sim$100 objects; nowadays, when orbital data are available for an even greater number of multiplanet systems and planetary systems of binary stars, period ratio histograms for them can be built analogously, so that to reveal possible resonant features. Such histograms for exoplanet systems were built in several studies; see Pichierri et al. (2019) and references therein.

In the analogy with asteroidal distributions, statistics of resonances in multiplanet systems, as well as in binary stellar systems with planets, were analysed by Popova and Shevchenko (2014). Orbital period ratios were calculated for each pair of objects in the whole sample of systems suitable for the analysis. Histograms of period ratios, using the exoplanet.eu database (Schneider 2018), were constructed separately for two possible configurations:

(a) The orbit of the dominating (in mass) body in the pair is inside the orbit of the smaller body in the pair. In the Solar system, this configuration is analogous to a "Neptune – TNO" configuration.
(b) The opposite case, analogous to a "main-belt asteroid – Jupiter" configuration.

The dominating body in the pair is considered to be the "perturber," analogously to orbital configurations in the Solar system, where Neptune and Jupiter are the main perturbers for TNOs and main-belt asteroids, respectively.

The obtained histograms are presented in Fig. 10.4. In the both cases (a) and (b), the outer body is indexed with "1" (for example, its orbital period is T_1), and the inner body is indexed with "2" (its orbital period is T_2).

As one may see, several peaks are prominent in the distributions. They are present at the nominal locations of resonances $3/2, 2/1, 5/2, 3/1$, and $4/1$. In Fig. 10.4, the histograms are fitted with the curves given by the multi-peaked function

$$f\left(\frac{T_1}{T_2}\right) = \sum_{i=1}^{5} f_i\left(\frac{T_1}{T_2}\right) + \phi\left(\frac{T_1}{T_2}\right), \tag{10.2}$$

where

$$f_i\left(\frac{T_1}{T_2}\right) = \frac{A_i}{\sigma_i\sqrt{2\pi}} e^{-\frac{(T_1/T_2-\mu_i)^2}{2\sigma_i^2}}, \tag{10.3}$$

$$\phi\left(\frac{T_1}{T_2}\right) = B\cdot\left(\frac{T_1}{T_2} - C\right)^5 e^{-(T_1/T_2-C)/D}. \tag{10.4}$$

Each fitting function f_i models nothing but the normal distribution of the period ratio values near a particular resonance. The function ϕ describes the wide bulge and decay in the distribution tail. The A_i, μ_i, σ_i, B, C, and D parameters were numerically evaluated using the Levenberg–Marquardt method. The numerical fits are shown in Fig. 10.4 by solid curves.

In this way, the model fitting of the period-ratio histograms allows one to obtain the best-fit locations of the resonant peaks; they are listed in Tables 10.1 and 10.2. It is obvious that the resonant peaks are systematically shifted "to the right." The shifts are less prominent in case (a) than in case (b). In case (a), almost no shift is present for the resonances of order greater than 1.

This is a known phenomenon in the statistics of exoplanetary systems, although discovered not long ago. Possible dynamical and cosmogonical mechanisms for the shifts of resonances were discussed in Lithwick and Wu (2012), Petrovich et al. (2013), Batygin and Morbidelli (2013b), Fabrycky et al. (2014), and Pichierri et al. (2019). For example, the shift may emerge, if the planetary masses in the pair

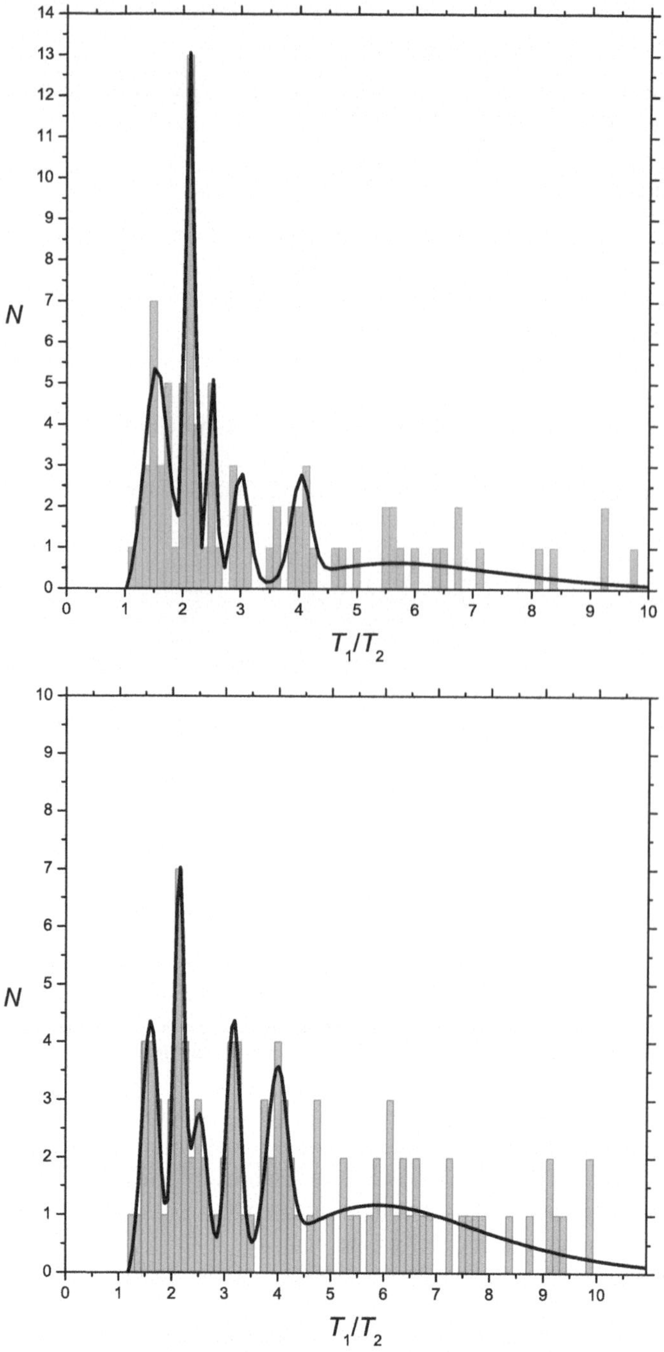

Fig. 10.4 The period ratio histograms. Upper panel: the inner perturber case (a). Lower panel: the outer perturber case (b) (Figures 1 and 2 from Popova and Shevchenko 2014)

Table 10.1 Locations of resonant peaks, case (a) (Popova and Shevchenko 2014)

Resonance	Peak position
2/1	2.122 ± 0.007
3/2	1.547 ± 0.024
3/1	3.000 ± 0.039
4/1	4.033 ± 0.046
5/2	2.500 ± 0.018

Table 10.2 Locations of resonant peaks, case (b) (Popova and Shevchenko 2014)

Resonance	Peak position
2/1	2.131 ± 0.018
3/2	1.599 ± 0.025
3/1	3.168 ± 0.024
4/1	3.991 ± 0.039
5/2	2.517 ± 0.056

increase with time (Petrovich et al. 2013). In Batygin and Morbidelli (2013b) and Lithwick and Wu (2012), the effect is interpreted as conditioned by tidal dissipation. Pichierri et al. (2019) argue that additional physical mechanisms, active during early stages of the dynamical evolution of planetary systems, are warranted to explain the observed statistics of close-to-resonant system architectures.

Chapter 11
Planetary Architecture: Stability, Packing and Ranging

In this chapter, we concentrate on resonances and manifestations of dynamical chaos in exoplanet systems of various types, and explore problems on their resonant and chaotic orbital dynamics. The analysed exosystems include multiplanet systems of single stars and planetary systems of binary stars. Theoretical methods and criteria for revealing the stability or instability in various planetary configurations are described. The presented topics include: classical results on stability criteria, Wisdom's criterion; the Kepler map criterion; the Holman–Wiegert criteria; packing and ranging criteria; the Moriwaki–Nakagawa criterion. The Kepler map, encounter map and its versions are introduced and their theory is used in analytical derivations of the stability criteria. We consider Lyapunov timescales, escape timescales, and linear and quadratic relationships between them. The presentation is partially based (in Sect. 11.1) on the papers by Shevchenko (2010), Shevchenko (2007a) (by permission from Cambridge University Press), Shevchenko (2011b) (by permission from Elsevier), and Shevchenko (2020).

11.1 Stability Criteria

11.1.1 Classical Results

De cette ardeur que vous dites,
où est le firmament?

Rainer Maria Rilke,
Vergers (1926)

In seventies of the eighteenth century, Pierre Simon Laplace and Joseph-Louis Lagrange explored secular variations of the semimajor axes of the Solar system planets under mutual perturbations and made first-ever rigorous conclusions on the

I. I. Shevchenko, *Dynamical Chaos in Planetary Systems*, Astrophysics and Space Science Library 463, https://doi.org/10.1007/978-3-030-52144-8_11

Solar system long-term stability. In the secular theory valid in the first order in planetary masses, the secular variations in the planetary semimajor axes were shown to be absent (Lagrange 1892; Laplace 1895). This result does not completely solve the stability problem, even in the first order in masses, because variations in the planetary eccentricities may obviously lead to close encounters between planets, but no limits on the variations of planetary eccentricities and inclinations were provided by the theory; for historical details see Laskar (2015).

In the twentieth century, a major breakthrough was achieved with the advent of Andrey Kolmogorov's theorem (Kolmogorov 1954). The theorem can be stated in the following formulation (Morbidelli 2002). Let a system with N degrees of freedom and the quasi-integrable analytical Hamiltonian

$$\mathcal{H}(\mathbf{p}, \mathbf{q}) = \mathcal{H}_0(\mathbf{p}) + \epsilon \mathcal{H}_1(\mathbf{p}, \mathbf{q})$$

be given, and also a point $\mathbf{p}_0$ in the action space, such that

1. $\boldsymbol{\omega}_0 = \mathrm{grad}_{\mathbf{p}} \mathcal{H}_0(\mathbf{p}_0)$ satisfies the diophantine condition

$$|\mathbf{k} \cdot \boldsymbol{\omega}_0| > \frac{\gamma}{|\mathbf{k}|^{\tau}}, \quad \forall \mathbf{k} \in \mathbf{Z}^N, \quad \mathbf{k} \neq \mathbf{0}, \tag{11.1}$$

   where $\gamma > 0$ and $\tau > 0$ are some constants;
2. $\mathcal{H}_0$ is locally non-degenerate in $\mathbf{p}_0$, i.e., the determinant of the $\partial^2 \mathcal{H}_0 / \partial p_i \partial p_k (\mathbf{p}_0)$ matrix is non-zero.

Then there exists an $\boldsymbol{\omega}_0$-depending threshold ϵ_{cr} such that at $\epsilon < \epsilon_{\mathrm{cr}}$ a canonical transform $(\mathbf{p}, \mathbf{q}) \to (\mathbf{J}, \boldsymbol{\phi})$ can render the original $\mathcal{H}$ in the form

$$\mathcal{H}(\mathbf{J}, \boldsymbol{\phi}) = \mathcal{H}_0(\mathbf{J}) + \mathcal{H}_1(\mathbf{J}, \boldsymbol{\phi}), \tag{11.2}$$

where $\|\mathcal{H}_1\| = O(\|\mathbf{J}\|^2)$ and $\mathrm{grad}_{\mathbf{J}} \mathcal{H}_0(\mathbf{0}) = \boldsymbol{\omega}_0$; $\mathbf{J}$ and $\boldsymbol{\phi}$ are conjugated action-angle variables.

At $\mathbf{J} = \mathbf{0}$, the equations of motion defined by the Hamiltonian (11.2) are

$$\dot{\mathbf{J}} = \mathbf{0}, \quad \dot{\boldsymbol{\phi}} = \mathrm{grad}_{\mathbf{J}} \mathcal{H}_0(\mathbf{0}), \tag{11.3}$$

and, for the initial conditions $(\mathbf{J}, \boldsymbol{\phi}) = (\mathbf{0}, \boldsymbol{\phi}_0)$, the solution of these equations is

$$\mathbf{J} = \mathbf{0},$$

$$\boldsymbol{\phi} = \mathrm{grad}_{\mathbf{J}} \mathcal{H}_0(\mathbf{0}) t + \boldsymbol{\phi}_0,$$

implying that the torus $\mathbf{J} = \mathbf{0}, \boldsymbol{\phi} \in \mathbf{T}^N$ is invariant with respect to the flow (11.2).

Therefore, at the small enough perturbations $\epsilon < \epsilon_{\mathrm{cr}}$, the quasi-integrable Hamiltonian systems possess invariant *KAM tori*, which carry motion with diophantine frequencies.

The Kolmogorov theorem was generalized in works of Vladimir Arnold and Jürgen Moser (Arnold 1963a,b; Moser 1962), and nowadays is usually referred to as the KAM theorem. For details of the mathematical treatment and a sketch of the theorem proof, see Morbidelli (2002).

The actual masses of planets in the Solar system (and in any other planetary system) are by many orders of magnitude too large for the original KAM theorem to be applicable to guarantee the stability. In 1966, Michel Hénon estimated that, if one tries to directly apply the KAM theorem to prove the Solar system stability, then one finds that the stability could be guaranteed if Jupiter's mass were less than 10^{-320} of the Solar mass (for details see Laskar 2014). This is 10^{260} times less than the electron mass. In other words, the original KAM theorem conditions are too stringent to be applicable in practice. However, in modern developments and refinements of the KAM theory (see Locatelli and Giorgilli 2000; Giorgilli et al. 2017, and references therein), the theoretical threshold perturbation values have been drastically raised, almost up to the actual Jovian case, but these refinements are model dependent.

The *Arnold diffusion*, universally taking place in multi-dimensional ($N > 2$) Hamiltonian systems, and the Nekhoroshev theory (Nekhoroshev 1977) describing the diffusion, is reviewed in Chirikov (1979). With an emphasis on applications in celestial mechanics and dynamical astronomy, the Arnold diffusion is reviewed in Cincotta (2002).

11.1.2 Wisdom's Criterion: The $\mu^{2/7}$ Law

Let us consider the first-order orbital resonances of a passively gravitating test particle and a perturber (say, a planet), both moving in initially circular close orbits around a primary (a host star). These resonances correspond to period ratios $(p+1)/p$, where integer $p \geq 1$. On increasing p, the resonances accumulate to the 1/1 commensurability, and they start to overlap at some critical p, because their widths do not decrease fast enough. Using Chirikov's overlap criterion, Wisdom (1980) inferred that the critical $p_{\rm cr}$ is given by

$$p_{\rm cr} \approx 0.51\mu^{-2/7}, \tag{11.4}$$

where $\mu = m_2/(m_1 + m_2)$ is the mass parameter, and m_1 and m_2 are the masses of the primary (the star) and the secondary (the planet), respectively.

From the third Kepler law, one directly finds that $p = p_{\rm cr}$ corresponds to the chaotic zone half-width

$$\Delta a_{\rm cr} \approx 1.31\mu^{2/7}a' \tag{11.5}$$

(Duncan et al. 1989; Murray and Dermott 1999), where a' is the orbital semimajor axis of the planet. The particles with the semimajor axis a within the interval

$a' \pm \Delta a_{cr}$ move chaotically (Duncan et al. 1989; Murray and Dermott 1999). They escape from this zone sooner or later, either by acquiring hyperbolic orbits, or by colliding with the finite-sized planet. In such a way, particle-free zones around planetary orbits can be formed.

At small initial eccentricities ($e < 0.15$) of the particle's orbit, in the planar circular restricted three-body problem, the radial half-width of the instability neighbourhood of the perturber's orbit can be well described by Eq. (11.5) (Murray and Dermott 1999). Mustill and Wyatt (2012) generalized the Wisdom law by deriving the dependence of the radial extent of the planetary chaotic zone on the initial eccentricity of a close-to-coorbital particle. Thus, the shapes of debris disc edges near a planet can be predicted when the disc particles are initially eccentric.

An insight in details of close-to-coorbital dynamics is provided by the so-called *encounter map*.

11.1.3 The Encounter Map

Duncan et al. (1989) constructed an *encounter map* to describe the close-to-coorbital motion of a particle and a perturber (a planet) around the primary body (a star). In the map's first version, it was assumed that the perturber is moving in a circular orbit, the orbital inclination of the particle with respect to the planet's orbital plane is zero, and the particle's orbital eccentricity is small or moderate. Initially, the particle is close to be coorbital with the planet. The encounter map is based on the assumption that the planet perturbs the particle's motion solely when the particle moves in a close vicinity of the planet. Otherwise, the particle moves in an unperturbed Kepler ellipse determined by the last encounter. Here we describe the derivation of the map in brief, following Duncan et al. (1989) and Murray and Dermott (1999).

The motion of the particle with zero mass orbiting around the central mass m_1 is subject to perturbations from the planet $m_2 \ll m_1$ moving in a coplanar circular orbit with the semimajor axis a_2. It is natural to consider the particle's motion in the frames centred on the planet. The x axis is directed from m_1 to m_2, and the y axis is orthogonal to x and is co-directed with the perturber's motion, see Fig. 11.1.

The particle's motion near the planet is described by Hill's equations

$$\ddot{x} - 2n_2\dot{y} - 3n_2^2 x = -\frac{\mathcal{G}m_2 x}{(x^2 + y^2)^{3/2}}, \tag{11.6}$$

$$\ddot{y} + 2n_2\dot{x} = -\frac{\mathcal{G}m_2 y}{(x^2 + y^2)^{3/2}} \tag{11.7}$$

(Hill 1878; Murray and Dermott 1999), where $n_2 = \left(\mathcal{G}m_1/a_2^3\right)^{1/2}$ is the mean motion of the planet m_2.

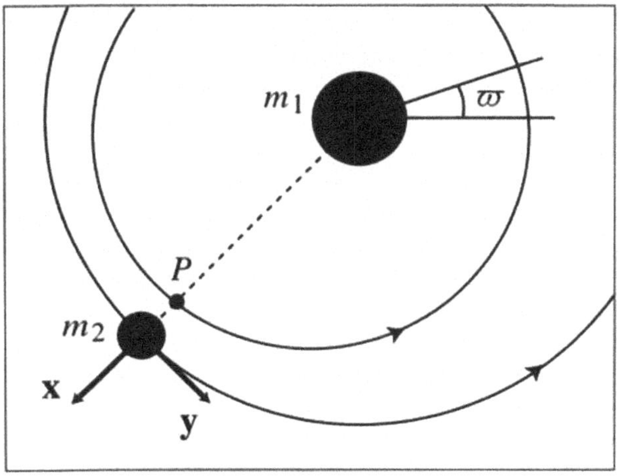

Fig. 11.1 A conjunction of the particle P, moving in an elliptic orbit, with the perturbing body in a coplanar circular orbit (Figure 9.19 from Murray and Dermott 1999, by permission from © Cambridge University Press)

In the limit $m_2 \to 0$, Hill's equations have the solution

$$x = D_1 \cos n_2 t + D_2 \sin n_2 t + D_3, \tag{11.8}$$

$$y = -2D_1 \sin n_2 t + 2D_2 \cos n_2 t - \frac{3}{2} D_3 n_2 t + D_4 \tag{11.9}$$

(Hénon and Petit 1986), where D_1, D_2, D_3, and D_4 are integration constants.

The motion is assumed to be unperturbed between approximately instant perturbations in the conjunctions. Therefore, the problem is reduced to calculating increments of D_1, D_2, D_3, and D_4 during a conjunction. D_4 can be set to zero, thus setting $t = 0$ when starting to consider each consequent conjunction. Duncan et al. (1989) calculated the increments of D_1, D_2, D_3 with the accuracy to the lowest orders in the particle's eccentricity e and in the quantity

$$\varepsilon = \frac{a - a_2}{a_2},$$

where a and a_2 are the semimajor axes of the particle and the planet, respectively. The most notable change is in D_2; the change in D_1 is zero.

Let e and ϖ be the eccentricity and the longitude of pericentre of the particle's orbit, and define the *complex eccentricity*

$$z = e \exp(i\varpi), \tag{11.10}$$

where i is the imaginary unit. Let λ_n be the mean longitude of the nth conjunction, z_n and ε_n be the values of z and ε just before the nth conjunction. The encounter map is given by the equations

$$z_{n+1} = z_n + \mathrm{sign}(\varepsilon_1)\frac{ig}{\varepsilon_1^2} \cdot \frac{m_2}{m_1} \exp(i\lambda_n), \tag{11.11}$$

$$\varepsilon_{n+1} = \varepsilon_n \left(1 + \frac{4(|z_{n+1}|^2 - |z_n|^2)}{3\varepsilon_n^2}\right)^{1/2}, \tag{11.12}$$

$$\lambda_{n+1} = \lambda_n + 2\pi \left|(1 + \varepsilon_{n+1})^{-3/2} - 1\right|^{-1} \tag{11.13}$$

(Duncan et al. 1989). The constant

$$g = \frac{8}{9}[2K_0(2/3) + K_1(2/3)] = 2.239566674\ldots,$$

where K_0 and K_1 are modified Bessel functions of the 2nd kind. Not that, although the map, as presented here, consists of three equations, it is actually four-dimensional, because the first equation is for a complex variable.

Another, more convenient, representation for the encounter map was derived in Duncan et al. (1989) by introducing the variable $\Delta\varepsilon_n = \varepsilon_n - \varepsilon_1$ (for $\varepsilon_1 > 0$) and expanding Eqs. (11.12) and (11.13) in power series; this gives

$$z_{n+1} = z_n + \frac{ig}{\varepsilon_1^2} \cdot \frac{m_2}{m_1} \exp(i\lambda_n), \tag{11.14}$$

$$\Delta\varepsilon_{n+1} = \frac{2(|z_{n+1}|^2 - |z_n|^2)}{3\varepsilon_1}, \tag{11.15}$$

$$\lambda_{n+1} = \lambda_n + \frac{4\pi}{3\varepsilon_1} - \frac{4\pi\,\Delta\varepsilon_{n+1}}{3\varepsilon_1^2}. \tag{11.16}$$

The *synodic period* (the time interval between consecutive conjunctions) is given by

$$T_{\mathrm{s}} = \frac{2\pi}{|n_2 - n|} = \frac{2\pi}{n_2} \left|\left(\frac{a_2}{a}\right)^{3/2} - 1\right|^{-1}. \tag{11.17}$$

It gives the encounter map period, i.e., the time interval corresponding to one iteration of the map. Here n is the particle's mean motion. As in cases of the separatrix and Kepler maps, it varies from iteration to iteration, i.e., depends on n.

Duncan et al. (1989) modified and generalized the encounter map for a more general case, in which the perturber's orbit is elliptic. Namouni et al. (1996)

provided an improved version, accounting for higher order terms in the particle's eccentricity and inclination.

The encounter map theory can be used to find analytically the radial width of the chaotic zone around the perturber orbit (Duncan et al. 1989; Murray and Dermott 1999). The main idea is analogous to a heuristic method used in Zaslavsky and Filonenko (1968), Zaslavsky et al. (1991) for estimating the chaotic layer width in the perturbed pendulum model. Namely, it is assumed that the motion is chaotic, being completely randomized in the phase angle variable, if this angle varies per one map iteration by more than π. Note that this chaos condition is completely heuristic and not sufficient, as, for example, a consideration of emergence of global chaos in the standard map dynamics can show. This condition can be shown to be partly analogous to the resonance overlap criterion in the perturbed pendulum model or in the standard map. It means that correlations between consecutive (at consecutive encounters) values of the mean longitude λ are lost, and the longitudes of consecutive conjunctions are completely randomized. Using Eq. (11.16), one may write down the criterion in the form

$$\frac{4\pi\,\Delta\varepsilon_n}{3\varepsilon_1^2} \gtrsim \pi, \tag{11.18}$$

where $\varepsilon = (a - a_2)/a_2$, $\Delta\varepsilon_n = \varepsilon_n - \varepsilon_1$. Equivalently,

$$|\Delta\varepsilon| \gtrsim \frac{3}{4}\varepsilon_1^2. \tag{11.19}$$

Let the initial eccentricity be zero, then $|z_1| = 0$, and from Eq. (11.14) one has

$$|z_2| = \frac{g}{\varepsilon_1^2}\frac{m_2}{m_1}. \tag{11.20}$$

On the other hand, Eq. (11.15) gives

$$\varepsilon\Delta\varepsilon = \frac{2}{3}|z_2|^2. \tag{11.21}$$

Hence

$$\Delta\varepsilon = \frac{2}{3}\frac{g^2}{\varepsilon^5}\left(\frac{m_2}{m_1}\right)^2, \tag{11.22}$$

and the condition (11.19) reduces to

$$\varepsilon_1 \lesssim \left(\frac{8}{9}g^2\right)^{1/7}\left(\frac{m_2}{m_1}\right)^{2/7}. \tag{11.23}$$

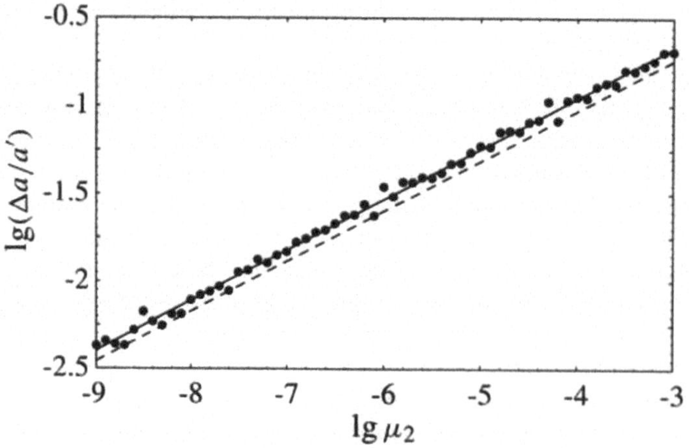

Fig. 11.2 The encounter map numerical-experimental data on $\Delta a_{\rm cr}/a'$ in the planar circular restricted three-body problem (Murray and Dermott 1999). Solid line: the best fit to the data. Dashed line: Wisdom's theoretical law (11.24). μ_2 is the mass parameter equivalent to μ in our notation (Figure 9.23 from Murray and Dermott 1999, by permission from © Cambridge University Press)

Evaluating g and setting $\varepsilon_1 = \Delta a_{\rm cr}/a'$, one arrives at

$$\Delta a_{\rm cr} \approx 1.24\mu^{2/7}a', \tag{11.24}$$

in accord with Wisdom's original law (11.5).

Note, once again, that the increment of λ in the chaos condition is set equal to π on solely heuristic grounds, that is why the perfect agreement in the numerical coefficient is largely by chance.

In Duncan et al. (1989) and Murray and Dermott (1999), the validity of the theoretical dependence (11.5) was checked in numerical experiments with the encounter map, and a close agreement of the theoretical power-law index in Eq. (11.5) with obtained numerical-experimental estimates was found. Murray and Dermott (1999) estimated that, in the adopted μ interval (which was rather broad: $10^{-9} \leq \mu \leq 10^{-3}$), the numerically measured power-law index was equal to 0.286, practically coinciding with its theoretical value 2/7,[1] but the numerically found coefficient at the scaling was found to be some 20% greater:

$$\varepsilon_{\rm cr} = \frac{\Delta a_{\rm cr}}{a'} = 1.57\mu^{0.286} \simeq 1.57\mu^{2/7}, \tag{11.25}$$

see Fig. 11.2. Therefore, the actual chaotic zone is about 20% broader than that given by Eq. (11.5), regardless of the mass parameter μ value.

[1] $2/7 = 0.2857\ldots$.

In Fig. 11.2, the numerical-experimental results of Murray and Dermott (1999) are represented by dots. The solid line shows the best linear fit in the logarithmic coordinates, it is given by Eq. (11.25). Wisdom's theoretical law (11.24) is shown by the dashed line. Obviously Wisdom's law serves as a lower bound for the numerical data of Murray and Dermott (1999), providing a perfect fit for the slope of the observed dependence.

It is important to outline that, if the initial conditions are inside the chaotic layer, the encounter map cannot be iterated infinitely, because sooner or later the particle's eccentricity is raised high enough, and one of the conditions for the map validity (namely, the low particle's eccentricity) breaks down. This is not at all a fault of the map, but rather a manifestation of the physical reality: the perturber "cleans up" in such a way the radial neighbourhood of its orbit. This close-to-orbit *clearing* is just one of the conditions set by the IAU General Assembly in Prague (IAU General Assembly 2006) for a celestial body to be called a planet.

11.1.4 The Two-Dimensional Encounter Map

To describe the motion inside the planetary chaotic zone, a two-dimensional kind of an encounter map can be derived using an adaptation of the *tokamap technique*, introduced in Balescu (1998), Balescu et al. (1998).

In the planar circular restricted three-body problem, the Hamiltonian of a particle's motion in the vicinity of a mean motion first-order $(p+1)/p$ resonance $(p \geq 1)$ with a planet can be written as

$$H = -\frac{1}{2\Phi^2} - \Phi + \left(1 + p - \frac{p}{\Phi^3}\right)\Psi - \frac{3p^2}{2\Phi^4}\Psi^2 - 2^{1/2}\mu B_p \Psi^{1/2}\cos\psi \qquad (11.26)$$

(Wisdom 1980, Equation (42)), where

$$\psi = -pl + (p+1)(t-g) \qquad (11.27)$$

is the resonant argument, and the Delaunay angles l and g are the mean anomaly and the argument of pericenter, respectively; t is time;

$$\Phi \simeq [(1-\mu)a)]^{1/2}\left(1 + \frac{p}{2}e^2\right), \quad \Psi \simeq [(1-\mu)a]^{1/2}\frac{e^2}{2} \qquad (11.28)$$

are the canonical momenta. The resonant argument ψ is conjugate to the momentum Ψ; a is the semimajor axis, e is the eccentricity. The canonical angle variable $\phi = l + g - t$, conjugate to the momentum Φ, is absent in the representation (11.26); therefore Φ is constant in this approximation.

The unit system is defined in such a way: $\mathcal{G}(m_1 + m_2) = 1$ (where $\mathcal{G}$ is the gravitational constant), and the distance between m_1 and m_2 is unity. Thus, the orbital period of the binary is equal to 2π and the binary's orbital frequency (mean motion) is unity.

The mass parameter $\mu = m_2/(m_1 + m_2)$ (where $m_1 \gg m_2$ are the masses of the binary's components) is assumed to be small. The numerical coefficient B_p is given by

$$B_p \simeq -\frac{p+1}{\pi}[2K_0(2/3) + K_1(2/3)] \simeq -0.802(p+1) \tag{11.29}$$

(Wisdom 1980; Malhotra 1998), where K_0 and K_1 are modified Bessel functions.

Let us consider the motion in the vicinity of a *guiding* resonance $(p+1)/p$, taking into account its interaction with the multitude of *all* other first order mean-motion resonances. Therefore, we incorporate all first-order resonances in the Hamiltonian (11.26). We set $\psi_k = k(-l+t-g)+t-g$ for any $k \geq 1$; for the guiding resonance, then, according to Eq. (11.27), $\psi \equiv \psi_p$. Therefore, $\psi_{p+j} = \psi_p - j\phi$ for any $j \geq 1 - p$, and

$$\begin{aligned} H &= -\frac{1}{2\Phi^2} - \Phi + \left(1 + p - \frac{p}{\Phi^3}\right)\Psi - \frac{3p^2}{2\Phi^4}\Psi^2 - 2^{1/2}\mu\Psi^{1/2} \cdot \sum_{k=1}^{\infty} B_k \cos\psi_k \\ &= -\frac{1}{2\Phi^2} - \Phi + \left(1 + p - \frac{p}{\Phi^3}\right)\Psi - \frac{3p^2}{2\Phi^4}\Psi^2 - 2^{1/2}\mu\Psi^{1/2} \cdot \sum_{j=1-p}^{\infty} B_{p+j}\cos(\psi - j\phi). \end{aligned} \tag{11.30}$$

The guiding resonance corresponds to the term with $j = 0$ in the sum.

We see that the canonical angle variable $\phi = l + g - t$ reappears in the Hamiltonian. If $p \gg 1$, in the vicinity of the guiding resonance in the case of particle's inner (with respect to the perturber) orbits one has

$$\dot{\phi} \approx n - n_{\mathrm{pl}} = \frac{p+1}{p} - 1 = \frac{1}{p}, \tag{11.31}$$

where $n = (p+1)/p$ and $n_{\mathrm{pl}} = 1$ are the mean motions of the particle and the planet, respectively. Defining the frequency $\Omega \equiv \dot{\phi}$, we rewrite Eq. (11.30) in the form

$$H = -\frac{1}{2\Phi^2} - \Phi + \left(1 + p - \frac{p}{\Phi^3}\right)\Psi - \frac{3p^2}{2\Phi^4}\Psi^2 - 2^{1/2}\mu\Psi^{1/2} \cdot \sum_{j=1-p}^{\infty} B_{p+j}\cos(\psi_p - j\Omega t). \tag{11.32}$$

Equation (11.29) for the B_p coefficients is valid in the vicinity of the guiding resonance. As soon as the resonances closest to the guiding one are most important, henceforth we treat the amplitudes B_{p+j} of the perturbing resonances in the $(p+1)/p$ "staircase" as having approximately the same value.

Henceforth we consider the dynamics in the specific angular momentum, Ψ, and its conjugate angle, ψ, assuming that the specific energy Φ is approximately constant. In reality, neither energy nor angular momentum are conserved here, i.e., in the restricted three-body problem. Instead, in the restricted circular three-body problem there exists the Jacobi constant, which, in some approximation, can be reduced to the *Tisserand relation*. The approximation implies that $\mu \ll 1$ and that the particle's motion is considered far from close encounters with the planet, i.e., the particle–planet distance is not much less than the particle–star distance; see, e.g., Murray and Dermott (1999). During the encounters, the relation is subject to deviations.

Introducing the *Tisserand parameter* Ti, we write down the Tisserand relation in the planar problem in the form

$$Ti = \frac{1}{2a} + \left[(1-e^2)a\right]^{1/2} \simeq \text{const}, \tag{11.33}$$

where a is the semimajor axis of the particle, measured in units of the semimajor axis a_{pl} of the planet, e is the particle's eccentricity.

If $e \sim 0$, and $a \sim 1$, as in our case, any relative (with respect to the initial value) variations in e are much greater than the corresponding relative variations in a; and the latter can be set relatively constant. Therefore, one may consider variations in the specific angular momentum assuming the energy to be constant. This situation is opposite to that in the Kepler map formalism, describing the particle's highly eccentric motion (Petrosky 1986; Petrosky and Broucke 1988), where the particle's pericentric distance is set constant and variations in the energy are considered; see Sect. 11.1.5.

Now let us see how an appropriate map can be constructed to describe the close-to-coorbital evolution. For this purpose, we use a *tokamap* technique. In Balescu (1998); Balescu et al. (1998), a specific area-preserving two-dimensional map, now known as *tokamap*, was derived to describe the dynamical behavior of the magnetic field lines in tokamaks. Its derivation was based on the Wobig–Mendonça formalism (Wobig 1987; Mendonça 1991). Nowadays, the tokamap technique is broadly used in physical applications (Caldas et al. 2017).

Note that the tokamap derivation itself is quite standard: its key step consists in representing the perturbation as a periodic delta-function (a sequence of "kicks"); see Chirikov (1979), Lichtenberg and Lieberman (1992), Murray and Dermott (1999), Wisdom (2018). However, as soon as we use the map equations as obtained in Wobig (1987), Mendonça (1991) for a given Hamiltonian, we use the tokamap terminology here.

The tokamap Hamiltonian is given by

$$H = \int \frac{dy}{q(y)} + \epsilon H_1(y) \sum_{k=-M}^{k=M} \cos(x - kt) \overset{M\to\infty}{=} \int \frac{dy}{q(y)} + \epsilon H_1(y) \cos x \sum_{k=-\infty}^{k=\infty} \delta(t-k), \tag{11.34}$$

see (Abdullaev 2006, equation (10.78)); x and y are conjugate canonical variables, t is time. The $q(y)$ function is called the *safety factor*. Based on Eq. (11.34), the general form of a tokamap can be written as

$$\begin{aligned} y_{k+1} &= y_k + 2\pi f(y_{k+1}) \sin x_k, \\ x_{k+1} &= x_k + \frac{2\pi}{q(y_{k+1})} + 2\pi f'(y_{k+1}) \cos x_k \end{aligned} \tag{11.35}$$

(Balescu 1998; Balescu et al. 1998), where $f(y) \equiv \epsilon H_1(y)$, $f'(y_{k+1}) = \frac{\partial f'(y_{k+1})}{\partial y_{k+1}}$. The map period is equal to the period of perturbation.

We do not reproduce here $q(y)$ and $\epsilon H_1(y)$ specific for tokamaks. Note that Wobig's map, derived originally in a specific tokamak model (Wobig 1987), can be shown to describe the motion in an infinite multiplet of equal-sized equal-spaced *parametric resonances* (considered, in particular, in Chirikov 1979). Analogously, the standard map describes the motion in an infinite multiplet of equal-sized equal-spaced pendulum-like resonances, i.e., the *first fundamental model* resonances; see Sect. 1.2.

Here we derive a map that describes the motion in an infinite multiplet of equal-sized equal-spaced *second fundamental model* resonances. We adopt the second fundamental model as defined in Henrard (1982, 1983), Henrard and Lemaître (1983). Its Hamiltonian is given by formula (11.34), if one sets

$$q(y) = \frac{1}{A + By}, \quad f(y) = Cy^{1/2}, \tag{11.36}$$

where A, B, C are arbitrary constants. Solving for y_{k+1} in the first equation of map (11.35), we derive the tokamap:

$$\begin{aligned} y_{k+1} &= \left[\pi C \sin x_k + \left(\pi^2 C^2 \sin^2 x_k + y_k \right)^{1/2} \right]^2, \\ x_{k+1} &= x_k + 2\pi (A + By_{k+1}) + \pi C y_{k+1}^{-1/2} \cos x_k, \end{aligned} \tag{11.37}$$

where it is implied that $y > 0$.

To reduce Hamiltonian (11.32) to paradigm (11.34), we rescale the time in Hamiltonian (11.32): $\Omega t \to t$; thus, the Hamiltonian itself is rescaled: $H \to H/\Omega$. In Eq. (11.32), we assume that the sum of the harmonic terms starts from minus infinity, as soon as $p \gg 1$. Therefore, we may apply the resulting Hamiltonian as a local approximation of the motion, valid in the vicinity of any guiding resonance $(p + 1)/p$ with $p \gg 1$. Using definitions (11.36) and setting $\Psi \equiv y$, $\psi \equiv x$, we compare Hamiltonians (11.32) and (11.34), and find

$$A = \Omega^{-1} \left(1 + p - \frac{p}{\Phi^3} \right), \quad B = -\Omega^{-1} \frac{3p^2}{\Phi^4}, \quad C = -2^{1/2} \Omega^{-1} \mu B_p, \tag{11.38}$$

where

$$B_p \simeq -0.802(p+1), \quad \Omega \simeq \frac{1}{p}. \tag{11.39}$$

The variables Φ and Ψ are expressed through μ, p, a, e by formulas (11.28). From Eqs. (11.32, 11.38), it is straightforward to see that at $p \gg 1$ and $e \sim 0$ one has

$$A \simeq 0, \qquad B \simeq -3p^3, \qquad C \simeq 1.134\mu p^2. \tag{11.40}$$

Making the substitution $y = z/(2\pi B)$ in Eqs. (11.37), we arrive at

$$\begin{aligned} z_{k+1} &= \left[J \sin x_k + \left(J^2 \sin^2 x_k + z_k \right)^{1/2} \right]^2, \\ x_{k+1} &= x_k + z_{k+1} + J z_{k+1}^{-1/2} \cos x_k + 2\pi A, \end{aligned} \tag{11.41}$$

where the new parameter

$$J = 2^{1/2} \pi^{3/2} |B|^{1/2} C \simeq 15.5 \mu p^{7/2}. \tag{11.42}$$

We see that the parameter A is mostly non-important, as it affects only local winding numbers of the map, but not the local Lyapunov exponents and diffusion rates, except at the border of the chaotic layer. At the border, its value controls the prominence of marginal resonances, analogously to the case of the separatrix map, as presented in Shevchenko (2012). Therefore, the derived tokamap has only one important parameter, J.

Let us consider the map (11.35) behavior in the standard map approximation in the vicinity of a fixed (integer-resonant) point y_{fp}. The conditions at which the standard map approximation is valid can be estimated using the original map in form (11.41), and also Eqs. (11.40). Indeed, the inequality $J z^{-1/2} \ll 1$ is required; and this condition can be reduced to

$$y \gg \pi^2 C^2 \simeq 12.7 \mu^2 p^4 \simeq 2.51 \mu^2 \varepsilon^{-4}. \tag{11.43}$$

In other words, the standard map approximation is valid if y is much greater than the "forced" y value, given by $y_{\mathrm{f}} = e_{\mathrm{f}}^2/2 \simeq 2.51 \mu^2 \varepsilon^{-4}$, where the forced eccentricity e_{f} is given below by Eq. (11.51). Note that one iteration of the map (11.37) just provides the forced y value, if one sets the initial y to zero.

Ignoring negligible (on conditions specified below) terms, one obtains an approximating map

$$\begin{aligned} \zeta_{k+1} &= \zeta_k - K \sin x_k, \\ x_{k+1} &= x_k + \zeta_{k+1} + \mathrm{const}, \end{aligned} \tag{11.44}$$

where $\zeta = 2\pi B(y - y_{\mathrm{fp}})$, $K(y_{\mathrm{fp}}) = 4\pi^2 |B| C y_{\mathrm{fp}}^{1/2}$.

The map (11.44) is identical to the standard map (considered in Sect. 1.5), except that here const $\neq 0$ generally. For the standard map's stochasticity parameter K, setting $y_{\rm fp} = e^2/2$, one has

$$K(y_{\rm fp}) = 4\pi^2|B|Cy_{\rm fp}^{1/2} \simeq 95.0 \cdot \mu e p^5. \qquad (11.45)$$

If $K > K_{\rm G} = 0.971635406\ldots$, the chaotic diffusion in the momentum I is non-bounded; see Sect. 1.5. In other words, on increasing the K parameter, at $K = K_{\rm G} \approx 1$ the last rotational invariant curve is destroyed and the motion becomes non-bounded in I. According to Eq. (11.45), the critical value of $y_{\rm fp}$ that marks the border of the chaotic layer and thus corresponds to $K = K_{\rm G}$ in map (11.44), is given by

$$y_{\rm cr} = \frac{K_{\rm G}^2}{16\pi^4 B^2 C^2} \simeq 5.23 \cdot 10^{-5} \mu^{-2} p^{-10}. \qquad (11.46)$$

Another important y value, which corresponds to $K = 4$ and which we designate as y_4, roughly separates a chaotic layer's mostly "non-porous" part (that with a negligible total measure of regular islands) and its mostly "porous" part (that with many large and small regular islands). In the "porous" part, trajectories may stick to fractal borders of the islands and therefore the diffusion is generally anomalous. The y_4 value is given by

$$y_4 = \frac{1}{\pi^4 B^2 C^2} \simeq 8.87 \cdot 10^{-4} \mu^{-2} p^{-10}. \qquad (11.47)$$

The ratio $y_{\rm cr}/y_4 = K_{\rm G}^2/16 \approx 0.059$. The fraction of the "porous part" in the whole extent of the layer in y depends on the values of p and μ. Recall that

$$y = \Psi \simeq [(1-\mu)a]^{1/2} \frac{e^2}{2}, \qquad (11.48)$$

according to Eq. (11.28). Therefore y may vary in the interval from 0 to $(1-\mu)/2 \approx 1/2$ (corresponding to the variation in eccentricity e from 0 to 1). If $p \gg 1$, then the porous part is negligible. For example, if $\mu > 10^{-3}$ and $p > 10$, one has $y_4 < 10^{-7}$.

Let us use the tokamap (11.41) to estimate the width of the Wisdom gap. From Eq. (11.45), at $K = K_{\rm G}$, as soon as $y \simeq e^2/2$ and B and C are given by Eqs. (11.40), one has

$$p \simeq 0.400 \cdot (\mu e)^{-1/5}, \qquad (11.49)$$

and the gap's half-width is given by

$$\varepsilon_{\rm cr} \equiv \frac{\Delta a_{\rm cr}}{a_{\rm pl}} \simeq \frac{2}{3p} \simeq 1.67 \cdot (\mu e)^{1/5}. \qquad (11.50)$$

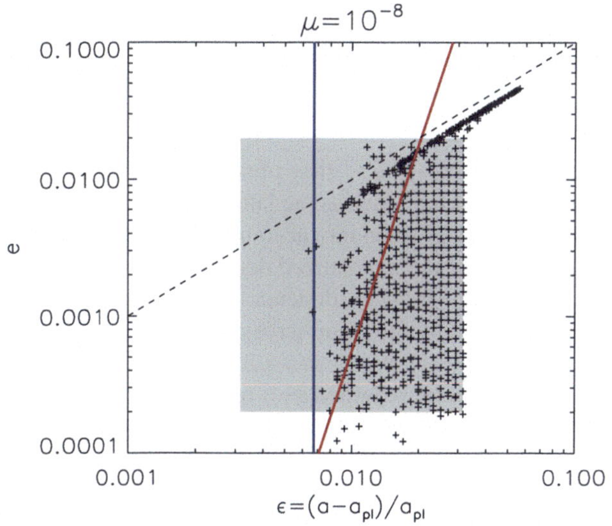

Fig. 11.3 The clearing of the planetary chaotic zone at $\mu = 10^{-8}$. Vertical solid blue line: the classical border of the Wisdom gap, as given by formula $\varepsilon_{cr} = 1.3\mu^{2/7}$. Solid red line: the border of the chaotic zone, as given by the generalized dependence $\varepsilon_{cr} = 1.8(\mu e)^{1/5}$. Dashed black line: the planet-crossing limit $e_{pc} \simeq \varepsilon$. Crosses: evolved semimajor axes and eccentricities of ~1000 particles, as obtained by direct numerical integrations over 10 Myr, in the three-body problem, with initial conditions uniformly distributed over the grey rectangle. The solid red and dashed black lines obviously delineate the cleared area (Figure 2 from Mustill and Wyatt 2012, © Oxford University Press)

This scaling reproduces formulas obtained in a different way by Mustill and Wyatt (2012) and Deck et al. (2013). However, the pre-law coefficient is slightly different. In Mustill and Wyatt (2012), the coefficient is 1.8, and in Deck et al. (2013) it is 1.38. The intermediate value 1.67 seems to be in a better agreement with numerical results presented in Mustill and Wyatt (2012); see Fig. 11.3.

The close-to-coorbital chaotic gap phenomenon was originally revealed in Wisdom (1980) and Duncan et al. (1989) in assumption that the particles are initially in circular orbits around the primary. However, the perturber (secondary) induces eccentricity in particle's motion. The forced eccentricity (the eccentricity raised from zero by the perturber in a single conjunction) of a particle initially placed in a circular orbit with radius $a = a_{pl} \pm \Delta a$ is given, as can be deduced from (Hénon and Petit 1986, Equations (33) and (127)), by

$$e_f = g\mu\varepsilon^{-2} \simeq 2.24\mu\varepsilon^{-2}, \tag{11.51}$$

where, as defined above, $\varepsilon \equiv \frac{\Delta a}{a_{pl}}$, and, according to (Murray and Dermott 1999, Equation (9.85)),

$$g = \frac{8}{9}\left[2K_0(2/3) + K_1(2/3)\right] \approx 2.23957, \tag{11.52}$$

where K_0 and K_1 are modified Bessel functions, as in Eq. (11.29).

Substituting the forced eccentricity in Eq. (11.50), one arrives at

$$\varepsilon_{\rm cr} \simeq 1.62\mu^{2/7}. \tag{11.53}$$

Therefore, the scaling $\varepsilon_{\rm cr} \propto (\mu e)^{1/5}$ first obtained in Mustill and Wyatt (2012) directly implies the classical $\mu^{2/7}$ law, as the latter directly follows from the former, when the forced eccentricity is taken into account.

A graphical illustration of the generalized dependence $\varepsilon_{\rm cr} \propto (\mu e)^{1/5}$, along with relevant numerical data, confirming the dependence, is given in Fig. 11.3. Obviously, the cleared area is delineated by the planet-crossing limit $e_{\rm pc} \simeq \varepsilon = \Delta a/a_{\rm pl}$ and the generalized dependence.

The tokamap (11.41) can be regarded as a reduced two-dimensional version of the four-dimensional *encounter map*, presented above in Sect. 11.1.3 to describe the close-to-coorbital motion both in the energy and angular momentum variables and their two conjugate angles. As soon as the diffusion in the eccentricity is dynamically more important than that in the semimajor axis, as justified above for the considered problem, the reduction to the two-dimensional case is appropriate. However, the two-dimensional map cannot be straightforwardly derived from the existing encounter map versions.

Comparing the tokamap (11.41) with the Kepler map (8.35), one should outline a certain complementary symmetry between the two maps: the Kepler map describes the motion in the energy variable, setting the angular momentum approximately constant, as justified in Sect. 11.1.5, whereby the tokamap describes the motion in the angular momentum variable, setting the energy approximately constant, as justified above.

11.1.5 The Kepler Map Criterion: The $\mu^{2/5}$ Law

The Kepler map is a two-dimensional area-preserving map, introduced by Petrosky (1986) and Chirikov and Vecheslavov (1986) to describe the highly-eccentric circumbinary motion of a massless (passively gravitating) particle in terms of energy and time. The model scheme is given in Fig. 11.4. The map's second equation is based on Kepler's third law, hence the title of the map. Since 1980s the Kepler map has become paradigmatic in applications in celestial mechanics and atomic physics.

The map was introduced by physicists, but in application to dynamical astronomy. The first publications, inspired by the return of the Halley comet in 1986–1987, were very soon followed by many other contributions, where the map theory was further developed and generalized. The generalizations concerned various problems in celestial mechanics (Petrosky and Broucke 1988; Vecheslavov and Chirikov 1988; Chirikov and Vecheslavov 1989; Emelyanenko 1990) and atomic physics (Casati et al. 1987; Gontis and Kaulakys 1987; Casati et al. 1988; Borgonovi et al. 1988; Jensen et al. 1988). In celestial mechanics, the Kepler map is broadly used

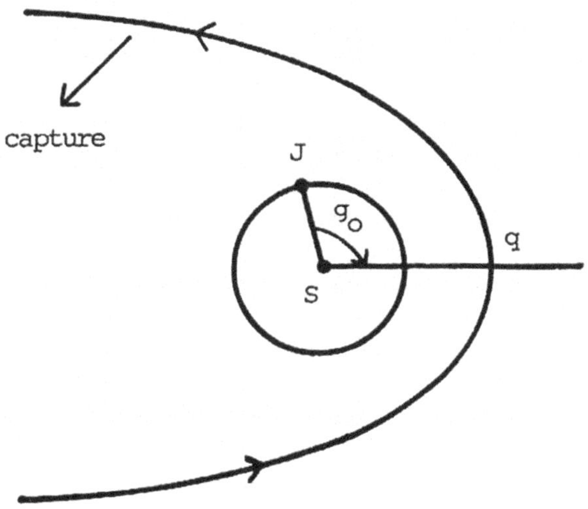

Fig. 11.4 Near-parabolic circumbinary motion of a particle around a "star–planet" binary (Figure 1 from Petrosky 1986, with permission from Elsevier)

nowadays to describe and analyse long-term dynamics of various kinds of cosmic objects in different settings of the hierarchical three-body and many-body problems.

Kepler's third law, which provides the second equation of the map, was formulated by Kepler in 1619 in the fifth book of *Harmonices Mundi* (Kepler 1619), a sequel to *Mysterium Cosmographicum*. A brief mention of the law appeared already in 1618, in *A short summary of Copernican astronomy* (*Epitome Astronomiae Copernicanae*) (Kepler 1618). Therefore, one may say that the second equation of the Kepler map was ready for derivation already some 400 years ago. It took quite a long time, almost the same 400 years, for a scientific community to become prepared to derive the first equation, although mathematically it does not look more complicated than the second one.

The major problem in constructing the Kepler map analytically is to derive the energy increment per one cometary orbital revolution. In the pioneering analytical works on the Kepler map (Petrosky 1986; Petrosky and Broucke 1988), a number of refined methods of mathematical physics were used to achieve this aim. The methods included, in particular, the Lie algebraic formalism (the Hori method), the Kolmogorov–Arnold–Moser (KAM) theory elements, a method of reduction of a Fourier series with a small denominator to the Fourier integral in the Cauchy integral form, a method of embedding the small denominator in an analytic function through a suitable analytic continuation, analogies with scattering theory in quantum mechanics. These methods became available in the twentieth century, and mostly in the sixties of the twentieth century. Using these methods could be convenient, but they are not necessary to derive the map. In principle, an analytical derivation of the energy increment could have been accomplished since 1836, when the Jacobi integral formalism became available; for historical details, see Shevchenko (2011b). This formalism is used below to demonstrate how the map is derived.

Modern domains of applications of the Kepler map and its generalizations in the dynamical astronomy include: studies in the dynamics of a passively gravitating particle in highly eccentric orbits in the planar restricted three-body problem without crossings of planetary orbits (Petrosky 1986; Petrosky and Broucke 1988), in the non-planar restricted three-body and four-body problems with crossings of planetary orbits (Chirikov and Vecheslavov 1986; Vecheslavov and Chirikov 1988; Chirikov and Vecheslavov 1989), studies of dynamical chaos in the perturbed highly-eccentric motion (Petrosky 1986; Malyshkin and Tremaine 1999; Pan and Sari 2004), chaotic diffusion in the dynamics of comets and meteor streams (Emelyanenko 1990, 1992; Liu and Sun 1994; Zhou and Sun 2001; Zhou et al. 2000, 2002; Malyshkin and Tremaine 1999), studies in the Sitnikov problem (Urminsky and Heggie 2008), exploration of dynamical environments of rotating contact binaries, such as cometary nuclei, trans-Neptunian objects, and other similar-shaped small bodies of the Solar system (Lages et al. 2017, 2018b).

Modern domains of developments and applications of the Kepler map in physics include studies in classical chaotic ionization processes in hydrogen atoms in microwave fields (Gontis and Kaulakys 1987; Casati et al. 1988; Jensen et al. 1988, 1991), generalizations of the Kepler map for cases of multi-frequency fields (Kaulakys and Vilutis 1999), studies in physics of hydrogen atoms driven by microwaves with arbitrary polarization (Pakoński and Zakrzewski 2001), elaboration of *synchronized* Kepler maps (Nauenberg 1990; Pakoński and Zakrzewski 2001).

Note that the encounter map, considered above in Sect. 11.1.3, and the Kepler map, considered here, are self-complementary: the first one describes the motion of a particle in trajectories that are low-eccentric prograde near-coorbital with the perturber (say, a planet), and the second one describes highly-eccentric dynamics of particles in circumbinary orbits.

The Kepler map, including analytical formulae for its parameter, can be derived by quite elementary methods (Shevchenko 2011b; Lages et al. 2018a). Let us consider the motion of a passively gravitating particle in the planar circular restricted three-body problem, say, a star–planet–comet problem, in an inertial Cartesian frame with the origin at the barycentre. For the distance unit, we take the constant star–planet distance, for the mass unit—the sum of the star's and planet's masses, and for the time unit—the $1/(2\pi)$th part of the planet's orbital period.

The particle's motion in the (x, y) coordinates is described by the differential equations

$$\ddot{x} = \nu \frac{x_S - x}{r_{13}^3} + \mu \frac{x_J - x}{r_{23}^3},$$
$$\ddot{y} = \nu \frac{y_S - y}{r_{13}^3} + \mu \frac{y_J - y}{r_{23}^3} \qquad (11.54)$$

(see, for example, Szebehely 1967), where

$$r_{13}^2 = (x_S - x)^2 + (y_S - y)^2,$$
$$r_{23}^2 = (x_J - x)^2 + (y_J - y)^2, \tag{11.55}$$

$$x_S = -\mu \cos(t - t_0),$$
$$y_S = -\mu \sin(t - t_0), \tag{11.56}$$

$$x_J = \nu \cos(t - t_0),$$
$$y_J = \nu \sin(t - t_0), \tag{11.57}$$

where r_{13} and r_{23} are, respectively, the star–comet and planet–comet distances; (x_S, y_S) and (x_J, y_J) are the coordinates of the star and planet (say, Jupiter), respectively; μ is the planet's mass, $\nu = 1 - \mu$ is the star's mass.

Expanding the right-hand sides of Eqs. (11.54) in power series of μ and retaining the first-order terms, one has

$$\ddot{x} = -\frac{x}{r^3} + \mu F(x, y, t, t_0),$$
$$\ddot{y} = -\frac{y}{r^3} + \mu G(x, y, t, t_0), \tag{11.58}$$

where $r = (x^2 + y^2)^{1/2}$,

$$F(x, y, t, t_0) = [x - \cos(t - t_0)]r^{-3} + 3x[x \cos(t - t_0) + y \sin(t - t_0)]r^{-5} + \\ + [\cos(t - t_0) - x]\{[x - \cos(t - t_0)]^2 + [y - \sin(t - t_0)]^2\}^{-3/2}, \tag{11.59}$$

$$G(x, y, t, t_0) = [y - \sin(t - t_0)]r^{-3} + 3y[x \cos(t - t_0) + y \sin(t - t_0)]r^{-5} + \\ + [\sin(t - t_0) - y]\{[x - \cos(t - t_0)]^2 + [y - \sin(t - t_0)]^2\}^{-3/2}; \tag{11.60}$$

see, for example, Liu and Sun (1994), Zhou et al. (2000). The initial epoch t_0 is chosen in such a way that the comet is at the pericentre when $t = 0$. Let the planet's phase at $t = 0$ be designated as $g = g_0$, then $t_0 = -g$; see Fig. 11.4.

The energy E of the unperturbed particle's motion is

$$E = \frac{1}{2}(\dot{x}^2 + \dot{y}^2) - \frac{1}{r} = -\frac{1}{2a}, \tag{11.61}$$

where a is the semimajor axis of the cometary orbit. If the planetary perturbation is switched on, the energy is no more conserved and is given by

$$E = \frac{1}{2}(\dot{x}^2 + \dot{y}^2) - \frac{1 - \mu}{r_{13}} - \frac{\mu}{r_{23}} = -\frac{1}{2a} \tag{11.62}$$

(Szebehely 1967). Equations (11.58) give

$$\dot{E} = \mu[\dot{x}(t)F(x(t), y(t), t, t_0) + \dot{y}(t)G(x(t), y(t), t, t_0)]. \quad (11.63)$$

The increment of E per one cometary orbital revolution is equal to the integral

$$\Delta E = \mu \int_{-\infty}^{+\infty} [\dot{x}(t)F(t, g) + \dot{y}(t)G(t, g)]\mathrm{d}t, \quad (11.64)$$

where $g = -t_0$. The integral is analogous to the Melnikov–Arnold integral in the separatrix map theory, see Sect. 1.6 and Appendix C. The function $\Delta E(gk)$ is 2π-periodic and anti-symmetric with respect to $g = \pi$.

In our inertial frame with the origin at the barycentre, the Jacobi integral is given by

$$\dot{x}^2 + \dot{y}^2 - \frac{2(1-\mu)}{r_{13}} - \frac{2\mu}{r_{23}} - 2(x\dot{y} - y\dot{x}) = E - D = \text{const} \quad (11.65)$$

(Szebehely 1967), where D is the angular momentum (note that, in our unit system, the planet's orbital frequency is set to unity). For the time derivatives of E and D we arrive at the equality

$$\dot{E} = \dot{D}.$$

Therefore, the increments in E and D are equal. In the following, we derive the increment in D, because in case of E the analytical calculation is much more complicated. However, the result should be the same. The angular momentum is

$$D = x\dot{y} - y\dot{x}, \quad (11.66)$$

and its time derivative

$$\dot{D} = x\ddot{y} - y\ddot{x}. \quad (11.67)$$

Using expressions (11.54) for $\ddot{y}$ and $\ddot{x}$, one has

$$\dot{D} = \nu\frac{xy_\mathrm{S} - x_\mathrm{S}y}{r_{13}^3} + \mu\frac{xy_\mathrm{J} - x_\mathrm{J}y}{r_{23}^3}, \quad (11.68)$$

where

$$r_{13}^2 = \mu^2 + r^2 - 2(x_\mathrm{S}x + y_\mathrm{S}y),$$
$$r_{23}^2 = \nu^2 + r^2 - 2(x_\mathrm{J}x + y_\mathrm{J}y). \quad (11.69)$$

Thus, $\dot{D}$ is the sum of four terms:

$$\dot{D} = \mathcal{A} + \mathcal{B} + \mathcal{C} + \mathcal{D}, \tag{11.70}$$

where

$$\mathcal{A} = \nu \frac{x y_S}{r_{13}^3}, \ \mathcal{B} = -\nu \frac{x_S y}{r_{13}^3}, \ \mathcal{C} = \mu \frac{x y_J}{r_{23}^3}, \ \mathcal{D} = -\mu \frac{x_J y}{r_{23}^3}. \tag{11.71}$$

It is sufficient to evaluate $\mathcal{A}$ and $\mathcal{B}$, because

$$\mathcal{C} = -\mathcal{A}(\nu \to -\mu), \ \mathcal{D} = -\mathcal{B}(\nu \to -\mu). \tag{11.72}$$

The motion on the unperturbed parabolic separatrix is described by

$$r = q(1+u^2), \ x = q\,(1-u)\,, \ y = 2qu, \ t = \kappa \left(u + \frac{u^3}{3} \right), \tag{11.73}$$

where

$$u = \left(\tau + (1+\tau^2)^{1/2} \right)^{1/3} + \left(\tau - (1+\tau^2)^{1/2} \right)^{1/3}, \ \tau = \frac{3}{2\kappa} t \tag{11.74}$$

(see, e.g., Petrosky 1986), where

$$\kappa = (2q^3)^{1/2},$$

the eccentric anomaly

$$u = \tan \frac{f}{2},$$

and q and f are the pericentric distance and the true anomaly, respectively. Note that we consider the prograde orbits here; the retrograde case can be analysed in a similar way.

Using formulas for the unperturbed separatrices, we follow a standard approach for estimating the Melnikov–Arnold integrals (Chirikov 1979). Inserting Eqs. (11.73) in Eqs. (11.71), we find

$$\mathcal{A} = -\mu\nu q \frac{(1-u^2)\sin\left[\kappa\left(u + \frac{u^3}{3}\right) - t_0\right]}{r_{13}^3},$$

$$\mathcal{B} = 2\mu\nu q \frac{u\cos\left[\kappa\left(u + \frac{u^3}{3}\right) - t_0\right]}{r_{13}^3}. \tag{11.75}$$

Combining Eqs. (11.69), (11.56), and (11.57) and inserting Eqs. (11.73), one finds

$$r_{13}^2 = \mu^2 + q^2(1+u^2)^2 + 2\mu q \left\{(1-u^2)\cos\left[\kappa\left(u+\frac{u^3}{3}\right) - t_0\right]\right.$$
$$\left. +2u\sin\left[\kappa\left(u+\frac{u^3}{3}\right) - t_0\right]\right\}. \qquad (11.76)$$

Substituting for r_{13} in the denominators in Eqs. (11.75) and expanding its right-hand sides in power series in μ, at $q \gg 1$ in the first order of μ one has

$$\mathcal{A} + \mathcal{C} = -\frac{3\mu}{2q^4}\frac{(1-u^2)\sin\left[\kappa\left(u+\frac{u^3}{3}\right) - t_0\right]}{(1+u^2)^5},$$
$$\mathcal{B} + \mathcal{D} = \frac{3\mu}{q^4}\frac{u\cos\left[\kappa\left(u+\frac{u^3}{3}\right) - t_0\right]}{(1+u^2)^5}. \qquad (11.77)$$

The energy increment (equal to the angular momentum increment) is given by the integral

$$\Delta E = \kappa \int_{-\infty}^{+\infty} (\mathcal{A} + \mathcal{B} + \mathcal{C} + \mathcal{D})(1+u^2)du. \qquad (11.78)$$

To evaluate it, it is suitable to beforehand define the following functions

$$I_n^0(x) = \int_{-\infty}^{+\infty} \frac{1}{(1+u^2)^n}\cos\left[x\left(u+\frac{u^3}{3}\right)\right]du,$$
$$I_n^1(x) = \int_{-\infty}^{+\infty} \frac{u}{(1+u^2)^n}\sin\left[x\left(u+\frac{u^3}{3}\right)\right]du,$$
$$I_n^2(x) = \int_{-\infty}^{+\infty} \frac{u^2}{(1+u^2)^n}\cos\left[x\left(u+\frac{u^3}{3}\right)\right]du.$$
$$(11.79)$$

These functions satisfy the recurrent relations

$$I_{n+1}^1(x) = \frac{x}{2n} I_{n-1}^0(x),$$
$$2nI_{n+1}^0(x) = (2n-1)I_n^0(x) + xI_{n-1}^1(x),$$
$$I_n^2(x) = I_{n-1}^0(x) - I_n^0(x),$$

$$\frac{\mathrm{d}I_n^0(x)}{\mathrm{d}x} = -\frac{2}{3}I_n^1(x) - \frac{1}{3}I_{n-1}^1(x),$$
$$\frac{\mathrm{d}I_n^1(x)}{\mathrm{d}x} = -\frac{2}{3}I_n^0(x) + \frac{1}{3}I_{n-1}^0(x) + \frac{1}{3}I_{n-2}^0(x) \quad (11.80)$$

(Petrosky and Broucke 1988; Shevchenko 2011b). Equation (11.78) then gives

$$\Delta E = W(q)\sin t_0, \quad (11.81)$$

where

$$W(q) = \frac{3\mu}{2^{1/2}q^{5/2}}[I_4^0(\kappa) + 2I_4^1(\kappa) - I_4^2(\kappa)] =$$
$$= \frac{3\mu}{2^{1/2}q^{5/2}}[2I_4^0(\kappa) + 2I_4^1(\kappa) - I_3^0(\kappa)], \quad (11.82)$$

and $\kappa = (2q^3)^{1/2}$.

Some of the terms in Eq. (11.82) can be expressed through modified Bessel functions and the Airy functions, because

$$I_0^0(x) = 3^{-1/2}K_{1/3}\left(\frac{2}{3}x\right) = \pi x^{-1/3}\mathrm{Ai}\left(x^{2/3}\right), \quad I_0^1(x) = 3^{-1/2}K_{2/3}\left(\frac{2}{3}x\right), \quad (11.83)$$

where

$$K_\nu(x) = \sec\left(\frac{1}{2}\nu\pi\right)\int_0^\infty \cos(x\sinh t)\cosh(\nu t)\mathrm{d}t,$$
$$\mathrm{Ai}(x) = \frac{1}{\pi}\int_0^\infty \cos\left(xt + \frac{t^3}{3}\right)\mathrm{d}t \quad (11.84)$$

by definition, see Abramowitz and Stegun (1970), Petrosky and Broucke (1988), Shevchenko (2011b).

With the help of recurrent relations (11.80), Eq. (11.82) is reduced to

$$W(q) = \frac{3\mu}{2^{1/2}q^{5/2}\kappa}\left[2I_6^0(\kappa) + 36I_6^1(\kappa) - 18I_6^2(\kappa) + 24\frac{\mathrm{d}I_6^0(\kappa)}{\mathrm{d}\kappa}\right]. \quad (11.85)$$

Asymptotic expressions for $I_6^n(x)$ ($n = 0, 1, 2$) at $x \to \infty$ were derived in Heggie (1975), Roy and Haddow (2003):

$$I_6^0(x) \simeq I_6^1(x) \simeq -I_6^2(x) \simeq \frac{\pi^{1/2}}{120}x^{5/2}\exp\left(-\frac{2}{3}x\right). \quad (11.86)$$

Finally, we arrive at

$$W(q) \simeq 2^{1/4}\pi^{1/2}\mu q^{-1/4} \exp\left(-\frac{(2q)^{3/2}}{3}\right), \tag{11.87}$$

in accord with findings of Petrosky and Broucke (1988).

If one writes down the expression (11.81) for the energy increment together with the expression for the increment of the planet's phase g (following from Kepler's third law) per one orbital revolution of the particle, then one obtains the Kepler map:

$$\begin{aligned} E_{i+1} &= E_i + W(q)\sin g_i, \\ g_{i+1} &= g_i + 2\pi|2E_{i+1}|^{-3/2}, \end{aligned} \tag{11.88}$$

where the subscript i denotes the current number of the pericentre passage, $g_i = -t_0$. The coefficient $W(q)$ is given by Eq. (11.87), if $\mu \ll 1$ and $q \gg 1$.

From the derivation of the Kepler map, it follows that the increments in energy and time per orbital revolution are taken in this map non-simultaneously: in energy, the increment is taken as accumulated between two consecutive *apocentre* passages, and in time it is taken as accumulated between two consecutive *pericentre* passages. In atomic physics, this difference was pointed out by Nauenberg (1990). A complicated "synchronized" version of the Kepler map can be constructed (Nauenberg 1990). The asynchronism can be as well removed by using a procedure of synchronization, at each step of the original map; for the classical separatrix map, the synchronization procedure is described in Sect. 1.8.

By replacing $E = Wy$, $g = x$, map (11.88) is reducible to form (8.35), where

$$\lambda = 2^{-1/2}\pi W^{-3/2}. \tag{11.89}$$

Both forms of the Kepler map, (11.88) and (8.35), depend on a single parameter, W or λ, respectively, but the advantage of (8.35) over (11.88) is that the λ parameter is an analogue of the adiabaticity parameter for the classical separatrix map. Therefore, by its value, one can judge whether chaos is adiabatic (if $\lambda \ll 1$) or not (if $\lambda \gg 1$). The term "adiabatic chaos" concerns conservation of an adiabatic invariant; at low values of λ, it is conserved on long time intervals between crossings of the separatrix; see Chirikov and Vecheslavov (2000a), Shevchenko (2008b).

At $q \gg 1$, one has $W \ll 1$ (see Eq. (11.87)); therefore, $\lambda \gg 1$. This means that chaos in the motion of particles is non-adiabatic (Shevchenko 2007a), and therefore the Kepler map can be locally approximated by the standard map with good accuracy (Shevchenko 2010, 2011b).

If $q > 1$, as in the case considered above, then the particle does not cross the planet's orbit. If $q < 1$, the planet's orbit is crossed, and ΔE as a function of g has two singularities with $|\Delta E| \to \infty$; see Zhou et al. (2000, 2002). The Kepler map can be generalized by introducing a *multi-harmonic Kepler map*, in which the energy increment is represented by a truncated Fourier series in the phase variable (Liu and Sun 1994). In the planar circular restricted three-body problem "Sun–planet–comet," Liu and Sun (1994), Zhou et al. (2000) derived and developed a multi-harmonic Kepler map describing the chaotic dynamics of comets in near-parabolic orbits perturbed by a planet, even when $q < 1$. If $q \gg 1$, the higher order harmonics in ΔE are exponentially small with q with respect to the first harmonic (Petrosky and Broucke 1988).

The Kepler map is applicable if the particle's pericentric distance q is constant. Indeed, the Kepler map is derived in the assumption that q is constant. Why q may play the role of a quasi-constant of the motion in the given problem? Based on the Jacobi constant formalism, Shevchenko (2015) argued that the condition $q \gg 1$ naturally provides the approximate constancy of q. Indeed, in the planar circular restricted three-body problem, the Tisserand relation (which is derived from the expression for the Jacobi constant) is given by

$$\frac{1}{a} + 2\left[(1 - e^2)a\right]^{1/2} \approx \text{const}, \tag{11.90}$$

where a is the particle's semimajor axis, measured in units of the perturber's semimajor axis a', e is the particle's eccentricity; see Eq. (11.33). If $a \gg a'$, and $e \sim 1$, one has

$$\frac{1}{a} + 2\left[(1 + e)q\right]^{1/2} \approx 2^{3/2} q^{1/2} \approx \text{const}. \tag{11.91}$$

Therefore, q is approximately conserved.

Finally, it is important to outline that, if the initial conditions are inside the chaotic layer, the Kepler map is formally not defined on the infinite time interval, because sooner or later the particle escapes from the chaotic layer, due to an encounter with the separatrix. However, one easily overcomes this technical difficulty by changing the sign of the energy each time the separatrix is crossed; thus, the trajectory does not leave the layer, and one may explore the statistics of the chaotic motion on time intervals that are arbitrarily long.

The Kepler map can be straightforwardly used to find analytically the width of a chaotic band in the vicinity of the perturbed parabolic orbit. Consider the planar circular restricted three-body problem. For the highly-eccentric particle's orbits, in the sequence of the $p/1$ resonances, the resonances start to overlap sooner or later with increasing p (this increase means approaching the separatrix), because the widths of the resonances do not decrease fast enough, with increasing p; for a scheme of the emerging chaotic layer, see Fig. 11.5. Petrosky (1986), based on an analysis of the Kepler map, showed that the energy width of the chaotic layer scales as the power $2/5$ of the mass parameter $\mu = m_2/(m_1 + m_2)$:

$$\Delta E_{\text{cr}} \propto \mu^{2/5}. \tag{11.92}$$

Fig. 11.5 The chaotic layer at the parabolic separatrix (Figure 3 from Petrosky 1986, with permission from Elsevier)

The particles with E within the interval $-\Delta E_{\rm cr} < E < 0$ move chaotically. Due to the diffusion in E they escape from this zone sooner or later.

Let us calculate the coefficient at the power law in Eq. (11.92). We use the Kepler map in form (8.35), where $W(q)$ is given by Eq. (11.87). By linearizing the Kepler map (11.88) near its fixed point at the border of the map's chaotic layer, one finds $y_{\rm cr} = (3\lambda/(2K_{\rm G}))^{2/5}$ (Shevchenko 2007a). Using the formulas for λ and W, we arrive at

$$\Delta E_{\rm cr} \simeq A\mu^{2/5}q^{-1/10}\exp\left(-Bq^{3/2}\right), \tag{11.93}$$

where

$$A = 2^{-1/2}3^{2/5}\pi^{3/5}K_{\rm G}^{-2/5} = 2.2061\ldots, \qquad B = 2^{5/2}/15 = 0.3771\ldots,$$

and $K_{\rm G} = 0.971635406\ldots$; see Sect. 1.5.

We consider only prograde orbits here; the analysis for the retrograde ones is completely analogous, resulting in a slightly different formula for the coefficient A and a different power of q, namely $q^{-7/10}$ instead of $q^{-1/10}$, before the exponent.

We define the critical eccentricity $e_{\rm cr}$ using the relation

$$\Delta E_{\rm cr} = -E_{\rm cr} = \frac{1}{2a_{\rm cr}} = \frac{1-e_{\rm cr}}{2q},$$

i.e.,

$$e_{\rm cr} = 1 - 2q\Delta E_{\rm cr}, \tag{11.94}$$

where $\Delta E_{\rm cr}$ is given by Eq. (11.93), and the pericentric distance q is assumed to be conserved in the given problem. The orbits with $e > e_{\rm cr}$ for a given q are chaotic.

To be able to analytically estimate Lyapunov timescales of the motion described by the Kepler map, it is useful to know its Chirikov's constant, as shown for the ordinary separatrix map in Sect. 3.1. Chirikov's constant $C_{\rm h}^{\rm gen}$ for the general separatrix map (1.45) with an arbitrary value of γ is introduced analogously, as the least upper bound for the maximum Lyapunov exponent of the motion in the separatrix chaotic layer.

By linearizing the map (1.45) in y it is straightforward to find out that the y value corresponding to the critical value of the stochasticity parameter $K = K_{\rm G} = 0.9716\ldots$ of the approximating standard map is

$$y_{\rm b} = \left(\frac{\gamma\lambda}{K_{\rm G}}\right)^{\frac{1}{\gamma+1}},$$

and the y value corresponding to $K = 4$ is

$$y_4 = \left(\frac{\gamma\lambda}{4}\right)^{\frac{1}{\gamma+1}}.$$

The layer's border is situated at $y = y_{\rm b}$, and any significant regular islands seize to exist inside the layer at $y = y_4$.

Integrating over the layer, one has

$$C_{\rm h}^{\rm gen}(\gamma) = \frac{K_{\rm G}^{\frac{1}{\gamma+1}}}{(\gamma+1)\sigma(\gamma)} \int_{K_{\rm G}}^{\infty} L(K)\mu(K)\,\frac{dK}{K^{\frac{\gamma+2}{\gamma+1}}}, \tag{11.95}$$

where

$$\sigma(\gamma) = \frac{K_{\rm G}^{\frac{1}{\gamma+1}}}{\gamma+1} \int_{K_{\rm G}}^{\infty} \mu(K)\,\frac{dK}{K^{\frac{\gamma+2}{\gamma+1}}} \tag{11.96}$$

(Shevchenko 2004a; Shevchenko 2007a). Here the $L(K)$ and $\mu(K)$ functions are, respectively, the K dependences of the maximum Lyapunov exponent of the standard map and the measure of the chaotic component in phase space of the standard map. They were computed in Shevchenko (2004a). By estimating the integral (11.95) numerically, one obtains Chirikov's constant

$$C_{\rm K} \equiv C_{\rm h}^{\rm gen}(\gamma = 3/2) = 2.21\ldots.$$

Note that the ratio

$$\frac{y_b}{y_4} \approx 4^{\frac{1}{\gamma+1}};$$

hence, the layer's "porous" (containing regular islands) component measure tends to zero with γ increasing. Since the contribution of the porous component to the integrated value of Chirikov's constant is comparatively uncertain, the integrated constant (11.95) for the map (1.45) with $\gamma = 3/2$ (generally, with $\gamma > 0$) can be estimated with a greater precision, in comparison to the case of map (1.31).

The value of $C_K \approx 2.21$ can be verified versus a direct computation of the maximum Lyapunov exponent L in the chaotic layer of the map (8.35). In Fig. 8.24, the L value is plotted as a function of λ. The limit of L at $\lambda \to \infty$ gives Chirikov's constant. The computed dependence is in accord with the estimate $C_K \approx 2.21$: the numerical estimates are all below this limit. The dependence is described by the fitting function

$$L_K(\lambda) = C_K - \frac{3}{\lambda}, \tag{11.97}$$

where $C_K = 2.15$. Therefore, $C_K \approx 2.2$.

The corresponding Lyapunov time is given by

$$T_L \approx \frac{T_{orb}}{L_K(\lambda)}, \tag{11.98}$$

where T_{orb} is the average orbital period of the tertiary (particle). Its lower bound is just

$$T_L \approx \frac{T_{orb}}{C_K}. \tag{11.99}$$

This value of T_L corresponds to the motion with $\lambda \gg 1$.

11.1.6 Hill's Criterion: The $\mu^{1/3}$ Law

In the framework of the hierarchical three-body problem, the Hill sphere can be defined as the secondary's spherical neighbourhood inside which the secondary can host satellites (tertiaries) in stable orbits.

The Hill radius (i.e., the Hill sphere radius) R_H can be heuristically estimated from a consideration of orbital timescales of the motion around the primary and the secondary; equating these two timescales, in the circular restricted three-body problem, one just gets $R_H \sim \mu^{1/3} a_{pl}$, where μ is the mass parameter and a_{pl} is the radius of the planet's orbit.

More precisely, the Hill sphere is delimited by the L_1 and L_3 Lagrange libration points. In the circular restricted three-body problem, the Hill radius is given by

$$R_H \approx \left(\frac{\mu}{3}\right)^{1/3} a_{pl} \approx 0.693\mu^{1/3} a_{pl}, \tag{11.100}$$

see, for example, Murray and Dermott (1999).

If the secondary's orbit is non-circular, then the radius of the stability zone is estimated as equal to the Hill radius calculated at the secondary's pericentre:

$$R_H \approx (\mu/3)^{1/3} a_{pl}(1 - e_{pl}), \tag{11.101}$$

where e_{pl} is the planet's orbital eccentricity. This formula renders the so-called Hill sphere at pericentre scaling (Hamilton and Burns 1992).

However note that the chaos borders, delimiting the zone of stability around the secondary, are not at all smooth surfaces, as it may seem judging from the form of these analytical approximations. In the "tertiary's semimajor axis—secondary's eccentricity" diagrams, constructed in Mudryk and Wu (2006), the borders clearly demonstrate a ragged fractal structure, conditioned by interaction and overlap of mean-motion resonances in the three-body system. Tantalizingly, the Hill criterion, although stemming from a very different theory (namely, the Jacobi integral formalism), provides the stability border that coincides with the smoothed border of the overlapped resonances (Mudryk and Wu 2006).

The Hill criterion is asymptotical, as it is valid in the limit $\mu \ll 1$. In applications with $\mu \sim 0.01$–0.5, more precise estimates can be achieved by exploiting numerical-experimental fitting formulas, presented in the next subsection.

11.1.7 The Holman–Wiegert Criteria

By performing massive numerical experiments, Holman and Wiegert (1999) obtained fitting formulas allowing one to estimate the radial sizes of major zones of stability and instability of the tertiary's motion in hierarchical three-body systems. In the framework of the planar elliptic restricted three-body problem, they showed that the radius a_{cr} of the circum-companion stability zone (around the secondary) of the initially circular prograde orbits of a massless particle is well described (providing the precision with at least two significant digits) by the fitting polynomial function

$$a_{cr}/a_b = 0.464 - 0.380\mu - 0.631e_b + 0.150e_b^2 + 0.586\mu e_b - 0.198\mu e_b^2, \tag{11.102}$$

where $\mu = m_2/(m_1 + m_2)$ is the primary binary's mass parameter, a_b and e_b are the binary's semimajor axis and eccentricity. The fitting is valid at $0.0 \le e_b \le 0.7$ and $0.1 \le \mu \le 0.9$. For an equal-mass circular-orbit binary one has $a_{cr}/a_b =$

$0.464 - 0.380 \cdot 0.5 = 0.274$. At $\mu < 0.1$ one may use formulas for the Hill radius, presented in the previous section.

For the circumbinary zone of instability, the fitting relation, obtained by Holman and Wiegert (1999), is

$$a_{\rm cr}/a_{\rm b} = 1.60 + 4.12\mu + 5.10e_{\rm b} - 5.09\mu^2 - 2.22e_{\rm b}^2 - 4.27\mu e_{\rm b} + 4.61\mu^2 e_{\rm b}^2. \tag{11.103}$$

The fitting is valid at $0.0 \leq e_{\rm b} \leq 0.8$ and $0.1 \leq \mu \leq 0.5$. For an equal-mass circular-orbit binary one has $a_{\rm cr}/a_{\rm b} = 1.60 + 4.12 \cdot 0.5 - 5.09 \cdot 0.25 = 2.39$.

At $\mu \lesssim 0.05$ the circumbinary zone of chaos does not form; a more precise estimate for this threshold is considered on theoretical grounds in Sect. 14.2.3.

For the description of the global instability borders, the Holman–Wiegert criteria utilize polynomial fits over numerical data in $a_{\rm b}$ and $e_{\rm b}$; these fits are given by smooth functions. However note that these borders are, in reality, fractal; examples of their "ragged" appearance will be given below.

11.1.8 Lyapunov and Escape Times: Linear and Quadratic Relationships

When the first results of calculations of Lyapunov exponents were obtained in celestial mechanics in 1980s, it was soon conjectured in a number of works that the Lyapunov timescale $T_{\rm L}$ of any celestial body's orbit is somehow related to its characteristic time of transport $T_{\rm r}$, i.e., to the time needed for the orbit to change its character qualitatively (e.g., the time needed for a particle to escape). In this framework, power-law dependences of the transport timescales on the Lyapunov times were numerically revealed (Soper et al. 1990; Lecar et al. 1992; Murison et al. 1994; Levison and Duncan 1993; Ferraz-Mello 1997; Shevchenko 1998a; Tsiganis et al. 2005). These power-law relationships were interpreted in Shevchenko (1998a, 1999b) as a phenomenon of critical dynamics; this phenomenon is immanent to the motion near chaos borders in the divided phase space of Hamiltonian systems. Generally speaking, the observed relationships mostly arise due to the *second kind Hamiltonian intermittency*. Conversely, the *first kind Hamiltonian intermittency* leads to the quasilinear $T_{\rm L}$–$T_{\rm r}$ relationship (Shevchenko 2010).

As follows from the second equation of the Kepler map (8.35), the length of a Poincaré recurrence in real (original) time units is

$$T_{\rm r}^{\rm (ru)} = \sum_{i=1}^{n} \Delta x_i = \lambda \sum_{i=1}^{n} |y_i|^{-3/2}, \tag{11.104}$$

where $n = T_r$ is the recurrence duration in the map time units, i.e., iterations; the time increment per recurrence is $\Delta x_i = x_i - x_{i-1}$, and the recurrence is started with iteration $i = 1$.

The finite-time (corresponding to the time interval of any performed numerical integration) maximum Lyapunov exponent of the original dynamical system referred to a recurrence in original time units is calculated as the map's finite-time maximum Lyapunov exponent L divided by the averaged (on the recurrence) length of the map iterations in real time units:

$$\langle P_{\rm map}^{\rm (ru)} \rangle = T_{\rm r}^{\rm (ru)} / T_{\rm r}.$$

Hereafter, $\langle P_{\rm map}^{\rm (ru)} \rangle$ is denoted by q:

$$q = \frac{T_{\rm r}^{\rm (ru)}}{T_{\rm r}} = \frac{1}{n} \sum_{i=1}^{n} \Delta x_i = \frac{\lambda}{n} \sum_{i=1}^{n} |y_i|^{-3/2}. \tag{11.105}$$

The maximum Lyapunov exponent referred to real time units is

$$L^{\rm (ru)} = \frac{L}{q}, \tag{11.106}$$

and for the Lyapunov time one has

$$T_{\rm L}^{\rm (ru)} = q T_{\rm L}. \tag{11.107}$$

On the other hand,

$$T_{\rm r}^{\rm (ru)} = q T_{\rm r}. \tag{11.108}$$

The quantity $q = T_{\rm r}^{\rm (ru)} / T_{\rm r}$ is the ratio of two random variables. If the y value hits close to the separatrix, then, according to Eq. (11.104), there is a jump in $T_{\rm r}^{\rm (ru)}$, but there is no jump in $T_{\rm r}$. If the magnitude of the jumps is large enough, then the relationship in real time units is spread, due to the jumps, in the direction $T_{\rm L} = T_{\rm r}$ in the $(T_{\rm r}, T_{\rm L})$ plane. Therefore, when $T_{\rm L}$ are $T_{\rm r}$ are expressed in real time units and Lévy flights of the first kind (LF1s) dominate over Lévy flights of the second kind (LF2s) (see Sect. 8.3.2), then the $T_{\rm L}$–$T_{\rm r}$ relationship becomes quasilinear.

This inference is valid when LF1s dominate. For the classical separatrix map (1.44) and general maps (1.45) with $\gamma < \gamma_{\rm crit}$, where the singularity is weaker, LF2s dominate if λ is large enough, and the generic relationship is not spread in the $T_{\rm L} = T_{\rm r}$ direction. This is the cause why the close-to-quadratic relationships (Soper et al. 1990; Lecar et al. 1992; Murison et al. 1994; Levison and Duncan 1993; Ferraz-Mello 1997; Shevchenko 1998a; Tsiganis et al. 2005), and not the quasilinear one, were observed to be present in the dynamics of minor Solar system

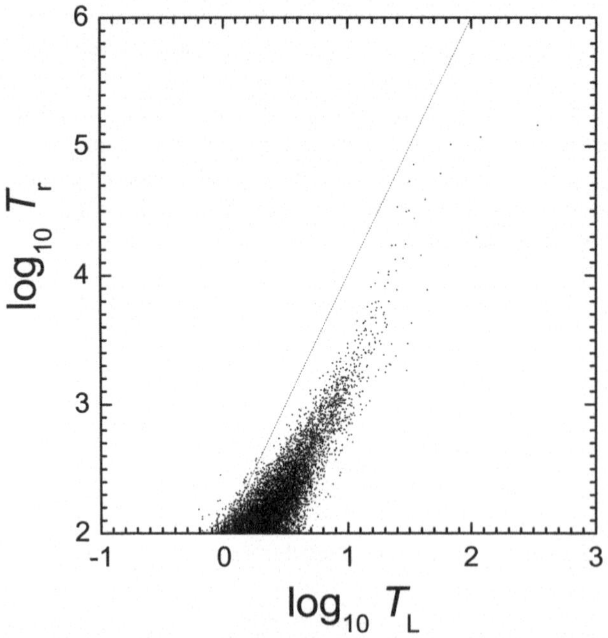

Fig. 11.6 A statistical "$\log_{10} T_L - \log_{10} T_r$" relationship, where T_L and T_r are expressed in the Kepler map's time units; $\lambda = 5$. Straight dotted line: the quadratic dependence. (Reprinted Figure 7 with permission from Shevchenko 2010. © 2010 by the American Physical Society)

bodies near resonances, where the classical separatrix map, but not the Kepler map, describes the relevant interacting non-linear resonances.

In Figs. 11.6 and 11.7, the T_L–T_r relationships are presented for the Kepler map with $\lambda = 5$, separately in the map's time units (iterations) and in the time units of the original system. For the both plots, the finite-time maximum Lyapunov exponent was computed for a recurrence; the total number of iterations $n_{it} = 10^7$ in the both cases.

In Fig. 11.6, one may see that, judging by the general slope in the log-log scale, the dependence is far from being linear. Its slope is much steeper: the power law index ~1.5–2, as predicted. The same dependence is shown in Fig. 11.7, but in real time units. Here the relationship looks obviously linear. One may even see how the "spreading" mechanism operates. Indeed, the cloud of points has a "V" shape, with the left wing being much shorter than the right one. This wing is nothing but a remnant (produced by spreading by the LF1s) of the generic close-to-quadratic relationship. Therefore, the general "composite" appearance of the T_L–T_r relationship in this Figure mimics the general character of the T_L–T_r relationships revealed in computations of the disruption process in three-body systems, such as presented in Figures 2 and 3 in Mikkola and Tanikawa (2007) and in Figures 3 and 7 in Urminsky and Heggie (2008).

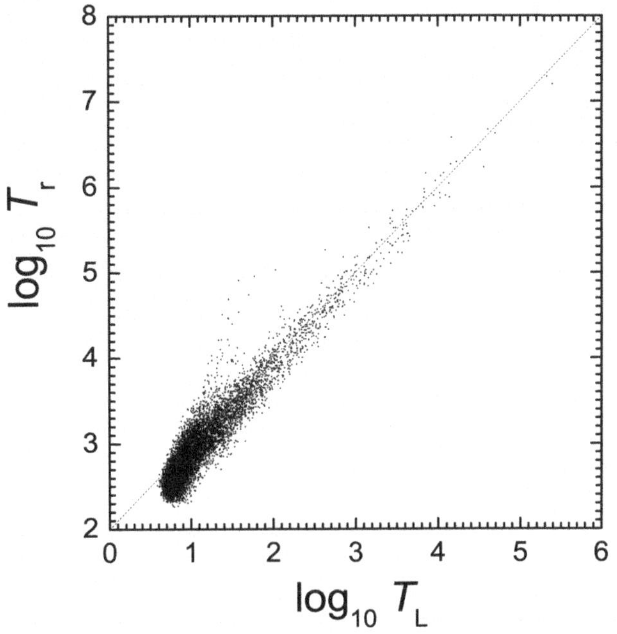

Fig. 11.7 The same as Fig. 11.6, but the recurrence times are expressed in real time units. Straight dotted line: the linear dependence. (Reprinted Figure 8 with permission from Shevchenko 2010. © 2010 by the American Physical Society)

The plot in Fig. 11.6 looks rather irregular, in comparison with Fig. 11.7. The nature of this irregularity can be clarified by means of construction of a *spectrum of winding numbers* (Shevchenko 1996, 2010). The winding number Q can be formally defined, for a recurrence, as $Q = \Delta x/n$, where n is the recurrence length measured in the map's iterations, $n = T_r$, and Δx is the total variation in x between crossings of the separatrix. The spectrum of winding numbers graphically demonstrates which of the resonant chains of islands produce the longest sticking events; see (Shevchenko 2010, Figure 9). The resonances produce obvious peaks and, therefore, are easily identified. The irregular structure of the cloud of points in Fig. 11.6 is produced by an overlay of individual relationships corresponding to several sticky island chains, corresponding to the peaks in the spectrum.

For the slope indices of the asymptotic power-law distributions of the Poincaré recurrence lengths T_r, already considered above in Sects. 4.3 and 8.3.2, there exists a sharp difference between the two kinds of Hamiltonian intermittency. This theoretical difference explains the emergence of the observed difference in the power-law indices of the distribution laws reported for chaotic transport times for Solar system minor bodies. Dones et al. (1996) reported on algebraic tails of the integral distributions, $\propto T_r^{-\alpha}$, with the slope index $\alpha = 0.8 \pm 0.2$, whereas Shevchenko and Scholl (1996, 1997) reported on the tails with $\alpha \approx 1.5$. In Dones et al. (1996), the escape times in the highly-eccentric chaotic cometary dynamics in the Solar system were studied (with the perturbations from the four giant planets taken into account), whereas in Shevchenko and Scholl (1996, 1997) the subject was the low-eccentricity intervals between the eccentricity jumps in the chaotic asteroidal dynamics in the Sun–Jupiter–asteroid restricted three-body problem. Judging by the values of the power-law index, the former statistics correspond to the Hamiltonian intermittency of the first kind, whereas the latter one to that of the second kind. The statistics are LF1-dominated and LF2-dominated, respectively, and the predicted α indices are equal to 2/3 and $\approx 3/2$, respectively. The evident inverse symmetry of the indices is a property of the gravitational dynamics, described by the Kepler map. For the general separatrix map (1.46) with $\gamma > 0$, the predicted α indices are equal to $1/\gamma$ and $\approx 3/2$, respectively; see Shevchenko (2010) for details.

Concerning the T_{L}–T_{r} relationship, the power-law index β in formula (4.15) changes from ≈ 2 to ≈ 1, in comparison with the numerical results of Soper et al. (1990), Lecar et al. (1992), Murison et al. (1994), Levison and Duncan (1993), Ferraz-Mello (1997), Shevchenko (1998a), Tsiganis et al. (2005). This change emerges because the singularity at crossing the separatrix is much stronger in the considered problem, which is described by the Kepler map instead of the classical separatrix map. Therefore, the first kind Hamiltonian intermittency dominates over the second kind Hamiltonian intermittency, and this domination defines the T_{L}–T_{r} relationship and the transport time distribution.

Generally, the typical way of disruption of a hierarchical three-body system, as described by the Kepler map, appears as a "Lévy unfolding" of the system in both time and space: at the edge of the system's disruption, the escaping body exhibits Lévy flights in its orbital period and semimajor axis, and in the course of this random process the orbital period and semimajor axis become arbitrarily large until the separatrix separating the bound and unbound states of the motion is crossed and the body escapes (Shevchenko 2010).

The considered statistics appear to be also valid in a more general "hierarchical resonant scattering" (Heggie 1975; Hut 1993) setting for a three-body interaction, in which the masses of three "stars" are arbitrary. Hut (1993) derived an heuristic $T_{\mathrm{r}}^{-2/3}$ law for the time decay of the system's survival probability, and showed it to describe well the tails of numerically found distributions. It is important to mention that the heavy-tailed character of the distribution is in accord with an early finding by Agekian et al. (1983) that the mean lifetime of a general isolated three-body system is infinite.

As we have seen above in Sect. 8.3.2, two basic conditions are necessary to be satisfied for the $T_{\mathrm{r}}^{-2/3}$ law to be valid; namely, (1) the distribution of ejection energies should be smooth in the neighbourhood of the energy threshold $E = 0$, and (2) the asymptotic decay of the survival probability should be the same as the tail of the distribution of the escaping body orbital periods. These two assumptions look rather plausible even in the general three-body problem. Therefore, the $T_{\mathrm{r}}^{-2/3}$ decay might be ubiquitous.

These theoretical asymptotics were indeed confirmed in numerical simulations in the equal-mass three-body problem (Orlov et al. 2010), as well as in the restricted three-body problem (Shevchenko 2010; Orlov et al. 2010). In Orlov et al. (2010), the statistics of the decay process in the equal-mass three-body problem with randomized initial conditions were investigated in extensive numerical experiments. The lifetime distributions obtained in Orlov et al. (2010) turned out to be heavy-tailed, i.e., algebraic. The calculated power-law index α for the integral distribution was found to be within a narrow range, from ≈ 0.4 to ≈ 0.7, depending on the virial coefficient; see Orlov et al. (2010). The theoretical value $\alpha = 2/3$ is within this narrow range.

Results of Orlov et al. (2010) were confirmed in Bogomolov et al. (2011), who studied triple systems with various mass ratios. Recall that for the first kind intermittency, the $T_r^{-2/3}$ law is for the integral distribution, and the $T_r^{-5/3}$ law is for the differential one. For the second kind intermittency, the $T_r^{-3/2}$ law is for the integral distribution, and the $T_r^{-5/2}$ law is for the differential one. The values of the power-law indices found in Bogomolov et al. (2011) for the decay time distribution tails fit in the range between the two theoretical values for the two kinds of Hamiltonian intermittency. These two values are 5/3 (1st kind) and 5/2 (2nd kind) for the differential distributions, and, respectively, 2/3 and 3/2 for the integral ones. The data on $\gamma \equiv \alpha + 1$ in Bogomolov et al. (2011) are generally much closer to 1.7 (characteristic to the 1st kind Hamiltonian intermittency), than to 2.5 (characteristic to the 2nd kind Hamiltonian intermittency); see Figures 2 and 3 and Table 1 in Bogomolov et al. (2011).

The proximity of the theoretical quasilinear T_L–T_r relationship to the numerical results by Mikkola and Tanikawa (2007) (obtained for an equal-mass three-body system) can be also naturally interpreted: the system just before disruption is hierarchical, with the outer body exhibiting final Lévy flights.

Concluding, there is no wonder that a simple one-parameter two-dimensional map, such as the Kepler map, is able to describe the essential dynamics of disruption of a system with several degrees of freedom (Shevchenko 2010). The matter is that we consider a specific stage of a three-body system evolution in a model subject to serious limitations: the orbit of the escaping body is assumed to be highly eccentric and its pericentre distance much greater than the size of the main binary. Generally speaking, the introduction of the Kepler map as a dynamical model is similar to the introduction of the separatrix map to describe a near-separatrix motion in general Hamiltonian systems. An important difference, however, is that the ordinary separatrix map has two parameters, and cannot be rendered, in contrast to the Kepler map case, in a one-parameter form.

11.2 Packing and Ranging Criteria

The heavens themselves, the planets, and this centre,
Observe degree, priority, and place

William Shakespeare,
Troilus and Cressida, I, iii (1602)

The *packing criteria* for the stability and structural properties of planetary systems are based on a simple idea that planetary orbits cannot be "packed" (placed with plausible radial intervals) too densely, because mutual perturbations may disrupt the system if the separations are not large enough. The critical separations can be derived either on a basis of the Hill criterion, considered above in Sect. 11.1.6, or employing Wisdom's notion of the close-to-coorbital band of instability, discussed above in Sect. 11.1.2.

11.2.1 Historical Background: Kepler's Goblet

The idea of packing was first ever introduced solely on the basis of geometrical harmony, as understood and interpreted by Johannes Kepler in 1593. *Kepler's goblet* represents a sequence of principal polyhedra inscribed into spherical surfaces with the radii approximately equal to the orbital radii of five planets known in Kepler's times; see Chap. 6 and Fig. 6.2. Although inadequate from the modern science viewpoint, Kepler's goblet was the first ever and therefore important landmark in the long-term and not yet finalized quest for understanding the architecture of our Solar system.

11.2.2 Dynamical Completeness

The dynamical completeness and ranging criteria stem from a common idea of hierarchy, namely, the idea that a hierarchical order should be initially set for any system to be stable, whatever the system may be.

In fact, the idea of hierarchy lies in the basis of the heuristic Titius–Bode "law," which played a major historical heuristic role in former studies of the Solar system architecture, long before exoplanets were discovered. However, nowadays this heuristic law is regarded as statistically insignificant, see arguments and discussion in Murray and Dermott (1999).

On the other hand, indeed, planetary orbits in any planetary system cannot be packed too densely, because, generally speaking, mutual perturbations may disrupt the system if the separations are not large enough. If two neighbouring planets strongly differ in masses, their critical separation can be approximately derived employing Wisdom's notion of the close-to-coorbital band of instability, Eq. (11.5), which we discussed above in Sect. 11.1.2.

The critical separations can be as well derived on the basis of the Hill criterion, considered in Sect. 11.1.6; this approach can be called the "mutual" Hill's criterion. By measuring the radial space between each two neighbouring planets in a planetary system in units called the "mutual Hill sphere" (see Pu and Wu 2015, Equation (1)), one may heuristically estimate the dynamical completeness of the system, i.e., whether the system possesses any free radial space for more planets, not yet observationally discovered.

On the other hand, resonant planetary systems can be hyper-closely packed, and it is not obligatory for them to obey simple packing or ranging criteria. For example, a large planet may possess "Trojan" smaller planets in 1/1 orbital resonance, obviously violating any literally applied "mutual" Hill's criterion.

In realistic cosmogonical situations, planets migrating in a primordial disc with a central circumstellar clearing eventually form and maintain a chain of mean motion resonances "anchored" at the disc inner edge. The cause of this phenomenon is clear: the innermost planet that is first to reach the edge is stalled there, because there is no more material to cause the migration, whereas all other (outer) planets continue to migrate inward and are eventually caught in the most pronounced (first-order) resonances with the inner bodies (Snellgrove et al. 2001; Lee and Peale 2002). In this way, a chain of mean motion resonances is "invariably organized" (Raymond et al. 2018). This process may provide formation of the known closely packed resonant exoplanet systems, such as *Kepler*-60 (Goździewski et al. 2016) or *Kepler*-223 (Lissauer et al. 2011a); see Sect. 10.4 for more details.

11.2.3 The Moriwaki–Nakagawa Criterion

Moriwaki and Nakagawa (2004) derived an analytical criterion for the possibility of effective planetesimal accretion in a circumbinary disc. If relative velocities between planetesimals exceed the escape velocity from the surface of a typical planetesimal, then it is impossible for planetesimals to accumulate and form planets. It follows then that the forced eccentricities of planetesimals must be small enough.

As shown on these grounds by Moriwaki and Nakagawa (2004), the inner boundary radius of the planetesimal accretion zone is given by

$$a_{\rm acc} \simeq \left[\frac{5}{2}(1-2\mu)a_{\rm b}e_{\rm b}\right]^{2/3}\left(\frac{3M^3}{32\pi m^2\rho}\right)^{1/9},$$

where $\mu = m_2/(m_1+m_2)$ is the binary's mass parameter; $M = m_1 + m_2$; $a_{\rm b}$ and $e_{\rm b}$ are the binary's semimajor axis and eccentricity; and the typical planetesimal has mass m and internal density ρ.

Chapter 12
Effects of Chaotic Clearing in Planetary Systems

In this Chapter, major types of the chaotic clearing effects in planetary systems are considered. Generally, the clearing effects consist in forming of orbital zones free from low-mass material (e.g., planetesimals, asteroids, particles, fragments), due to the dynamical chaotization in presence of perturbations. Among them, the following effects are known and well-studied:

1. Close-to-coorbital clearing—the forming of a ring-like clearance in the planetesimal disc (around a single or a binary star), due to accumulation and overlap of the "particle–planet" orbital (mean-motion) resonances of the first order in the radial neighbourhood of a planetary orbit (Wisdom 1980; Duncan et al. 1989; Morrison and Malhotra 2015; Demidova and Shevchenko 2016).
2. Circumbinary clearing—the forming of a total circular clearance in the planetesimal disc around a binary star, due to overlap of the "particle–binary" integer orbital resonances accumulating to the separatrix corresponding to the parabolic motion (Shevchenko 2015).
3. Circumnuclear clearing—the forming of a total circular clearance in the cloud of particles and fragments around a rotating irregularly-shaped body (e.g., a contact binary, such as cometary nuclei, many objects in the Kuiper belt, and asteroids), due to overlap of the "particle–body" integer orbital resonances, accumulating to the separatrix corresponding to the parabolic motion (Lages et al. 2018b).
4. Inner disintegration—the long-term chaotic decay of the inner "low-mass" region of a planetary system, due to overlap of secular resonances (Lithwick and Wu 2011; Batygin et al. 2015).

In this Chapter, analytical approaches that serve to characterize these effects, in particular, serve to estimate the cleared zone sizes and the clearing timescales, are described and discussed.

I. I. Shevchenko, *Dynamical Chaos in Planetary Systems*, Astrophysics and Space Science Library 463, https://doi.org/10.1007/978-3-030-52144-8_12

12.1 Close-to-Coorbital Clearing

The differential distributions (histograms) of known exoplanet systems in ratios of planetary periods, presented in Fig. 10.4, provide a graphical insight in the close-to-coorbital clearing effect. The histograms make evident that the planets avoid vicinities of the 1/1 resonance, i.e., they elude being close-to-coorbital. This avoidance is similar to what is observed in the asteroidal statistics in the Solar system, considered above in Sect. 8.2. However, in contrast to the asteroids, there are no observed "Trojan planets," i.e., any bodies residing in the exact 1/1 resonance. For the both classes, asteroids and exoplanets, the absence of objects (apart from the horseshoe and tadpole librators, which are in the exact 1/1 resonance with the perturber)[1] in orbits close to the perturber manifests one and the same dynamical effect, namely, that of the *close-to-coorbital clearing*. This demonstrates the robustness of the planet definition approved by IAU General Assembly (2006); see Sect. 10.2. This definition, in its point (c), is based on rigorous dynamical grounds, both theoretical and numerical-experimental; it states that any object that is called a planet should clear its orbital neighbourhood from low-mass material, e.g., planetesimals.

The formation of close-to-coorbital clearings was demonstrated in many massive numerical simulations on the dynamics of planetary systems with planetesimal discs. The motion of planets inside a planetesimal disc may produce patterns that are stable on secular time scales.

Prominent types of such patterns are revealed in a work by Demidova and Shevchenko (2016). In this research, the discs are modelled for two mass ratios of stars in the central binary: a circular binary with masses $m_1 = M_\odot$ and $m_2 = 0.2M_\odot$ (model 1) and a circular binary with masses $m_1 = m_2 = M_\odot$ (model 2). Also a model with a single central star with mass $m = 1.2M_\odot$ (model 3) is considered. The orbital periods of the binaries in models 1 and 2 are both set to $P_b = 0.2$ yr. The planet of Jovian mass is put initially in a circular orbit around the system's barycentre. Its orbital radius corresponds to mean motion resonances with the binary. The ratios of the orbital periods of the planet and the binary are set equal to 5/1, 11/2, 6/1, 13/2, 7/1, 15/2, and 8/1. These locations are chosen because actual *Kepler* circumbinary planets are mostly located in resonance cells at the outer border of the chaotic region around the central binary (see Popova and Shevchenko 2013, 2016a; in particular, Table 3 in Popova and Shevchenko 2016a). In model 3, in which the host star is single, the planet's orbit is started at the same grid of radial distances as in model 1. In each model, the disc consists of 20,000 massless (passively gravitating) planetesimals initially distributed from 0.3 to 5.3 AU in radial distance r from the barycentre in such a way that the surface density decreases with radial distance as $\propto r^{-1}$.

[1]The horseshoe and tadpole types of orbits are considered and discussed in detail in Murray and Dermott (1999).

The disc's surface density distribution along the radius from the barycentre is found by subdividing the planetesimal disc in a set of annular bands with a radial step 0.02 AU. The local surface density is thus found as a function of the radial distance by calculating the number of particles in each band and dividing this number by the band area.

The simulations make evident emerging ring-shaped patterns coorbital with the planet; see Fig. 12.1, where the case of the 8/1 planet resonance with the central binary is illustrated. The patterns are most pronounced in models 1 and 2. For a single host star (as in model 3), similar coorbital patterns were revealed by Ozernoy et al. (2000), Quillen and Thorndike (2002), and Küchner and Holman (2003). The patterns are formed by planetesimals in tadpole and horseshoe orbits.

The radial half-width $\Delta a_{\rm cr}$ of the Wisdom gap is given by Eq. (11.5). On the other hand, the radial half-width $\Delta a_{\rm H}$ of the coorbital (with the planet) band of the stable horseshoe and tadpole orbits is approximately equal to the Hill sphere's radius $R_{\rm H}$, as illustrated in Figure 3.28 in Murray and Dermott (1999). In the planar circular restricted three-body problem the Hill radius is given by Eq. (11.100). From Eqs. (11.5) and (11.100) one has

$$\frac{\Delta a_{\rm cr}}{\Delta a_{\rm H}} \approx 2.26\mu^{-1/21}. \tag{12.1}$$

From this formula it is clear that the ratio $\Delta a_{\rm cr}/\Delta a_{\rm H}$ is virtually insensitive to the mass parameter μ variations, in a broad range of μ.

The central one third part (in radial extent) of the Wisdom gap may thus contain stable horseshoe and tadpole material. From the theoretical viewpoint, therefore, the emergence of a ring-like pattern, surrounding the orbit of a planet embedded in a planetesimal disc, is natural. The pattern consists of at least three lanes: the populated central one and two components of the broader Wisdom gap. They are all illustrated in Fig. 12.1.

Modern astronomical observational techniques allow one to identify clearance patterns in actual systems with discs. The ALMA telescope provided images of

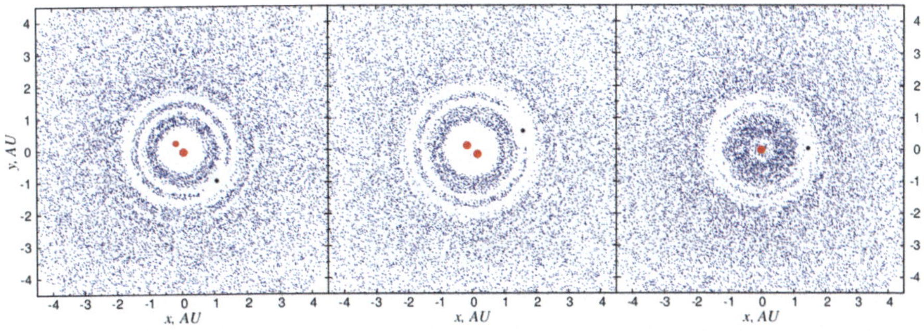

Fig. 12.1 Time-evolved distributions of planetesimals in models 1, 2, and 3 (from left to right) (Figure 1 from Demidova and Shevchenko (2016), by permission from © Oxford University Press)

the disc of HL Tau, where multiple clearings of resonant nature are obvious; see Fig. 12.2. The numerical-experimental resonant patterns (Fig. 12.1) and the observed structures (Fig. 12.2) look rather similar.

Various aspects of the theory of Wisdom's close-to-coorbital chaotic gap, along with results of massive numerical experiments on the close-to-coorbital chaotic dynamics, were considered in Wisdom (1980), Duncan et al. (1989), Murray and Holman (1997), Malhotra (1998, 2012), Murray and Dermott (1999), Quillen and Faber (2006), Mustill and Wyatt (2012), Deck et al. (2013), Ramos et al. (2015), Petit et al. (2017), and Morrison and Malhotra (2015). The phenomenon of the close-to-coorbital chaotic clearing was revealed: planets open coorbital gaps, free from low-mass material (Duncan et al. 1989; Murray and Holman 1997; Murray and Dermott 1999; Morbidelli 2002; Morrison and Malhotra 2015).

In the planetary chaotic zone, the chaotic diffusion in the particle's eccentricity and semimajor axis eventually leads the particle to escape, or to collide with the planet (Duncan et al. 1989; Murray and Holman 1997; Murray and Dermott 1999; Morbidelli 2002; Morrison and Malhotra 2015). In such a way the particle-free gap coorbital with the planet is formed. This is a manifestation of the physical reality: in such a way, the perturber cleans up the radial neighbourhood of its orbit from any low-mass material.

Characteristic timescales of opening the close-to-coorbital gap, as a function of the mass parameter μ, were numerically estimated in simulations by Morrison and Malhotra (2015). The simulations were performed at three fixed values of the ratio $R_{\rm pl}/R_{\rm H}$, where $R_{\rm pl}$ is the planet physical radius and $R_{\rm H}$ is its Hill radius. The test particles were regarded as removed if they either reached large values of their orbital semimajor axes (more than two times the planet's orbit radius) or collided with the finite-size planet. The resulting μ dependences of the removal timescale $T_{\rm r}$ are presented in Fig. 12.3.

To describe the μ–$T_{\rm r}$ relationships observed in the numerical experiments, Morrison and Malhotra (2015) adopted an empirical fitting composite function in

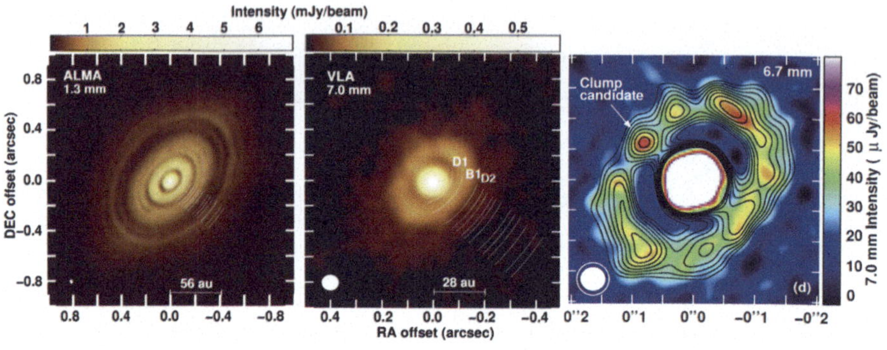

Fig. 12.2 The disc of HL Tau (Figure 1 from Carrasco-González et al. (2016), by permission of IOP Publishing/AAS)

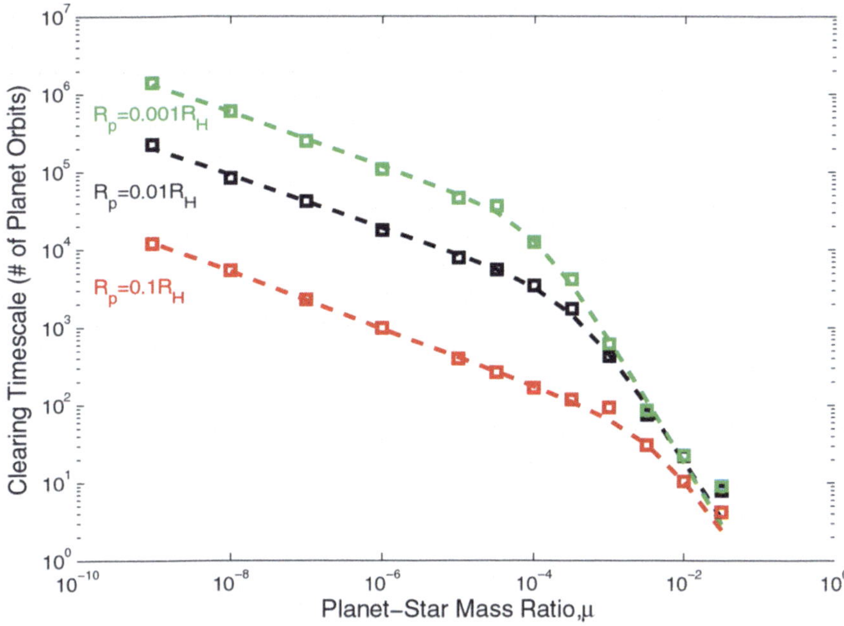

Fig. 12.3 The μ dependences of the clearing (removal) timescale T_r, at three fixed values of R_{pl}/R_H. The timescale T_r is in units of the planet's orbit period. Dashed: the best-fit functions (12.2) (Figure 7 from Morrison and Malhotra (2015), by permission of IOP Publishing/AAS)

the form of two power laws conjugated at some transition μ value (designated as μ_b); namely,

$$T_r(\mu) = 2T_b \left[(\mu/\mu_b)^{-\alpha_1} + (\mu/\mu_b)^{-\alpha_2}\right]^{-1}, \tag{12.2}$$

where T_b is the T_r transition value: if $\mu = \mu_b$, then $T_r = T_b$. Asymptotic dependences at $\mu \ll \mu_b$ and $\mu \gg \mu_b$ are represented by the power laws $T_r(\mu) = 2T_b(\mu/\mu_b)^{\alpha_1}$ and $T_r(\mu) = 2T_b(\mu/\mu_b)^{\alpha_2}$ (where $\alpha_1 < 0, \alpha_2 < 0$), respectively. The power-law index α_1 turned out to be in the range from -0.37 to -0.34, and α_2 in the range from -1.59 to -1.48.

The power-law index approximately equal to $-1/3$ is expected in the particle-planet collisional regime, and it naturally takes place at smaller μ values. At larger μ values, the scattering regime starts to statistically dominate. When the particle's eccentricity is inflated to high values, the chaotic diffusion in the semimajor axis can be described by the Kepler map (8.35), and a semi-analytical Kepler-map result for the diffusion timescale can be used:

$$T_r \propto \mu^{-2} \tag{12.3}$$

(Shevchenko 2020). The scaling (12.3) seems to be in general accord with the data in Fig. 12.3 at $\mu > 10^{-4}$–10^{-3}; indeed, at the smallest planet size adopted in the simulations ($R_{\rm pl}/R_{\rm H} = 1/1000$), Morrison and Malhotra (2015) found that $\alpha_2 = -1.59 \pm 0.18$. This is not far from the theoretical inverse-quadratic slope.

12.2 Circumbinary Clearing

More than a half of the main-sequence stars in our Galaxy belong to multiple, mostly binary, star systems (Duquennoy and Mayor 1991; Mathieu et al. 2000). To date, planets have been discovered to be present in more than a hundred multiple, mostly binary, stars. Most of the planets discovered in binary systems are the so-called *S-type* planets (*satellite-type*, i.e., orbiting one of the components of a stellar binary), and others are *P-type* planets (*planet-type*, i.e., orbiting around the both components of a binary); see Fig. 12.4. The S-type planets are also called inner or *circumstellar* planets, and the P-type planets are called outer or *circumbinary* planets.

Theoretical studies of the stability of hypothetical planetary systems in binary stellar systems had been initiated already in sixties of the twentieth century (Huang 1960), long in advance of the first discoveries of planets in binary star systems. The theory was thoroughly developed in the eighties, in application to some binary stars in the Solar neighbourhood (Benest 1988a,b, 1989).

The problem of stability of planetary orbits in binary and multiple stellar systems is important first of all due to its relevance to the problem of potential habitability of planets in such systems. Indeed, at a first glance, it may seem that multiple stars

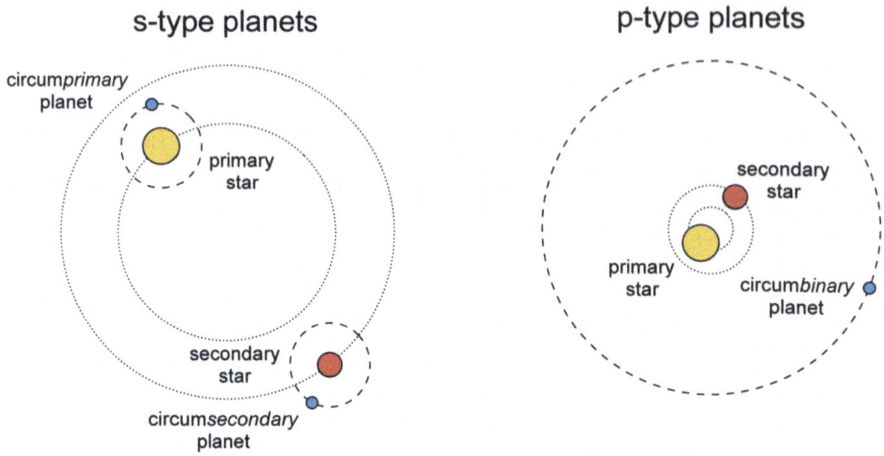

Fig. 12.4 Orbital schemes of planetary systems of binary stars: S-type systems (left panel) and P-type systems (right panel) (Figure 2 from Martin (2018), by permission from Springer Nature, © 2018)

are less suitable (in comparison with single stars) for hosting habitable planets, because their dynamical environments seem to be generally unstable. Combined with some observational data that Solar-like stars are formed mostly in pairs, this argument, among others, even served to validate the concept of practical uniqueness of our technological civilization in the Galaxy (Shklovsky 1979). Nowadays, based on modern theoretical and observational data, one may argue quite conversely that circumbinary planetary systems might be even much more suitable for life than systems of single stars; for details see Shevchenko (2017b) and references therein.

Certain conditions must be satisfied for a planet to be potentially habitable. First of all, the insolation level of its surface should be appropriate, long-term stable, and does not vary too much in amplitude (Huang 1960). Observational data on sizes of orbits of actual circumbinary planets of main-sequence binaries, combined with numerical-experimental estimates of sizes of the chaotic zones around the host binaries (see Sect. 11.1.7), show that these actual planets are orbiting close to the chaotic zone outer boundary (Doyle et al. 2011; Welsh et al. 2012, 2014; Popova and Shevchenko 2013). What is more, their orbits are either inside or quite close to the potential habitability zone (Welsh et al. 2014).

For a circumbinary planet, the stability zone starts at some large enough distance from the host binary's barycenter; see Sect. 11.1.7. The habitability zone, where the insolation level allows water on planetary surfaces to be in liquid state, is not at all obliged to overlap with the circumbinary stability zone. However, in the known circumbinary planetary systems they mostly overlap (Welsh et al. 2014). Apart from the binaries with known planets, a number of binaries in the Solar neighbourhood are identified (e.g., EZ Aqr) that have the circumbinary zones of stability and potential habitability overlapping (Popova and Shevchenko 2016a; Shevchenko et al. 2019).

It is curious that notwithstanding the former scientific doubts on the suitability of binary stars for hosting habitable worlds, circumbinary habitable planets had long ago emerged in fiction (most notably in *Solaris*, written by Stanisław Lem in 1959) and also in cinema and painting. Due to the popularity of the *Star Wars* saga, the newly-discovered circumbinary planets of Solar-like stars, such as their prototype *Kepler*-16b, discovered in 2011, are nicknamed *tatooines*; see Paardekooper et al. (2012).

A relevant painting, *The world of two Suns* by Andrey Sokolov, is reproduced in Fig. 12.5. Note that the depicted world is obviously circumbinary. The painting exists also in a version named *The world of three Suns*, where a circumtriple world is depicted. In the latter case, the stability of the world (not to mention the stability of the non-hierarchical stellar triple) is indeed questionable.

Generally, planetary formation scenarios and observed planetary dynamics (often at the "brink of stability") in binary star systems present a number of theoretical challenges, especially concerning circumbinary systems (Meschiari 2012, 2014; Paardekooper et al. 2012).

Until the launch of the *Kepler* space observatory mission, a few circumbinary planetary systems had been known to exist (belonging to the binary stars HW Vir, NN Ser, UZ For, DP Leo, FS Aur, SZ Her, among others). However, none of them

Fig. 12.5 Andrey Sokolov's picture *The world of two Suns*, on a Cuban postal stamp issued in 1974

belonged to a main-sequence star. Due to the *Kepler* mission, several circumbinary planetary systems of main-sequence stars were first ever discovered: *Kepler*-16, 34, 35, 38, and 47 (Doyle et al. 2011; Welsh et al. 2012; Orosz et al. 2012a,b). Among them, the *Kepler*-47 system is multiplanet, as it has at least two discovered planets.

The planets in the *Kepler* circumbinary systems are all coplanar with the parent binaries; but this does not say anything on the real inclination statistics, because the planets were discovered using the transit method, implying the practical coplanarity of the orbits of the eclipsing stellar binary and the transiting planet. *Kepler*-16 represents a paradigmatic example of circumbinary systems, sketched in Fig. 12.4 (right panel). The orbital parameters of the *Kepler*-16b planet are exhaustively retrieved by the TTV analysis; i.e., we know its current dynamics almost completely.

A characteristic feature of almost all known circumbinary systems of main-sequence stars is that they possess planets orbiting on the brink of stability, within 15% of the border of the chaotic zone around the central binary. The existence of the central zone of instability around a gravitating binary was earlier revealed by Holman and Wiegert (1999) in massive numerical experiments; for the zone radial size, see Eq. (11.103). From this formula one may easily estimate, setting $m_1 = m_2$ and $e_b = 0$, that the radius of the chaotic zone around a twin circular binary is about 2.4 times greater than the binary size.

A MEGNO chart of global dynamics of the *Kepler*-413 planetary system is shown in Fig. 12.6. In this diagram, constructed by Kostov et al. (2014) in the (a, e) (semimajor axis – eccentricity) space of initial conditions, dark (purple) regions correspond to stable dynamics, whereas light (yellow) regions correspond to chaos.

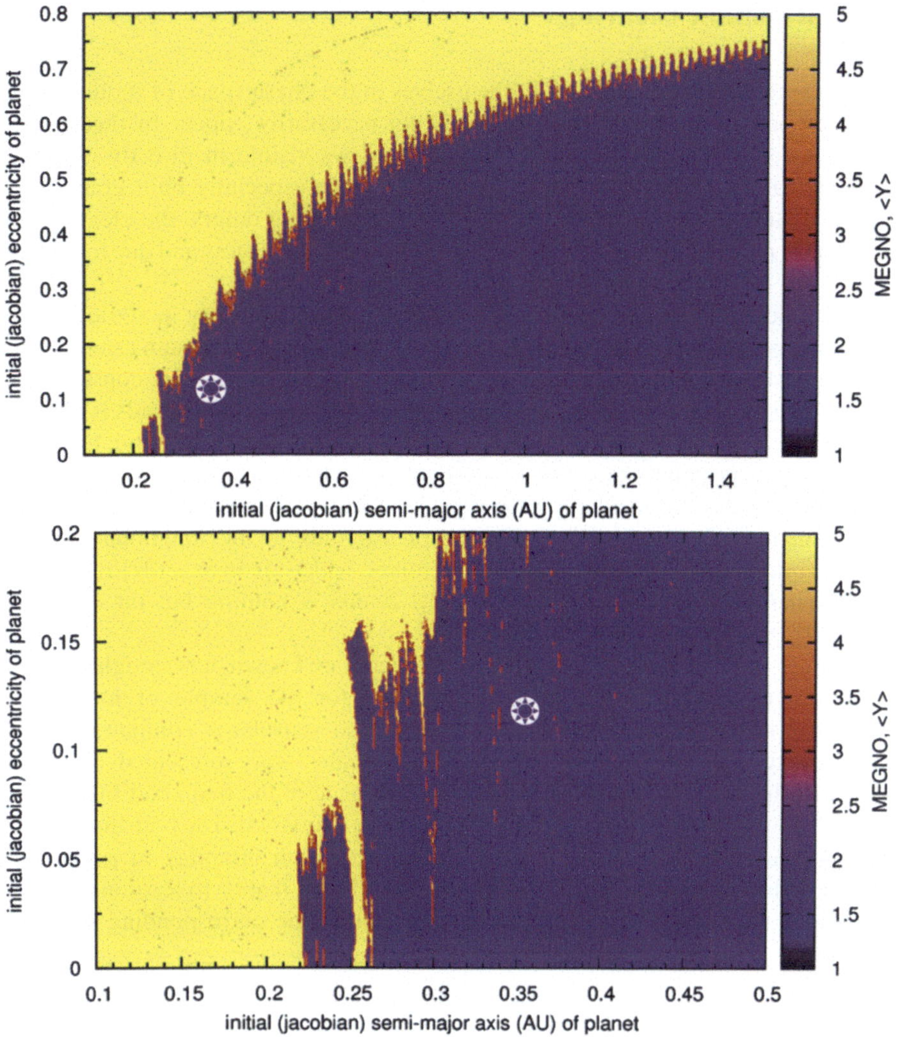

Fig. 12.6 A MEGNO chart for the *Kepler*-413 circumbinary system, in the plane of initial values of the planetary semimajor axis and eccentricity, with all other initial orbital elements fixed. The lower panel is a close-up of the upper panel. Yellow colour corresponds to chaos. The actual location of planet *Kepler*-413b is shown by a white circle (Figure 12 from Kostov et al. (2014), by permission of IOP Publishing/AAS)

The actual location of planet *Kepler*-413b is shown by a white circle. Since the planet is deeply in the purple region, its motion is stable (Kostov et al. 2014).

An interplay of clearing effects in circumbinary systems is graphically illustrated in Fig. 12.1, where the circumbinary clearings, Wisdom gaps, and, in contrast, horse-show coorbital accumulations, are all clearly visible.

12.3 Resonant Clearings

Clearing effects first of all manifest themselves in the phase space of motion. Any clearance present in the phase space does not necessarily appear in the configuration space. Accordingly, clearings are not always visible in directly obtained astronomical images, but may manifest themselves in especially built graphs and diagrams. Such "indirect" clearing effects include, among others, the clearings in discs due to mean motion resonances of higher orders with planets and due to secular resonances.

Therefore, the history of chaotic clearing effects starts already in sixties of the nineteenth century, when Daniel Kirkwood discovered gaps in the main asteroid belt (Kirkwood 1867), now called the Kirkwood gaps; see Sect. 8.2.1. He considered a sample of about 100 asteroids; today, the number of discovered asteroids is about a million; therefore, the histogram can be built with much greater resolution. It reveals many gaps, corresponding to various two-body resonances; see Fig. 8.5.

Apart from the spectacular gaps and peaks, Kirkwood's histogram has another prominent feature: the complete absence of asteroids in the neighbourhood of Jupiter's orbit (see Fig. 8.5), apart from the Trojans, which are in exact 1/1 resonance with Jupiter. As we have seen in Sect. 11.1.2, this is nothing but the close-to-coorbital clearing, namely, the Wisdom gap.

Nowadays, when thousands of exoplanet systems are known, it is straightforward to construct analogous "Kirkwood histograms" for the sample of all known two-planet configurations, taking into account also star-planet configurations in planetary systems of binary stars. These histograms were presented above in Sect. 10.5. In Fig. 10.4, the lower panel is an analogue of the main belt histogram, and the upper panel is an analogue of the Kuiper belt histogram. The both histograms show a characteristic resonant structure, representing an interplay of peaks and troughs. Whether any feature (peak or trough) is due in reality to a resonance can be verified only by analysing the time behaviour of the corresponding resonant arguments.

12.4 Inner Disintegration

Our Sun is on the half-way of its main-sequence evolution: its age is about 4.5 Gyr, while about 5 Gyr is left until it leaves the main sequence and becomes a red giant; see, e.g., Ribas (2010). This metamorphosis will have catastrophic consequences for the inner zone of the Solar system, because the atmosphere of the red giant will partly engulf it, possibly up to the Earth's orbit, and, what is more, the Sun's luminosity will rise hundreds of times, disastrously affecting even those planets that will not be engulfed by the Solar atmosphere. In Fig. 12.7, the luminosity, radius, and effective temperature of the Sun are shown as functions of time, during the time interval of 12 Gyr, starting with the Sun's birth. The rapid transformation of the Sun

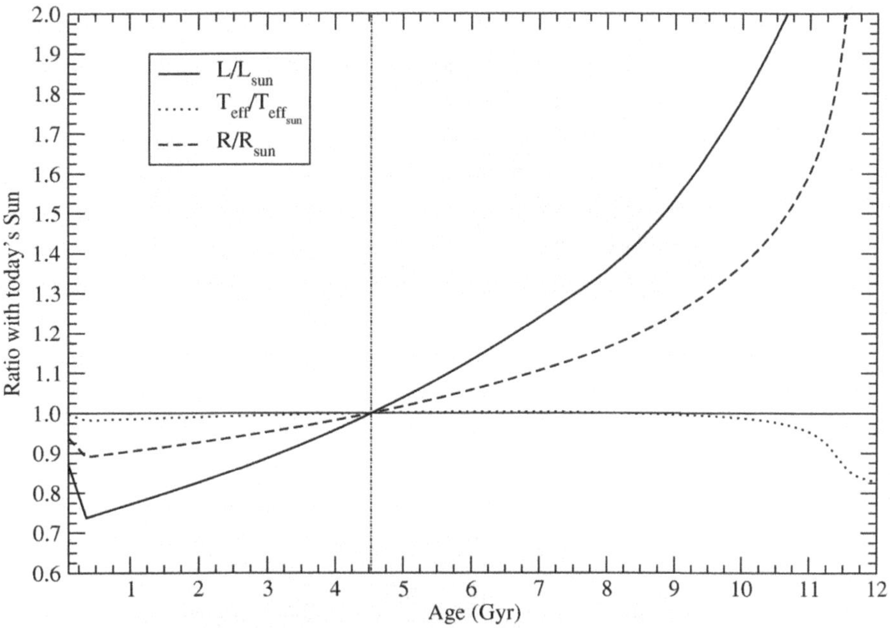

Fig. 12.7 Luminosity, radius, and effective temperature of the Sun, as functions of time (Figure 1 from Ribas (2010), by permission from © Cambridge University Press)

at ~5 Gyr from present is evident. Therefore, any analysis of the future dynamical evolution of the inner Solar system on any timescales greater than ~5 Gyr has but a solely academical interest

However, the finite lifetime of the inner Solar system is not necessarily conditioned by the Sun's evolution: there exists a chance that the inner rocky-planet zone, including the "habitable" annular band, will be already cleared from any planets when the red-giant epoch arrives. First of all, this concerns the fate of Mercury: it may escape the first. Mercury's possible escape signalled itself already in a number of numerical experiments performed in nineties of the twentieth century (Laskar 1994); and the most spectacular demonstration of its vagabond fate was performed by Laskar and Gastineau (2009).

What is more, Laskar and Gastineau (2009) showed that not only Mercury may escape, but there also exist future mutually-collisional trajectories of Venus, Earth, and Mars. In other words, in a few Gigayears the whole inner Solar system may disintegrate.

In Fig. 12.8, Mercury's eccentricity is shown as a function of time, over 5 Gyr from present, for a large set of possible initial conditions (shown in colour gradation), differing by tiny shifts in initial semimajor axis. In the lower panel, the relativistic and Lunar corrections are taken into account in the integrations, and in the upper panel, they are ignored. It turns out that the general relativity greatly enhances Mercury's stability, while the Lunar corrections are not so important.

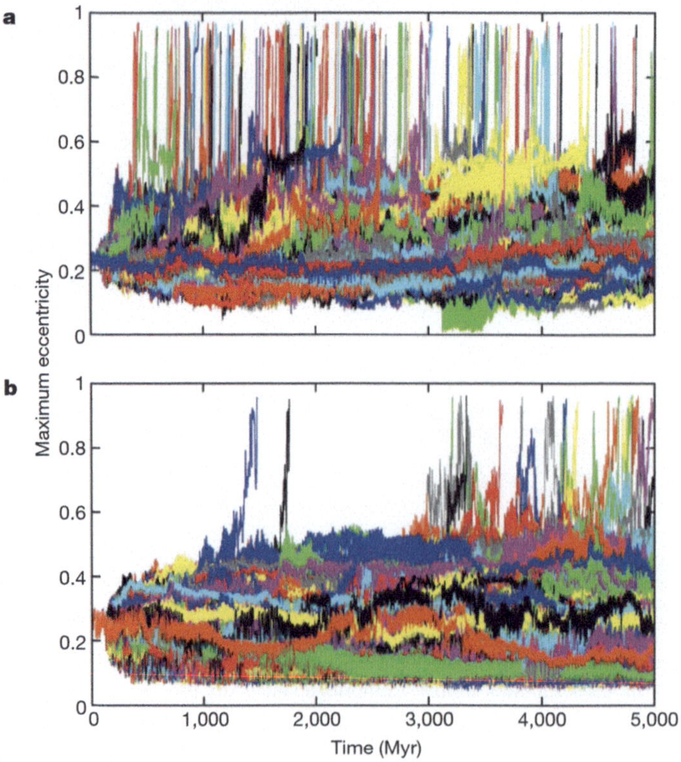

Fig. 12.8 Mercury's eccentricity, as a function of time, over 5 Gyr from present. Upper panel: the relativistic and Lunar corrections are ignored; ≈200 orbits are shown; they differ in the initial semimajor axis by tiny deviations ≈4 cm. Lower panel: the relativistic and Lunar corrections are taken into account; ≈2000 orbits are shown; they differ in the initial semimajor axis by ≈0.4 mm (Figure 1 from Laskar and Gastineau (2009), by permission from Springer Nature, © 2009)

To construct these plots, direct numerical integrations of the non-averaged equations of motion were performed (Laskar and Gastineau 2009). The integration model comprised eight major planets and Pluto and optionally included relativistic and averaged Lunar corrections. A symplectic integrator was used with the step size of 0.025 yr, which was decreased (conserving the integration accuracy) when the eccentricity of the planets raised above 0.4.

Figure 12.9 demonstrates a possibility of collisional trajectories of the Earth with Venus and Mars at ∼3 Gyr from present (Laskar and Gastineau 2009). In panels (a) and (b), the evolution of semimajor axes and eccentricities of all four inner planets is shown. Panel (c) shows the evolution of the Earth–Mars (green) and the Earth–Venus (violet) minimum distances (computed over 1000 yr consecutive time intervals); the critical (collisional) distance values are depicted as horizontal lines of the same colour. Panel (d) verifies the conservation of the system's total energy during the integrations.

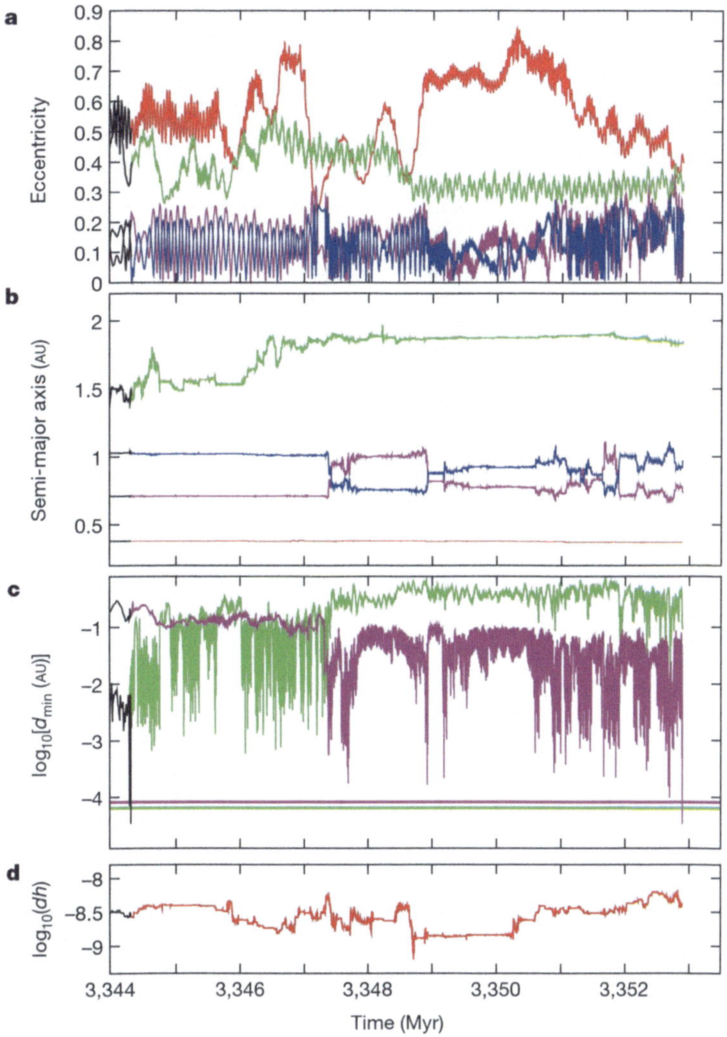

Fig. 12.9 Collisional trajectories: Earth with Venus and Earth with Mars at ~3 Gyr from present (Figure 3 from Laskar and Gastineau (2009), by permission from Springer Nature, © 2009)

The pumping of Mercury's eccentricity to a planet-crossing value is due to its eventual entrance into the ν_5 secular resonance, as revealed numerically in Laskar (2008). Recall that in the ν_5 resonance, the difference of Mercury's and Jupiter's longitudes of pericentre should librate. This is just the behaviour found in the integrations.

In Fig. 12.10, lower panel, Mercury's eccentricity is shown as a function of time, over 400 Myr (Boué et al. 2012). The corresponding Poincaré section, also presented in the Figure in its upper panel, demonstrates a typical chaotic pattern tracing the

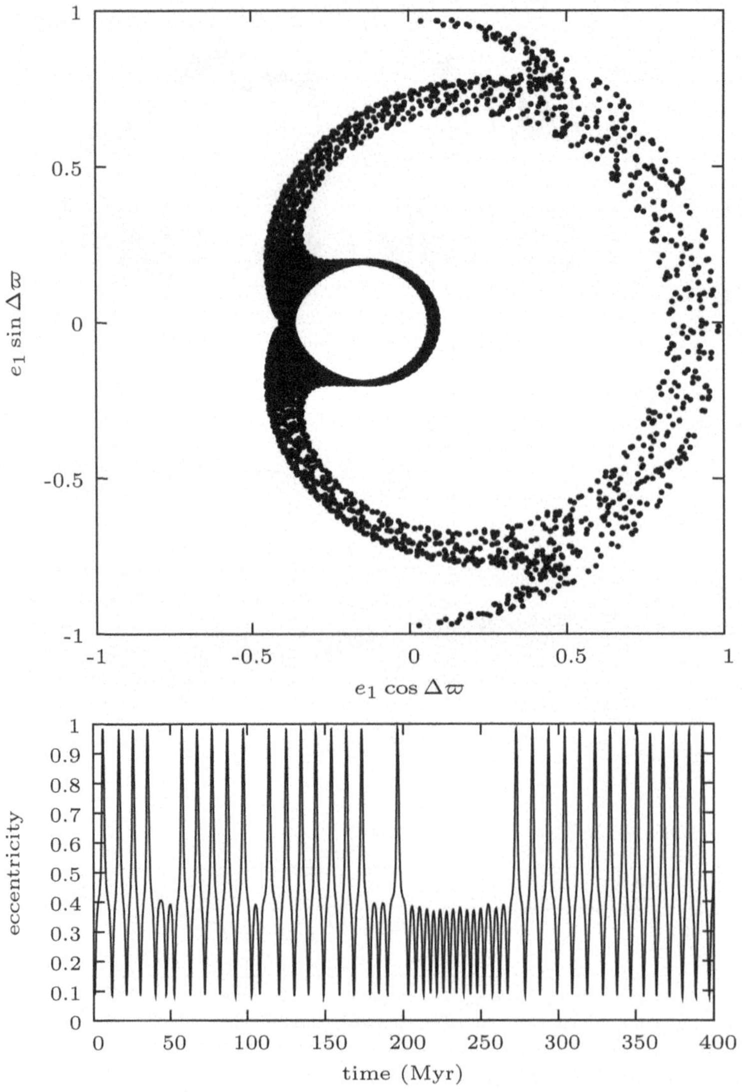

Fig. 12.10 Mercury's eccentricity, as a function of time over 400 Myr from present (lower panel) and the corresponding Poincaré section (upper panel) (Figure 6 from Boué et al. (2012), reproduced with permission © ESO)

perturbed separatrices of a non-linear resonance in the second fundamental model; see Sect. 1.2. Hamiltonian intermittency is evident in the eccentricity graph; this is its second kind, see classification in Sect. 4.4.

In a semi-analytical way, the future fate of Mercury was studied in Batygin et al. (2015). The authors succeeded in deriving a Hamiltonian model with 2 and

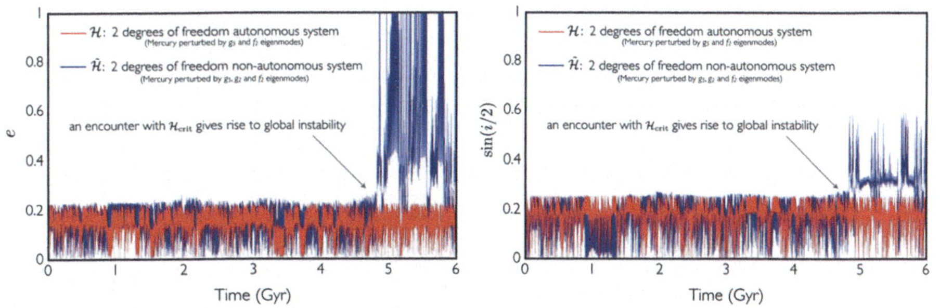

Fig. 12.11 Mercury's eccentricity (left panel) and inclination (right panel), as functions of time, over 6 Gyr from present, in two models of Batygin et al. (2015): the 2 degree of freedom system (red curve) and the 2 and 1/2 degree of freedom system (blue curve) (Figure 3 from Batygin et al. (2015), by permission of IOP Publishing/AAS)

1/2 degrees of freedom, perfectly representing the already numerically revealed qualitative behaviour of Mercury. In two models, an autonomous (with 2 degrees of freedom) one and a non-autonomous (with 2 and 1/2 degrees of freedom) one, Mercury's eccentricity and inclination, as functions of time on Gigayear scales, behave differently. A slow diffusion in the Hamiltonian value turns out to be the main factor leading to the orbit's entrance into a violent mode at time 4–5 Gyr from present (Fig. 12.11).

Two precession frequencies associated with the eccentricity and inclination of Mercury's orbit are close (within 25%) to two fundamental frequencies of the Solar system. These two latter eigenfrequencies are associated with Jupiter's eccentricity and Venus's inclination; therefore, at least two secular resonances may eventually arise. According to Lithwick and Wu (2011), this is just an eventual diffusive entrance of Mercury in a regime of overlap of these resonances that would cause Mercury's violent chaos.

The inner disintegration of a similar kind can be typical in many single-star planetary systems, in which rocky planets are in the system inner zone and giant planets are in the outer zone, as in our Solar system. In Fig. 12.12, it is demonstrated in a simplest model how secular resonances of a "particle" with two outer giant planets may interact and overlap, depending on parameters characterizing the overall architecture of the system (Lithwick and Wu 2011).

For an inner small planet's secular evolution, the Hamiltonian, in its lowest-order expansion, is given by

$$\mathcal{H} = -\frac{\mathcal{G} m m_{\mathrm{J}}}{a_{\mathrm{J}}} \left[f_{e,2}^{(1)} e^2 + f_{e,2}^{(2)} e e_{\mathrm{J}} \cos(\varpi - \varpi_{\mathrm{J}}) + f_{i,2}^{(1)} + f_{i,2}^{(2)} s s_{\mathrm{J}} \cos(\Omega - \Omega_{\mathrm{J}}) \right], \tag{12.4}$$

as rendered in orbital elements, see Murray and Dermott (1999) and Batygin et al. (2015). Here m is mass, and a, e, i, ϖ, and Ω are semimajor axis, eccentricity, inclination, longitude of pericentre, and longitude of ascending node, respectively;

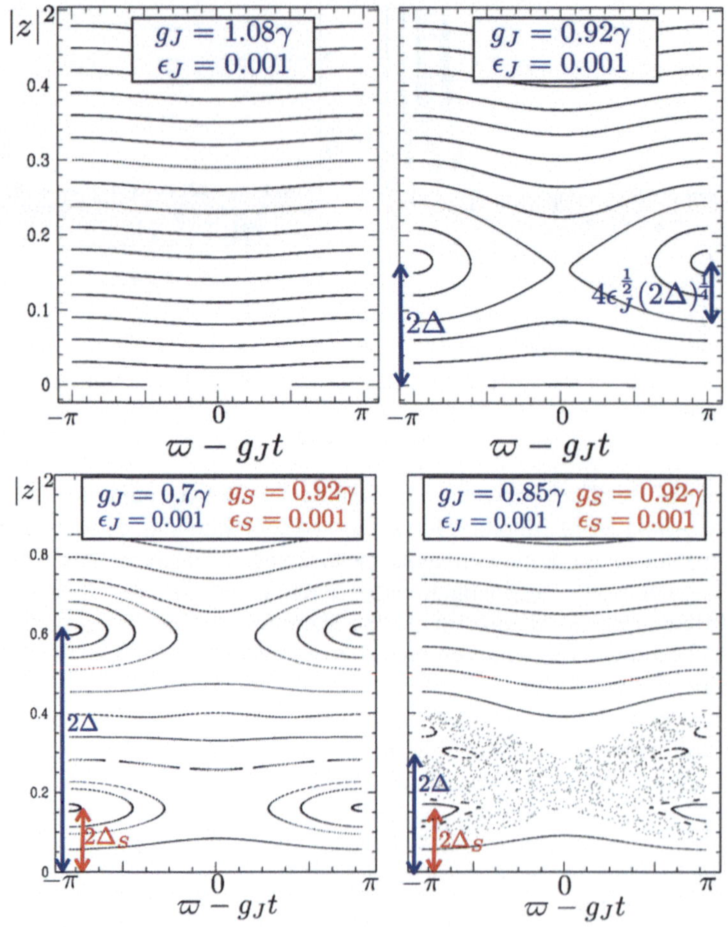

Fig. 12.12 Secular resonances and their interaction. No resonance (left upper panel), single resonance (right upper panel), two interacting resonances (left lower panel), two overlapping resonances (right lower panel) (Figures 1 and 2 from Lithwick and Wu (2011), by permission of IOP Publishing/AAS)

$s = \sin(i/2)$; the coefficients f are functions of $\alpha = a/a_J < 1$. The mass and orbital elements of the particle are non-indexed; for the perturber, they are indexed by "J" (a paradigmatic "Jupiter").

A model Hamiltonian of a single secular resonance, when the inclination terms are ignored, can be written down in the form

$$\frac{H}{\gamma} = -\frac{p^2}{4} + \Delta \cdot p - 2\varepsilon_J p^{1/2} \cos\varphi, \qquad (12.5)$$

where $\varphi = \varpi - g_J t$; and ϖ is the particle's longitude of pericentre, g_J is the perturber's fundamental frequency (approximately the rate of precession of the perturber's apsidal line); Δ is a detuning parameter; ε_J characterizes the strength of perturbation. The constant γ is the particle's apsidal rate of precession forced by the perturber. The action-like variable p is conjugated to the angle φ.

In the y axis of Fig. 12.12, $y \approx e^2$. In case of secular resonances with two planets ("J" and "S", for paradigmatic Jupiter and Saturn), a second resonance cell emerges in the phase space section. The phase space section (lower panel of Fig. 12.12) is taken at $\exp[i(g_J - g_S)t] = 1$; here i is the imaginary unit.

Generally, as presented in Fig. 12.12, in this three-planet planar model four basic dynamical situations may take place: no resonance (left upper panel), single resonance (right upper panel), two interacting resonances (left lower panel), two overlapping resonances (right lower panel). In the last case, global chaos is present, signalling an eventual destabilization of the system. This theoretical example demonstrates the intrinsic importance of secular resonances in defining the dynamical architectures of planetary systems; for more examples, see Barnes and Greenberg (2006) and Wu and Lithwick (2011).

Chapter 13
Multiplanet Systems of Single Stars

In planetary systems of single stars, mean motion resonances are possible if the system comprises two or more planets, i.e., if it is multiplanet. In this chapter, we consider secular, resonant, and chaotic dynamics of multiplanet systems of single stars. Location and interaction of resonances are characterized in various dynamical models, in particular, if the perturber is in an outer orbit in the system, and if it is in an inner orbit. We discuss actual resonant and formally chaotic exoplanet systems.

13.1 Secular Planetary Dynamics

In a hierarchical setting of the planar elliptic restricted three-body problem (R3BP), Heppenheimer (1978) derived a secular perturbation theory, providing analytical formulas for the time variation of the forced eccentricity e and the longitude of periastron ϖ of a passively gravitating body (tertiary) put initially in a circular orbit around one of the components of the primary binary (say, a star with a planet). In the hierarchical setting, the primary binary size is assumed to be much greater than the tertiary's orbit; in a more rigorous formulation, the perturbing companion's orbital pericentric distance, in astrocentric coordinates, is assumed to be much greater than the tertiary's apocentric distance. This problem setting can be called the *circum-companion* version of the hierarchical planar elliptic R3BP. The version can be either *circumprimary* or *circumsecondary*.

In Heppenheimer's (1978) equations, the leading terms in the powers of the semimajor axes ratio of the system bodies are kept, and higher order terms are neglected. The theory was initially intended for analytical descriptions of circumstellar planetesimal disc dynamics in binary star systems; therefore, it is useful in various exoplanetary applications, especially in systems with discs, and also in systems with a planet whose mass can be neglected with respect to the perturbing body. Whitmire et al. (1998) and Thébault et al. (2006) applied this

I. I. Shevchenko, *Dynamical Chaos in Planetary Systems*, Astrophysics and Space Science Library 463, https://doi.org/10.1007/978-3-030-52144-8_13

secular perturbation theory to analytically describe a stirring of the circumstellar disc of a young star by a distant stellar companion.

An alternative formulation of the R3BP hierarchical setting is the *circumbinary* one, in which the primary binary size is assumed to be much smaller than the tertiary's pericentric distance, in barycentric coordinates. Roughly speaking, the primary binary size is assumed to be much smaller than the tertiary's orbit; therefore, the tertiary moves around the primary binary in a weakly perturbed circumbinary orbit.

Moriwaki and Nakagawa (2004) considered this R3BP setting and derived equations of the secular circumbinary motion, keeping the leading terms in the powers of the semimajor axes ratio and powers of the eccentricities of the system bodies.

Demidova and Shevchenko (2015) combined the approaches of Heppenheimer (1978) and Moriwaki and Nakagawa (2004) and derived explicit analytical formulas for the secular evolution of the tertiary's eccentricity and longitude of pericentre in the both problem settings, the circum-companion and circumbinary ones.

Here we consider first the circumbinary setting. In fact, it is superfluous to consider a non-hierarchical setting, because, as we show further on in Sect. 14.2.3, the central chaotic circumbinary zone exists at all eccentricities of the tertiary if $\mu \gtrsim 0.05$, where $\mu = m_2/(m_1 + m_2)$ is the mass parameter of the primary binary and $m_1 \geq m_2$ are the masses of the primary binary components. This mass threshold has an important physical meaning (Shevchenko 2015): above it, the tertiary can diffuse, even starting from small eccentricities, following the sequence of the overlapping integer $p/1$ mean motion resonances between the primary binary and the particle, up to ejection from the system; any close encounters with other bodies of the system are not required to provide the escape. Note that, in the alternative circum-companion case, the motion inside the secondary's Hill sphere is generally stable.

We adopt the barycentric frame and the following designations: a_b and e_b are the primary binary's semimajor axis and eccentricity, respectively; a and e are the semimajor axis and eccentricity of the tertiary's orbit. All masses are measured in Solar units, distances in astronomical units (AU), and time in years. In this unit system, the gravitational constant $\mathcal{G}$ is equal to $4\pi^2$.

The averaged perturbing function in the circumbinary case is given in Moriwaki and Nakagawa (2004) in the form of a truncated power-law series in the ratio of the primary binary's and the tertiary's semimajor axes and in the eccentricities. The corresponding equations of the secular motion (Moriwaki and Nakagawa 2004, Equations (A7)) can be readily analytically integrated (Demidova and Shevchenko 2015) to provide formulas for the secular time evolution of the circumbinary tertiary's eccentricity e and longitude of pericentre ϖ:

$$e = e_{\max} \left| \sin \frac{ut}{2} \right|, \tag{13.1}$$

$$\tan \varpi = -\frac{\sin ut}{1 - \cos ut}, \tag{13.2}$$

where t is time,

$$u = \frac{3\pi}{2} \frac{m_1 m_2}{(m_1 + m_2)^{3/2}} \frac{a_\mathrm{b}^2}{a^{7/2}} \left(1 + \frac{3}{2} e_\mathrm{b}^2\right), \tag{13.3}$$

and

$$e_{\max} = 2e_\mathrm{f}, \tag{13.4}$$

where the forced eccentricity

$$e_\mathrm{f} = \frac{5}{4} \frac{(m_1 - m_2)}{(m_1 + m_2)} \frac{a_\mathrm{b}}{a} e_\mathrm{b} \frac{\left(1 + \frac{3}{4} e_\mathrm{b}^2\right)}{\left(1 + \frac{3}{2} e_\mathrm{b}^2\right)}. \tag{13.5}$$

By introducing the variable $y = ut/2$, Eq. (13.2) can written down as

$$\begin{aligned}
&\text{if } y \geq -\pi \text{ and } y \leq -\frac{\pi}{2}, \text{ then } \varpi = y + 5\frac{\pi}{2};\\
&\text{if } y \geq -\frac{\pi}{2} \text{ and } y \leq 0, \text{ then } \varpi = y + \frac{\pi}{2};\\
&\text{if } y \geq 0 \text{ and } y \leq \frac{\pi}{2}, \text{ then } \varpi = y + 3\frac{\pi}{2};\\
&\text{if } y \geq \frac{\pi}{2} \text{ and } y \leq \pi, \text{ then } \varpi = y - \frac{\pi}{2}.
\end{aligned} \tag{13.6}$$

In some sense, the u quantity can be regarded as a precession rate of the tertiary's line of apsides.

Numerical model integrations of the dynamical stirring of planetesimal discs on secular timescales, performed in Moriwaki and Nakagawa (2004), Meschiari (2012, 2014), and Paardekooper et al. (2012) for various sets of the problem parameters, provide plots of the eccentricity e and the longitude of pericentre ϖ of a circumbinary body in dependence on its semimajor axis a. The numerically obtained dependences are perfectly reproduced by the described above secular theory; see Demidova and Shevchenko (2015).

In their form, Eqs. (13.1) and (13.2) turn out to be rather similar to formulas derived in Heppenheimer (1978) for the same quantities (but in the circumbinary case the motion is described in the barycentric frame instead of the astrocentric one) in the circum-companion hierarchical setting of the problem. In the circumprimary case, the quantities are also given by Eqs. (13.1), (13.2), and (13.6), but the u and e_f parameters are expressed differently; see Heppenheimer (1978), Whitmire et al. (1998), and Thébault et al. (2006). With our notations, they are

$$u = \frac{3\pi}{2} \frac{m_2}{m_1^{1/2}} \frac{a^{3/2}}{a_\mathrm{b}^3} \left(1 - e_\mathrm{b}^2\right)^{-3/2} \tag{13.7}$$

and

$$e_{\rm f} = \frac{5}{4}\frac{a}{a_{\rm b}}\frac{e_{\rm b}}{\left(1-e_{\rm b}^2\right)}. \tag{13.8}$$

In the given circum-companion case, m_1 is the primary mass (around which the massless tertiary orbits) and m_2 is the secondary (perturbing) mass; $m_1 > m_2$.

13.2 Location and Interaction of Resonances

13.2.1 The Outer Perturber Case

Let us consider the circum-companion case: the perturber moves in an outer orbit with respect to the particle; i.e., the tertiary orbits around one of the components of the primary binary. Circum-companion mean motion resonances correspond to commensurabilities between the orbital frequencies of the particle and the outer perturber.

An approximate Hamiltonian of the particle's motion around the primary, in the vicinity of the $(k + j)/k$ mean motion resonance (where $k \geq 1$ and $j \geq 1$ are integers) with the primary binary, in the planar elliptic restricted three-body problem is given by

$$H = \frac{1}{2}\beta\Lambda^2 - \sum_{p=0}^{j} \phi_{k+j,k+p,k} \cos(\psi + p\varpi) \tag{13.9}$$

(Holman and Murray 1996; Murray and Holman 1997), where $\beta = 3k^2/a^2$, $\Lambda = \Psi - \Psi_{\rm res}$, $\Psi = (\mu_1 a)^{1/2}/k$, $\Psi_{\rm res} = (\mu_1^2/(k^2(k+j)n_{\rm b}))^{1/3}$, $\mu_1 = 1 - \mu$, $\mu = m_2/(m_1 + m_2)$ (we set $m_1 > m_2$); ϖ is the longitude of the particle's pericentre; a and e are the particle's semimajor axis and eccentricity; $\psi = kl - (j + k)l_{\rm b}$, where l and $l_{\rm b}$ are the mean longitudes of the particle and the primary binary, respectively.

Here the units are chosen in such a way that the gravitational constant, the total mass of the primary binary, the binary's semimajor axis $a_{\rm b}$ are all equal to one. The binary's mean longitude $l_{\rm b} = n_{\rm b}t$, and the binary's mean motion $n_{\rm b} = 1$, i.e., the time unit is the $\frac{1}{2\pi}$th part of the binary's orbital period.

Equation (13.9) demonstrates that any resonance $(k + j)/k$ is split in a cluster of $j + 1$ subresonances with $p = 0, 1, \ldots, j$. For each of them, the resonant argument is

$$\phi = \psi + p\varpi = kl - (j + k)l_{\rm b} + p\varpi, \tag{13.10}$$

where $p = 0, 1, \ldots, j$. The coefficients of the resonant terms are given by

$$|\phi_{k+j,k+p,k}| \approx \frac{\mu}{j\pi a_b} \binom{j}{p} \left(\frac{\epsilon}{2}\right)^p \left(\frac{\epsilon_b}{2}\right)^{j-p}, \tag{13.11}$$

where $\epsilon = ea_b/|a - a_b|$ and $\epsilon_b = e_b a_b/|a - a_b|$.

The model (13.9) is intended for description of resonances of relatively high order, $j \geq 2$, and the approximation (13.11) is adequate if $\epsilon j < 1$ (Holman and Murray 1996).

The frequency of small-amplitude oscillations on subresonance p is

$$\omega_0 = (\beta|\phi_{k+j,k+p,k}|)^{1/2} \approx \frac{a_b}{|a - a_b|} n_b \left[\mu_1 \mu \frac{4j}{3\pi} \binom{j}{p} \left(\frac{a}{a_b}\right) \left(\frac{\epsilon}{2}\right)^p \left(\frac{\epsilon_b}{2}\right)^{j-p}\right]^{1/2} \tag{13.12}$$

and for the perturbation frequency (the apsidal precession rate) one has

$$\Omega = \dot{\varpi} \approx -\frac{\mu_1 \mu}{2\pi} n_b \left(\frac{a}{a_b}\right)^{1/2} \left(\frac{a_b}{a - a_b}\right)^2. \tag{13.13}$$

13.2.2 The Inner Perturber Case

Consider the circumbinary case (in otherwise the same setting of the three-body problem): a particle orbits around the primary binary. Circumbinary mean motion resonances correspond to commensurabilities between the orbital frequencies of the central primary binary and the particle.

Using an analysis presented in Holman and Murray (1996); Murray and Holman (1997), the Hamiltonian of the circumbinary motion of a tertiary in the vicinity of a mean motion resonance in the planar elliptic restricted three-body problem can be written down as

$$H = \frac{1}{2}\beta\Lambda^2 - \sum_{p=0}^{j} \phi_{k+j,k+p,k} \cos(\psi + p\varpi), \tag{13.14}$$

where the designations are the same as in the previous Sect. 13.2.1. The variable $\psi = kl - (j + k)l_b$, where l and l_b are the mean longitudes of the particle and the primary binary.

For the circumbinary motion, one has $a > a_b$ and j is negative and p non-positive. The $(k + j)/k = (k - |j|)/k$ resonance is split in a cluster of $|j| + 1$ subresonances with $p = 0, -1, -2, \ldots, j$. For each of them, the resonant argument is

$$\phi = \psi + p\varpi = kl - (j + k)l_b + p\varpi = kl - (k - |j|)l_b - |p|\varpi, \tag{13.15}$$

where $p = 0, -1, -2, \ldots, j$.

For $k \geq 2$ and $j = 1 - k$, one has resonances $1/k$, each split in a cluster of k subresonances with the resonant arguments

$$\phi = kl - l_b + p\varpi \tag{13.16}$$

where $p = 0, -1, -2, \ldots, 1 - k$.

The coefficients of the resonant terms are

$$|\phi_{k+j,k+p,k}| \approx \frac{\mu}{|j|\pi a} \binom{|j|}{|p|} \left(\frac{\epsilon}{2}\right)^{|p|} \left(\frac{\epsilon_b}{2}\right)^{|j|-|p|}, \tag{13.17}$$

where $\epsilon = ea/|a - a_b|$, $\epsilon_b = e_b a/|a - a_b|$.

As in the described above model (13.9), the model (13.14) is intended for description of resonances of relatively high order, $|j| \geq 2$. The approximation (13.11) is adequate, if $\epsilon|j| < 1$ (Holman and Murray 1996).

The frequency of small-amplitude oscillations on subresonance $|p|$ is given by

$$\omega_0 = (\beta|\phi_{k+j,k+p,k}|)^{1/2} \approx \frac{a}{|a - a_b|} n_b \left[\mu_1\mu \frac{4|j|}{3\pi} \binom{|j|}{|p|} \left(\frac{a}{a_b}\right) \left(\frac{\epsilon}{2}\right)^{|p|} \left(\frac{\epsilon_b}{2}\right)^{|j|-|p|}\right]^{1/2}. \tag{13.18}$$

The perturbation frequency Ω is considered further on. In the pendulum model, the half-width of any subresonance is equal to $2\omega_0$.

In the hierarchical planar *general* three-body problem, an analytical description of resonant terms can be found in Mardling (2008a, Equations (3.51), (3.60), and (3.61)).

There is a number of mechanisms that may cause apsidal precession, among them the general relativity. Recall the famous case of apsidal precession of the innermost planet, Mercury, in our Solar system. The rate of Mercury's apsidal precession due to the perturbations from all planets beginning with Venus is equal to 532″ per century, and the general relativity adds 43″ per century; see Clemence (1947). The Solar oblateness and tidal effects are known to be negligible. The classical perturbations from planets, therefore, dominate in summoning Mercury's apsidal precession.

The secular evolution in the circumbinary and circum-companion cases is described above in Sect. 13.1. In each of these two cases, the classical non-relativistic apsidal precession rate is given by individual expressions. In the circumbinary hierarchical setting of the circular ($e_b = 0$) restricted three-body problem, where the barycentric frame is naturally adopted, for the classical precession rate one has

$$\omega_{cl} \equiv \dot{\varpi} \approx \frac{3\pi}{2} \frac{m_1 m_2}{(m_1 + m_2)^{3/2}} \frac{a_b^2}{a^{7/2}} = \frac{3\pi}{2} \frac{m_1 m_2}{(m_1 + m_2)^{3/2}} \frac{a_b^2(1-e)^{7/2}}{q^{7/2}}, \tag{13.19}$$

in radians per year. Here, the masses are measured in Solar units, distances in astronomical units, and time in years; the gravitational constant $\mathcal{G}$ is therefore equal to $4\pi^2$. As adopted above, $m_1 \geq m_2$ are the masses of the primary binary components, a_b and e_b are the binary's semimajor axis and eccentricity, a, q, and e are the particle's semimajor axis, pericentric distance, and eccentricity, respectively.

It is convenient to express the particle's apsidal precession rate in ratio to the particle's mean motion n:

$$\frac{\omega_{cl}}{n} \approx \frac{3}{4} \frac{m_1 m_2}{(m_1 + m_2)^2} \left(\frac{a_b}{a}\right)^2 = \frac{3}{4}\mu(1-\mu)\left(\frac{a_b}{q}\right)^2 (1-e)^2, \tag{13.20}$$

where $\mu = m_2/(m_1 + m_2)$.

13.3 Resonant and Formally Chaotic Exoplanet Systems

Mean motion resonances may emerge in systems of single stars that have two or more planets, as well as in systems of binary stars that have one or more planets. About one third of all discovered exoplanets belong to multiplanet systems (Rein 2012). Orbital resonances seem to be ubiquitous in planetary systems, as confirmed in computations of the behaviour of resonant arguments. However, orbital and physical parameters of the observed systems are usually not well-constrained, and this circumstance does not allow one to judge definitely whether any given resonant argument librates or circulates; perhaps solely in case of three-body resonances some definite conclusions can be made (Pichierri et al. 2019).

The current data on planetary masses and orbital elements indeed suffer from observational uncertainties; however, the occurrence of low-order resonances (such as 2/1 and 3/2) seems to be statistically significant (Wright et al. 2011; Fabrycky et al. 2012), especially in pairs of planets with similar masses. Modern classifications usually attribute any planetary system exhibiting a mean motion resonance to the first basic dynamical class of planetary systems (Ollivier et al. 2009). In application to exoplanetary systems, in the context of the unrestricted three-body problem, the planetary resonances are analysed and discussed by Batygin and Morbidelli (2013a).

The most widespread planetary resonances or close-to-resonance states in two-planet configurations correspond to the 2/1 and 3/2 orbital frequency commensurabilities, see Fig. 10.4. It is a natural outcome of the primordial dynamical evolution of planets migrating in primordial discs, as numerical simulations show; see Wang et al. (2012), Davies et al. (2014), Pichierri et al. (2019), and references therein. The migration may lead to either resonant or non-resonant final orbital configurations, in which the lines of apsides are aligned. This phenomenon is observed in a number of actual planetary systems and, therefore, serves as a confirmation that the migration took place indeed.

Three-body mean motion resonances may generate prominent chaos, due to overlap of subresonances in subresonance supermultiplets. There exist examples of such overlap in dynamics of the Solar system bodies, including giant planets, see Sect. 9.1; chaos due to three-body mean motion resonances in case of asteroids was considered above in Sect. 8.2.1. In exoplanetary studies, with the number of discovered multiplanet systems increasing, planetary three-body resonances became a perspective subject of analytical and numerical analysis (Quillen 2011; Pichierri et al. 2019).

Apart from the mean motion resonances, a major role in the dynamical evolution of exoplanet systems is played by secular resonances, as in our Solar system. Some systems are likely to exhibit both apsidal and mean motion resonances; a prototype (though not yet confirmed) is the planetary system of the star HD 12661 (Goździewski 2003).

Several outstanding examples of mean-motion resonant and close-to-resonant multiplanet systems are presented in what follows. Note that an interested reader may compare the main physical and orbital parameters of planets in the discussed exoplanetary systems with the analogous parameters of planetary class prototypes present in our Solar system; some of the latter ones are listed in Appendix B.

13.3.1 The Kepler-223 *System*

Kepler-223 (initially designated as KOI-730, where "KOI" means *Kepler's Object of Interest*) is a G5V dwarf star around which four super-Earth planets revolve. The physical radii of the four planets b, c, d, and e are equal to 1.8, 2.1, 2.8, and 2.4 R_{Earth}, respectively.[1] They were all identified by means of the TTV technique. As described above in Sect. 10.1, this technique provides virtually complete and accurate information on planetary orbital parameters and masses.

The *Kepler*-223 system is considered nowadays to be the most remarkable example of a *closely packed resonant system*. The orbital periods of its planets are equal to 7.4, 9.8, 14.8, and 19.7 days, forming pairs close to mutual integer commensurabilities; and the planetary quartet was indeed confirmed, by analysing the time behaviour of the corresponding resonant arguments, to reside in the multi-body mean motion resonance 8:6:4:3.

Lissauer et al. (2011a) state that "This resonant chain is potentially the missing link that explains how planets that are subject to migration in a gas or planetesimal disc can avoid close encounters with each other, being brought to a very closely packed, yet stable, configuration."

[1] Note that the literal nomenclature of planets in any exoplanetary system starts with letter "b," the letter "a" being implicitly attributed to the host star.

Fig. 13.1 *Kepler*-11, a triple transit. Artist's impression (Credits: NASA/Tim Pyle)

13.3.2 *The* Kepler-11 *System*

Kepler-11 is a Sun-like G6V dwarf star. It possesses a compact (with all planets' semimajor axes $a < 0.5$ AU) system of six low-mass transiting planets (Lissauer et al. 2011b, 2013); see an illustration in Fig. 13.1. The planets from b to f have densities 0.5–2.0 g cm^{-3} and masses from 2 to 8 $M_{\rm Earth}$. Such masses are typical for super-Earths, see Sect. 10.3.2, although the planets' densities are Saturn-like and Neptune-like. Five inner planets in this system orbit very close to the host star; the radial distances between their orbits are small both in relative (with respect to the inner planet's orbit size) and in absolute units. The orbit of the outermost 6th planet is only a little bit larger than the orbit of Mercury. The two innermost planets are close to the 5/4 mean motion resonance. Outer non-transit planets, of course, can be actually present, although they are non-observed now.

This planetary system is super-coplanar: the current mutual inclinations of the planetary orbits do not exceed $\sim 1°$ (Lissauer et al. 2011b; Migaszewski et al. 2012); for comparison, in our Solar system (with eight planets, Pluto excluded), the planetary mutual inclinations have values up to $\sim 2.3°$.

Stability MEGNO-charts,[2] constructed by Migaszewski et al. (2012) for neighbourhoods of observationally allowed sets of planetary parameters and initial conditions, demonstrate that the phase space of orbital dynamics of the *Kepler*-11 planetary system is mostly chaotic. Zones of regular motion in the diagrams are

[2]On the MEGNO technique, see Sect. 2.3.

shallow. Migaszewski et al. (2012) explain the observed chaos as due to overlap of three-body and four-body mean motion resonances, and argue that the *Kepler*-11 system can be stable and long-lived only if caught in particular three-body and four-body resonances, in close-to-coplanar system configurations. Migaszewski et al. (2012) conclude that the system is "extremely resonant," residing in a number of many-body resonances, and can therefore be long-term stable.

Concerning two-body resonances, Lissauer et al. (2011b) find that the system is non-resonant, with only one planet pair being close to the 5/4 mean motion resonance. Any presence of non-transiting giant planets in large outer orbits (if real but not yet observationally revealed) may significantly modify the inferences on the long-term dynamical behaviour of the *Kepler*-11 system.

13.3.3 The GJ-876 System

GJ-876[3] is an M4V dwarf star, which has at least four planets, one super-Earth (planet b), two Jupiters (planets c and d), and one Neptune (planet e). The masses of the b, c, d, and e planets are 6.83 $M_{\rm Earth}$, 0.71 $M_{\rm J}$, 2.28 $M_{\rm J}$, and 14.6 $M_{\rm Earth}$, respectively, and the orbital periods are, respectively, 1.9, 30, 61, and 124 days (Rivera et al. 2010). Planets c, d, and e are involved in the 4:2:1 Laplace mean motion resonance (Rivera et al. 2010). Along with the Io–Europa–Ganymede subsystem of Jovian moons, it is the second currently known, to be present anywhere in the Universe, case of the Laplace resonance.

A MEGNO chart for the d planet dynamics is shown in Fig. 13.2. The planet's location is marked with a black circle, situated in the green area at the border of regular and chaotic zones. According to this MEGNO diagram, the d planet seems to be located just at the chaos border.

13.3.4 The Kepler-36 *System*

Kepler-36 is a sub-giant G1IV star with radius $R = 1.63 R_{\rm Sun}$ and mass $M = 1.07 M_{\rm Sun}$, hosting two super-Earths with masses 4.1 and 7.5 $M_{\rm Earth}$ and orbital periods 13.8 and 16.2 d, respectively (Carter et al. 2012). Thus, the planet system has a close-to-coorbital compact configuration, with the planetary semimajor axes $a < 0.13$ AU.

By means of massive direct numerical integrations on grids of observationally allowed initial conditions, the long-term dynamics of the *Kepler*-36 system was explored by Deck et al. (2012). The TTV technique provides strict bounds on the planetary masses and possible domains of the initial conditions. Tantalizingly, most

[3]The "GJ" designation means that the star is from the Gliese–Jahreiss catalogue.

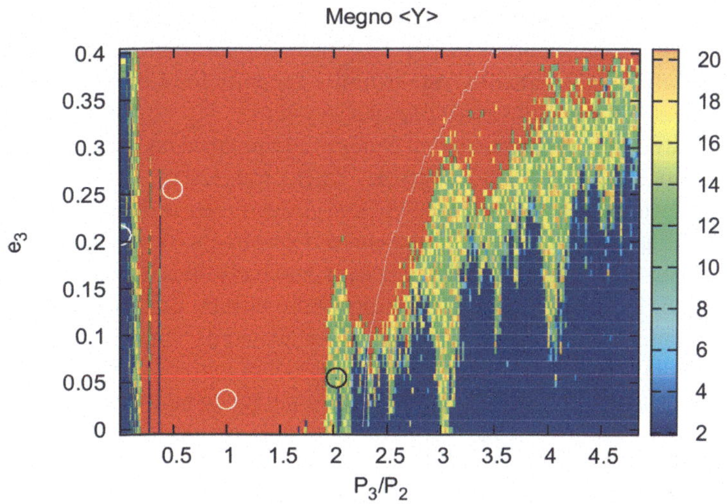

Fig. 13.2 A MEGNO chart for the d planet in the GJ-876 system, in the "period ratio of d and c planets—eccentricity" plane. An analytical stability limit (Gladman 1993) is shown by a white curve. The locations of planets b and c are shown for reference (white circles); the d planet location is marked with a black circle (Figure 2a from Martí et al. 2013, © Oxford University Press)

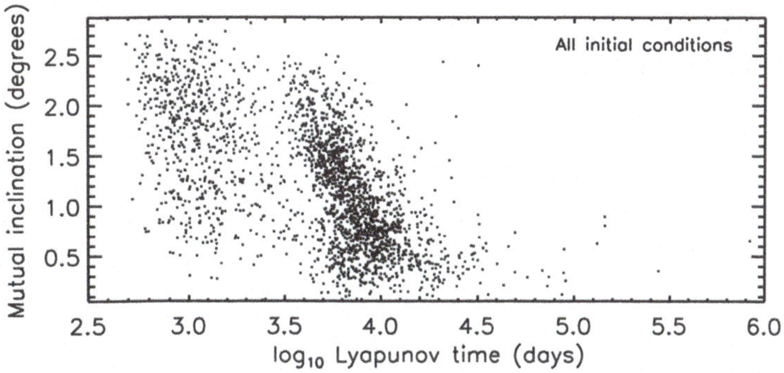

Fig. 13.3 Distributions of the Lyapunov times for *Kepler*-36, with respect to the planetary mutual inclination. Higher initial mutual inclinations imply stronger dynamical chaos, i.e., shorter Lyapunov times (Figure 1a from Deck et al. 2012, by permission of IOP Publishing/AAS)

of the allowed initial conditions ($\approx$95%) turns out to imply chaotic behaviour with extremely short Lyapunov times. The Lyapunov times can be as small as $\sim$10 yr; see Fig. 13.3. From this diagram, it follows that any initial mutual planetary inclinations higher than nominal ones imply stronger chaos (chaos with shorter Lyapunov times).

However, the remaining ≈5% of the possible initial conditions imply regular long-lived orbits. The diagram in Fig. 13.3 provides an impressive example how a stability analysis can radically refine limits for initial conditions (and also for planetary masses) for any observed planetary system.

What is the cause for the dynamical chaos predominance on the plane of possible initial conditions in the *Kepler*-36 system? Deck et al. (2012) argue that the chaos originates due to interaction of the 7/6 first-order mean motion resonance with the nearby (much weaker) 34/29 resonance. Conversely, one may observe that, if the less massive planet were treated as a passively gravitating body in some approximation, it would belong to the close-to-coorbital chaotic zone formed in the radial vicinity of the massive planet. In other words, the less massive planet moves inside the planetary chaotic zone (the Wisdom gap) of the larger planet; see Sect. 11.1.2. Therefore, the dynamical chaos emergence here can be attributed to the overlap of the $(p + 1)/p$ first order mean motion resonances; the neighbouring 8/7, 7/6, and 6/5 mean motion resonances, among them.

13.3.5 The HD 12661 System

HD 12661 is a Solar-like G6V dwarf star with radius $R = 1.10R_{\rm Sun}$ and mass $M = 1.07M_{\rm Sun}$. It hosts a pair of Jupiters with masses $m \sin i = 2.3$ and 1.56 $M_{\rm J}$, where i is the angle between the line of sight and the normal to the orbit plane; the planets' orbital periods $P = 263.3$ and 1444.5 days, and eccentricities $e \simeq 0.35$ and 0.20, respectively (Goździewski 2003). (For different estimates of masses and orbital parameters, see Wright et al. 2009.) The planets were discovered by the RV-method; hence, only lower bounds for their masses are known. As follows from the period ratio in the planetary pair, this two-planet system is close to the 11/2 mean motion resonance.

Goździewski (2003) explored the stability of the HD 12661 two-planet system in massive numerical N-body simulations; a MEGNO chart is presented in Fig. 13.4. For the given above nominal orbital initial conditions, the Lyapunov time turns out to be about 1300 yr (Goździewski 2003), i.e., the planetary system is nominally chaotic.

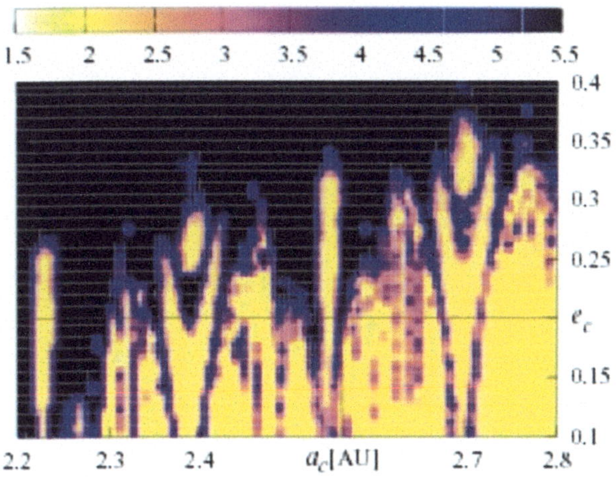

Fig. 13.4 A MEGNO chart of the HD 12661 system in the "semimajor axis—eccentricity" plane for the outer planet c, which is less massive than planet b. Planet c is nominally situated at the intersection of the black solid lines (Figure 6a from Goździewski 2003, reproduced with permission © ESO)

Chapter 14
Planetary Systems of Multiple Stars

Mean motion resonances, apart from their general significance, evoke several important effects specific to the hierarchical circum-companion and circumbinary planetary systems of binary stars. In the circumbinary case, their overlap is responsible for the formation of a continuous chaotic zone around the stellar binary, if the binary's mass parameter is large enough. In the circum-companion case, their overlap forms the outer cut-off of any circum-companion planetary system. Circumbinary planetary systems may contain planets captured in resonances with the central stellar binary; such resonant planets may even form many-body resonance sequences, "anchored" to the innermost resonance. The circumbinary resonances may form observable patterns in circumbinary planetesimal discs, in case the disc is present. Similar effects can be active in circum-companion systems. We consider potential and actual examples of circum-companion systems and circumbinary systems. Concepts of migration and chaos as system architects, chaotic orbital zones, structure of chaos borders, and the mass parameter threshold for forming the circumbinary chaotic zone are described and discussed. The presentation in this chapter is partially based (in Sect. 14.2) on the papers by Popova and Shevchenko (2013) and Shevchenko (2015).

14.1 Circum-Companion Systems

14.1.1 The α Centauri System

The α Centauri system, being the binary star closest to our Sun, naturally attracts a lot of attention on the subject how stable planetary configurations in such a system may look like (Benest 1988a; Popova and Shevchenko 2012; Quarles and Lissauer 2016; Quarles et al. 2018). According to Anosova et al. (1994), Proxima Centauri, which is some 0.02 pc closer to us, is not gravitationally bound to the α Centauri

I. I. Shevchenko, *Dynamical Chaos in Planetary Systems*, Astrophysics and Space Science Library 463, https://doi.org/10.1007/978-3-030-52144-8_14

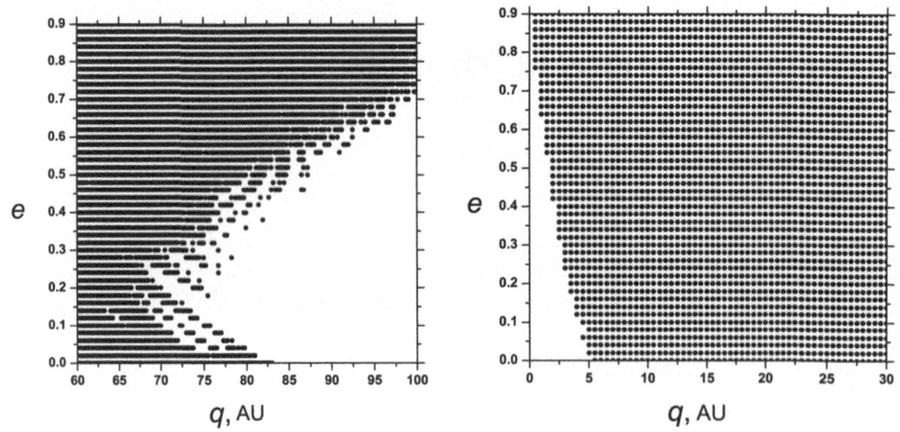

Fig. 14.1 Stability diagrams in the "pericentric distance q – eccentricity e" plane of initial orbital elements for hypothetical planets in the α Centauri A–B system. Areas of instability are shaded. Left panel: outer (circumbinary) planetary orbits. Right panel: inner (circumstellar) planetary orbits (Figures 2 and 8 from Popova and Shevchenko (2012). With permission from Pleiades Publishing Inc.)

binary star; therefore, the planetary systems of the single star Proxima Centauri and the binary star α Centauri A–B can be considered in isolation.

A numerical study, conducted in Popova and Shevchenko (2012) by means of massive numerical integrations along with calculations of Lyapunov exponents, in the planar problem, made it possible to identify zones of orbital stability of hypothetical circumstellar and circumbinary planets in the α Centauri A–B system in the space of planetary initial orbital elements (Fig. 14.1).

In the right panel of Fig. 14.1, the hypothetical (not yet confirmed) circumstellar planet, reported by Dumusque et al. (2012), would have the q and e coordinates close to the frame origin; therefore, the planet would be deeply inside the triangular zone of stability in the panel. At the planet's initial eccentricity equal to zero, the outer boundary radius of the stability domain is $\sim$5 AU.

In case of the circumbinary planetary motion (Fig. 14.1, left panel), the outer boundary radius (in the barycentric frame) of the circumbinary chaos zone is $\sim$80 AU. Planets with smaller barycentric radii of initially circular orbits cannot survive.

In Fig. 14.1, left panel, the chaos boundaries obviously exhibit a "ragged" pattern, which is explained by the emergence of major mean motion resonances with the central stellar binary. An analogous resonant pattern is considered in detail in Sect. 12.2 for the *Kepler*-16 circumbinary system.

According to Popova and Shevchenko (2012), for the chaotic circumbinary motion of planets in the α Centauri A–B system, the Lyapunov times turn out to be $\sim$500 yr; and for the chaotic circumstellar motion around any of the stellar components, they are $\sim$60 yr.

14.1.2 The υ Andromedae System

υ And (Upsilon Andromedae) is a binary star, comprising a yellow F8V dwarf and a red M4.5V dwarf. The binary size is 750 AU. At least four giant planets orbit around the yellow companion (Butler et al. 1999; McArthur et al. 2010; Ligi et al. 2012). According to McArthur et al. (2010); Ligi et al. (2012), the b, c, and d planets' masses are 0.62, 14.0, and 10.3 M_{J}, orbital semimajor axes are 0.060, 0.83, and 2.53 AU, and orbital periods are 4.6, 241, and 1290 d, respectively. The system is not planar: the orbital planes of the c and d planets are mutually inclined by 30°. The c and d planets are super-Jupiters. The orbit of the fourth discovered planet (υ And e) is larger than the orbits of the b, c, and d planets; its semimajor axis is 5.25 AU, orbital period 3849 d, and mass 0.96 M_{J} (Ligi et al. 2012).

The c and d planets are close to the mean motion 11/2 resonance; the four-planet system as a whole is long-term stable (Michtchenko and Malhotra 2004). The close-to-resonant configuration of the c–d pair is similar to that observed in the circumbinary planetary system *Kepler*-16, which is considered below in Sect. 14.2.

14.1.3 The 55 Cancri System

The main (most massive) component of the 55 Cancri binary star is a yellow G8 dwarf star. It is orbited by at least five planets with masses ranging from 0.034 to 3.84 M_{J}; their orbital periods range from 2.8 to 5200 days. Planets b and c are close to the 3/1 mean motion resonance, or may even reside in it. The system as a whole is long-term stable (Gayon et al. 2008). It may contain also other planets, since it is far from being tightly packed.

Gayon et al. (2008) identified the 55 Cnc planetary system to probably reside in a state of stable (bounded) chaos. Recall that there exist such chaotic objects in our Solar system; notably, the asteroid Helga, see Sect. 8.2.3. However, in the 55 Cnc planetary system, planetary masses and mutual inclinations of planets are not well enough constrained observationally. Therefore, the conclusion on the presence of chaos, bounded or not, in this system is still uncertain.

14.1.4 The 16 Cygni System

The Lidov-Kozai effect, considered in detail further on in Chap. 15, is often present in hierarchical circum-companion systems. It can be regarded as a kind of resonance, with the resonance argument (critical angle) equal to the argument of the particle's pericentre $\omega = \varpi - \Omega$, where ϖ is the longitude of the particle's pericentre and Ω is the longitude of the particle's ascending node (Morbidelli 2002; Shevchenko 2017a; Sidorenko 2018). The effect might be widespread in planetary systems of binary stars with highly inclined (with respect to the stellar binary's orbit plane) planetary orbits; in particular, it was shown to be possibly present in the 16 Cygni system (Holman et al. 1997) and in the γ Cephei system (Haghighipour et al. 2010). Moreover, the Lidov-Kozai effect was invoked to provide a general mechanism of production of hot Jupiters (Lithwick and Naoz 2011).

Holman et al. (1997) suggested that the motion of the planet 16 Cyg Bb orbiting the B component of the star 16 Cyg A–B might be chaotic. This planet's orbit is prone to the Lidov–Kozai oscillations, and the amplitude of ω-libration (libration of the argument of pericentre) turns out to be rather large. Consequently, the planet's motion might be close to the separatrix of the Lidov–Kozai resonance; see Fig. 14.2.

Conversely, Melnikov (2016) performed massive computations of the Lyapunov exponents of the planetary motion in this system on representative sets of observationally allowed initial conditions, and showed that the presence of chaos is unlikely; for chaos to emerge in this system, the planetary motion should take place closer to the Lidov–Kozai separatrix. Indeed, in comparison with the Lidov–Kozai resonance size, the near-separatrix chaotic layer in the Lidov–Kozai diagram (Fig. 14.2) is very narrow.

In more detail, chaotic implications of the Lidov–Kozai effect in exoplanetary systems are considered in Chap. 15.

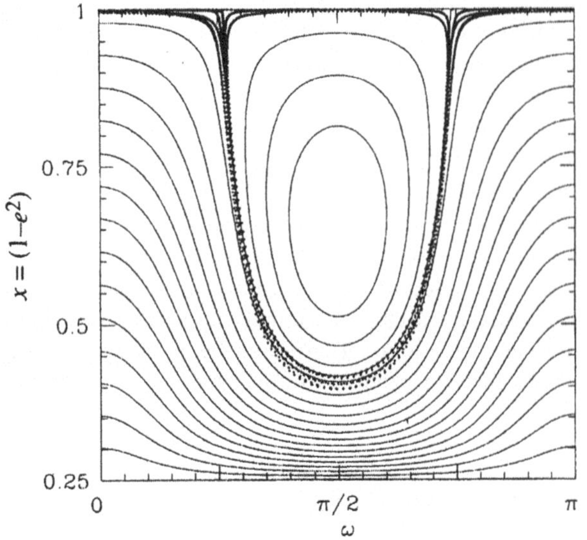

Fig. 14.2 The long-term Lidov–Kozai oscillations (dotted curves) in the motion of the planet 16 Cyg Bb. The integral curves and the Lidov–Kozai separatrix are superimposed (solid curves) (Figure 3 from Holman et al. 1997, by permission from Springer Nature, © 1997)

14.2 Circumbinary Systems

I send two Sunsets –

Emily Dickinson, Poem 308 (1862)

Amongst a couple of hundreds of exoplanets discovered up to now in systems of multiple stars, about two dozens are circumbinary ones, that is, they orbit around stellar binaries. Most (more than a half) of the circumbinary planets are hosted by evolved stellar binaries, consisting of white and red dwarfs in very close configurations. Conversely, the other half is formed by the circumbinary planets belonging to main-sequence binaries, whose components are Sun-like dwarfs; the first ever known such planets were discovered by the *Kepler* team in 2011–2012 (Doyle et al. 2011; Orosz et al. 2012a,b; Welsh et al. 2012).

14.2.1 Kepler*'s Circumbinary Planets*

The Neptune-like planet *Kepler*-16b is known to follow a circumbinary orbit around a system of two main-sequence stars (Doyle et al. 2011; Welsh et al. 2012). The TTV analysis allowed one to specify all orbital parameters of the planet with remarkable accuracy, see Table 14.1.

The empirical numerical-experimental criterion (11.103) for the stability of planetary circumbinary orbits identifies this system as stable, although not far ($\lesssim$10–20% in the orbital radius) from the outer border of the circumbinary chaotic zone. This estimate was confirmed in Doyle et al. (2011); Welsh et al. (2012) in long-term numerical integrations of the *Kepler*-16b's orbit.

Table 14.1 presents data on ten *Kepler*'s circumbinary planetary systems. The data is compiled from Doyle et al. (2011), Welsh et al. (2012), Orosz et al. (2012b), Orosz et al. (2012a), Schwamb et al. (2013), Kostov et al. (2013), Kostov et al.

Table 14.1 *Kepler*'s circumbinary planets

System	m_1, M_{Sun}	m_2, M_{Sun}	m_{pl}, M_J	a_b, AU	e_b	a_{pl}, AU	e_{pl}
Kepler-16	0.69	0.20	0.33	0.22	0.16	0.71	0.007
Kepler-34	1.05	1.02	0.22	0.23	0.52	1.09	0.18
Kepler-35	0.89	0.81	0.13	0.18	0.14	0.60	0.042
Kepler-38	0.95	0.25	0.38	0.15	0.10	0.46	<0.03
Kepler-47(b)	1.04	0.36	–	0.08	0.02	0.30	<0.04
Kepler-47(c)	1.04	0.36	–	0.08	0.02	0.99	–
Kepler-64	1.53	0.41	0.53	0.16	0.22	0.63	0.054
Kepler-413	0.82	0.54	0.21	0.10	0.04	0.36	0.12
Kepler-453	0.94	0.20	0.20	0.19	0.05	0.79	0.04
Kepler-1647	1.21	0.98	1.52	0.13	0.16	2.72	0.06

(2014), Welsh (2015), and Kostov et al. (2016). The masses of the stars are given in Solar units, and the masses of planets in Jovian units.

In Popova and Shevchenko (2013), stability charts for *Kepler*-16 in the "pericentric distance—eccentricity" plane were constructed, using a general approach analogous to that described in Sect. 14.1.1 on the α Centauri system; i.e., the planetary orbits were numerically integrated on a fine grid of initial values of the orbital pericentric distance and eccentricity, while other orbital elements were fixed at their values for a particular epoch. To assess the stability, two criteria were used. The first one is the value of the maximum Lyapunov exponent, and the second one is as follows: the orbit is stable if the planet does not escape (passing to hyperbolic orbit) from the system, or does not encounter with any of the two host stars. The resulting stability charts are shown in Fig. 14.3.

A direct inspection of Fig. 14.3b demonstrates that the Lyapunov exponent criterion provides a more clear-cut picture of the chaotic domain borders, in comparison with the escape-collision criterion, when equal time intervals of the integrations are used. To identify chaotic orbits at high pericentric distances and eccentricity, the escape-collision criterion can be used only if the time intervals of integration are appropriately increased.

The actual position of the *Kepler*-16b planet is marked in Fig. 14.3 by a dot. The planet turns out to reside almost at the edge of the chaos domain, in a hazardous vicinity to it, just between the instability "teeth." Linear extrapolation of these "teeth" to the $e = 0$ axis reveals that they are conditioned by integer resonances between the orbital periods of the planet and the central stellar binary. The two

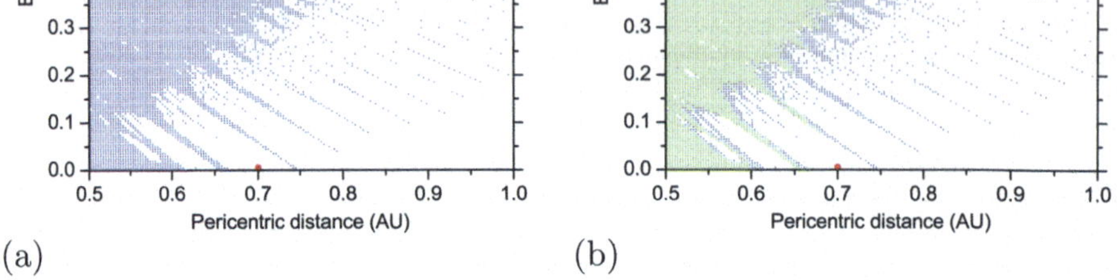

Fig. 14.3 Stability charts for *Kepler*-16b. (**a**) Left panel: the Lyapunov exponent criterion is used. (**b**) Right panel: both the Lyapunov exponent and escape-collision criteria are used. The initial conditions resulting in regular orbits are left white, those resulting in chaos are shown in blue colour, and those resulting in chaos and escapes and collisions are shown in green colour. The dot marks the location of *Kepler*-16b (Figure 2 from Popova and Shevchenko 2013, by permission of IOP Publishing/AAS)

teeth that closely surround *Kepler*-16b correspond to the 5/1 and 6/1 resonances. The smaller teeth centred between the "integer" teeth correspond to half-integer resonances.

Although in a dangerous vicinity to the chaos zone around the central binary, the *Kepler*-16b planet survives, because its orbit is deeply inside the cell between the chaotic bands conditioned by the mentioned integer resonances with the central binary (Popova and Shevchenko 2013). In fact, the planet is close to the half-integer 11/2 orbital resonance (which is stable at such large distance from the barycentre): the resonant tooth almost pointing at *Kepler*-16b corresponds to this resonance. However, there is no chaos where *Kepler*-16b resides, whereas the neighbouring "integer" teeth extend down to the $e = 0$ axis.

The resonant argument for a circumbinary mean motion resonance is defined by the formula

$$\sigma = (k + q)\lambda_{\text{pl}} - k\lambda_{\text{s}} - l\varpi_{\text{pl}} \tag{14.1}$$

(Murray and Dermott 1999; Morbidelli 2002), where λ_{s} and λ_{pl} are the mean longitudes of the secondary and the tertiary, respectively; ϖ_{pl} is the longitude of pericentre of the tertiary; here k, q and l are integer numbers (q is the resonance order). For the resonance, the ratio of orbital periods of a planet and the central binary is equal to $(k + q)/k$. In Eq. (14.1), the longitude of pericentre of the secondary is ignored, because it is practically constant.

Mean motion resonances split in multiplets of subresonances, corresponding to the l integer sequence. The coefficients of the subresonant terms in the perturbing function expansion are proportional to the orbital eccentricities of the particle and perturber in some powers depending on q; in particular, the coefficients of the first and last subresonant terms in the multiplet are proportional, respectively, to the eccentricities of the perturber and particle in the power equal to q, see Sect. 13.2.2. In the subresonance pendulum model, the subresonance width is proportional to the square root of this coefficient (Chirikov 1979; Holman and Murray 1996). Therefore, the resonance order q controls the subresonance width.

Consider two neighbouring outer integer resonances $(q + 1)/1$ and $(q + 2)/1$; their orders are q and $q+1$, respectively. The half-integer resonance between them is $(2q+3)/2$, and its order is $2q+1$. We see that, in comparison with its neighbouring integer resonances, a high-order half-integer resonance have the power-law indices in the subresonant term coefficients much greater. Therefore, the strengths of the subresonances are accordingly much less, and their interaction is much weaker. On increasing e, the subresonances start to overlap much later than in the neighbouring integer resonances. This explains why the 11/2 resonance at the diagram location of *Kepler*-16b is stable, whereas the neighbouring 5/1 and 6/1 resonances are unstable. Thus, the planet is safe inside the resonance cell bounded by the unstable 5/1 and 6/1 resonances.

In the Solar system, this phenomenon is analogous to the survival of Pluto and Plutinos (Popova and Shevchenko 2013). They are in the 3/2 outer orbital resonance with Neptune; therefore, the order of the occupied high-integer resonance is much smaller than in the *Kepler*-16 system. The mass parameter μ of the Sun–Neptune binary is much less and, therefore, the chaos border in the stability diagram radically shifts to smaller pericentric distances.

The resonant structure of the Kuiper belt was discussed above in Sect. 8.2.2. One may see that the analogy with the case of resonant trans-Neptunian objects is striking. According to Gladman et al. (2012), the population of TNOs in the next half-integer resonance (5/2) with Neptune is estimated to be as large as in the 3/2 resonance, or even greater, whereas other (non-half-integer) resonant populations are radically smaller.

14.2.2 Migration and Chaos as System Architects

It might seem from the dynamical chart in Fig. 14.3 that no radial migration was possible for *Kepler*-16b since its formation epoch, because otherwise it would cross several chaotic bands corresponding to integer resonances with the central binary, and, therefore, it could be removed. On the other hand, any in situ formation of *Kepler*-16b is a theoretical challenge (Meschiari 2012; Paardekooper et al. 2012).

However, the presence of zones of instability on the migration path does not necessarily mean catastrophic consequences. From Sect. 7.1.1, one may recall that the planetary satellites in the Solar system are known to be slowly tidally despun, until they reach the 1/1 synchronous spin-orbit resonance. In the course of despinning, the satellites cross a number of chaotic layers in the phase space of motion (Wisdom 1987a); the broadest layer is at the separatrix of the synchronous resonance. Nevertheless, all tidally-evolved satellites (with a possible exception of no more than three of them, see Kouprianov and Shevchenko 2005; Melnikov and Shevchenko 2010) at present reside in the final stable synchronous resonance, as the Moon does. Clearly, most of the satellites were able to cross the chaotic layers without being caught in chaos forever. The cause is that the timescales for developing the gross instability are usually much longer in comparison with the times needed for crossing the chaotic zones (Wisdom 1987a; Kouprianov and Shevchenko 2005). In the given planetary migration case, the interplay of timescales can be analogous (Popova and Shevchenko 2013).

The currently favoured scenario for the formation of circumbinary planets, within the planet accretion framework, is as follows: the planetary core forms farther out in the protoplanetary disc (in its accretion-friendly zone) and then migrates inward, until the migration is stalled at the border of the disc's inner cavity cleared by the central stellar binary (Pierens and Nelson 2007; Meschiari 2012; Paardekooper et al. 2012). The cavity is comparable in size to the circumbinary chaos zone. Paardekooper et al. (2012) estimate the final locations of *Kepler*-16b, 34b and 35b to be close to the truncation radii of the gas discs. Although the in situ formation of *Kepler*-16b is still not ruled out (Meschiari 2012), it is less likely, owing to the hostile conditions for planetesimal accretion in such close vicinities to the stellar binary. The hostile conditions comprise, in particular, high encounter velocities of planetesimals and low planetesimal spatial density (Meschiari 2012; Paardekooper et al. 2012).

An important role of orbital resonances in the formation scenarios was outlined in Moriwaki and Nakagawa (2004) and Pierens and Nelson (2007, 2008). Moriwaki and Nakagawa (2004) pointed out that the pumped eccentricities of planetesimals, as a function of the semimajor axis, show "interesting behaviour such as somewhat resonant features;" see Figure 1 in Moriwaki and Nakagawa (2004). Such resonances with the central binary as the 5/1 resonance affect the formation and orbital evolution of giant Saturn-mass planets embedded in a circumbinary disc, as data of hydrodynamic simulations by Pierens and Nelson (2007, 2008) show.

At least two astrophysical processes are known to date that may lead to matter being trapped in high-order outer resonances in circumbinary systems. In a dynamical (e.g., any planetary) system, whose parameters are slowly (adiabatically) varying, captures in resonances may occur; in particular, Plutinos are believed to have been trapped in the 3/2 resonance with Neptune in the Kuiper belt due to the outward migration of this planet. Generally, the outward migration of a planet may lead to capture of outer particles in circumbinary mean motion resonances with the inner star-planet binary; see Quillen (2006) and references therein. Another relevant astrophysical process takes place in dusty debris discs around stars with planets: when, due to dissipation forces, dust spirals inward, it can be efficiently captured in resonances; see Deller and Maddison (2005); Quillen (2006) and references therein.

Deller and Maddison (2005) performed simulations of the debris disc evolution in the Fomalhaut system with a hypothetical giant planet; notable examples of matter trapping in high-order circumbinary resonances were demonstrated: see, for example, Figure 14 in Deller and Maddison (2005), where integer (such as 4/1 and 5/1) and half-integer (5/2, 7/2, and 9/2) resonances dominate or are prominent in the "semimajor axis—resonance occupation" plot.

Why there is only one sole resonance cell in the *Kepler*-16 system that is occupied by a planet? At least for the cells neighbouring to the *Kepler*-16b planet the answer is straightforward (Popova and Shevchenko 2013). Again, it concerns resonances and their interaction, but the resonances are different from those considered above: these are the first-order orbital resonances $(p+1)/p$ of the test particles with the planet. Equation (11.5) gives the radial half-size of the planetary chaotic zone emerging due to overlap of these resonances. Using Eq. (11.5) and data on $a_{\rm pl}$ and masses from Doyle et al. (2011), and setting the total stellar mass $M_{\rm s} = M_1 + M_2$, one finds $\mu_{\rm pl} = 3.56 \cdot 10^{-4}$ and $\Delta a_{\rm cr} \approx 0.095$ AU for the *Kepler*-16b's planetary chaotic zone. Therefore, at least two neighbouring resonance cells, namely those centred at the 9/2 and 13/2 circumbinary resonances, are purged by *Kepler*-16b, residing in the 11/2 resonance cell, because these two resonances are within the $\Delta a \approx 0.1$ AU radial distance; see Fig. 14.3.

Dynamical properties of circumbinary planets in the *Kepler*-34 and 35 systems are qualitatively similar to those of the *Kepler*-16 system; see Popova and Shevchenko (2013). The multiplanet circumbinary system *Kepler*-47 has at least two planets, with the planet c moving in a much larger orbit than the planet b. The ratios of the orbital periods of the innermost planet and the binary star, as follows from data given in Orosz et al. (2012a,b), are equal to 5.62 and 6.65 in the systems *Kepler*-38 and *Kepler*-47, respectively. Therefore, the over-all planetary dynamical configurations and states (including planets not yet discovered) in these systems might be similar to those in the *Kepler*-16 and *Kepler*-35 systems, where the period ratios are 5.57 and 6.34, respectively.

14.2.3 Chaotic Zones

Zones of orbital instability are known to exist around binary stars. In particular, such zones are known to be present in circumbinary protoplanetary discs. The latter may contain planetesimals, dust, and gas. Gas is present during the initial stages of the disc evolution; it dissipates later on. Numerical simulations show that, irrespective of the gas content, a free-from-matter central cavity always forms in the planetesimal disc around a gravitationally bound pair of stars, whose masses are comparable; see Moriwaki and Nakagawa (2004), Pierens and Nelson (2007, 2008), Meschiari (2012), and Paardekooper et al. (2012).

The existence and possible characteristics of central cavities in gaseous circumbinary discs were first ever considered theoretically in Artymowicz and Lubow (1994, 1996), in view of the observational data on the discs around stars of GW Ori type. Artymowicz and Lubow outlined the role of Lindblad resonances in the cavity formation.

In a quite separate field of study, namely, on the dynamics of triple stars, it is well known that the stability of hierarchical triple stars is mostly determined by the pericentric distance of the tertiary (Mikkola 2008; Valtonen et al. 2008; Saito et al. 2012, 2013): if this distance can decrease below a critical value, the system is unstable. A number of heuristic semi-analytical criteria for the critical pericentric distance were proposed; see reviews by Mikkola (2008) and Valtonen et al. (2008). Most of these criteria do not appeal to resonant phenomena, with a few exceptions.

Mudryk and Wu (2006) explored local chaos borders in the phase space of motion of planets in binary star systems, and revealed the overlap of subresonances of mean motion resonances, responsible for the borders' emergence. In the general three-body problem framework, Mardling (2008a) considered the role of overlap of mean motion resonances for the stability of triples and developed an algorithm for constructing the chaos composite border as an envelope of a set of local resonance borders. For hierarchical triples, Mardling (2008a) revealed the major role (for the border formation) of the overlap of orbital resonances $p/1$ (where p is integer) between the tertiary and the central binary.

If the masses of the gravitating binary components are not comparable, i.e., one of them is much less than the other one, then a narrow annular chaotic band exists, surrounding the orbit of the secondary, with the radial width $\propto \mu^{2/7}$, see Eq. (11.5). Extrapolating the validity range of formula (11.5) from small values of the mass parameter μ to $\mu = 1/2$ (the equal-mass case), one finds $\Delta a_{\rm cr}/a' \sim 1.3$, i.e., the zone inside the binary is expected to be totally continuously chaotic, and the outer border of this continuous chaotic zone is expected to be situated at $\sim(0.5 + 1.3)a' \approx 1.8a'$ from the barycentre. Further on we shall see that the actual size of the chaos zone is severely underestimated here; in reality, it is $\approx 2.8a'$. The cause for the discrepancy is not just a numerical uncertainty of the extrapolation. It is conceptual: the class of resonances responsible for the formation of the circumbinary chaotic zone is different from that responsible for the formation of the annular chaotic band. To show this, we use the Kepler map theory (Shevchenko 2015).

As discussed above in connection with Eq. (11.92), the energy width of a one-sided chaotic band in the vicinity of the perturbed parabolic orbit scales as the power 2/5 of the mass parameter:

$$\Delta E_{\rm cr} = |E_{\rm cr}| = -E_{\rm cr} \propto \mu^{2/5}, \tag{14.2}$$

if $\mu \ll 1$. The particles with $E \in (-\Delta E_{\rm cr}, 0)$ move chaotically.

It is curious that the power-law index in the scaling $R_{\rm H} \propto \mu^{1/3}$ for the radius of the *regular* zone (the Hill sphere, see Sect. 11.1.6) around the secondary is intermediate between the indices in the Wisdom and "Kepler-map" scalings, given by Eqs. (11.5) and (11.92), respectively. Thus, the indices form the "2/5, 2/6, 2/7" sequence.

We linearize the Kepler map (8.35) in the y variable near the fixed point at the border of the map's chaotic layer. In this way, the Kepler map is locally approximated by the standard map, which represents a mathematical model of a multiplet of equally-spaced equally-sized resonances, see Sect. 3.4. Then, for the layer's border location, one finds

$$y_{\rm cr} = \left(\frac{3\lambda}{2K_{\rm G}}\right)^{2/5}, \tag{14.3}$$

where $K_{\rm G} = 0.971635406\ldots$, see Sect. 1.5. Using Eqs. (11.87) and (11.89) for the map parameters W and λ, one arrives at Eqs. (11.93) and (11.94) (Shevchenko 2007a).

What if $\mu \approx 1/2$, i.e., the binary is approximately equal-mass? This is quite common in stellar binaries. According to (Roy and Haddow 2003, formula (26)), the energy increment in the $\mu = 1/2$ case in the restricted three-body problem limit is given by

$$\delta E \simeq -2^{7/4}\pi^{1/2}q^{3/4} \exp\left(-\frac{2^{5/2}q^{3/2}}{3}\right) \sin 2g_i. \tag{14.4}$$

The possibility to use the Kepler map at moderate and high values of μ (i.e., at $\mu \sim 1/2$) was discussed in Shevchenko (2010). Note that, in Eq. (11.88), the harmonic term in the first line ($\propto \sin g_i$) is just the first most prominent one in the

Fourier expansion of the energy increment, if $\mu \ll 1$ (Petrosky 1986; Shevchenko 2011b). If one increases μ, the second harmonic ($\propto \sin 2g_i$) becomes more and more important. If $\mu = 1/2$, the first harmonic ($\propto \sin g_i$) disappears, whereas the second one ($\propto \sin 2g_i$) becomes the largest one, in amplitude, in the series expansion. Indeed, due to the equality of the primaries' masses, the perturbation frequency is effectively doubled. Thus, the Kepler map, formally, takes the form

$$E_{i+1} = E_i + W_{1/2} \sin 2g_i,$$
$$g_{i+1} = g_i + 2\pi |2E_{i+1}|^{-3/2} \qquad (14.5)$$

(Shevchenko 2015), where $W_{1/2} = \delta E$, given by formula (14.4). By substituting $E = W_{1/2} y$ and $g = x/2$, map (14.5) is reducible to map (8.35) with

$$\lambda = 2^{1/2} \pi W_{1/2}^{-3/2}. \qquad (14.6)$$

The critical $y_{\rm cr}$ is given by Eq. (14.3). Therefore, in the given $\mu = 1/2$ case one has

$$\Delta E_{\rm cr} = \left| W_{1/2} y_{\rm cr} \right| \simeq A_{1/2} q^{3/10} \exp\left(-B_{1/2} q^{3/2}\right), \qquad (14.7)$$

where

$$A_{1/2} = 2^{1/2} 3^{2/5} \pi^{3/5} K_{\rm G}^{-2/5} = 4.4122\ldots, \qquad B_{1/2} = 2^{7/2}/15 = 0.7542\ldots$$

and $K_{\rm G} = 0.971635406\ldots$, see Sect. 1.5. The critical eccentricity $e_{\rm cr}$ is given by formula (11.94).

Naturally, the critical curve given by Eq. (14.7) is different from the limiting curve that results from extrapolating Eq. (11.93) from small μ values to $\mu \to 1/2$, although they are rather similar. This difference is due to the fact that the actual energy increment, in the both cases, is not given by a single harmonic term, but represents a Fourier series of harmonic terms (Petrosky 1986; Liu and Sun 1994), while we took only the leading terms of these series. These are the terms $\propto \sin g_i$ (in case of $\mu \ll 1$) and $\propto \sin 2g_i$ (in case of $\mu \approx 1/2$). Other harmonic terms can be no less important, when q is close to unity (Liu and Sun 1994); leaving apart the case of crossing orbits.

In Fig. 14.4, theoretical "pericentric distance q—critical eccentricity $e_{\rm cr}$" relationships are graphically presented for several values of the mass parameter μ. The just derived formulas for the critical curves are used. In these plots, any theoretical curve with a fixed μ separates the chaos domain (situated to the left of the curve) from the regular domain (situated to the right of the curve), at the given μ value.

Extrapolating the critical curves to the $e = 0$ axis, one may see that, if $\mu = 0.1$, the curve hits the horizontal axis at $q = 2.3$, and, from Kepler's third law, this corresponds to the ratio $\approx$3.5 between the orbital periods of the tertiary and the central binary. Another relevant value, $q = 2.8$, where the critical curve with

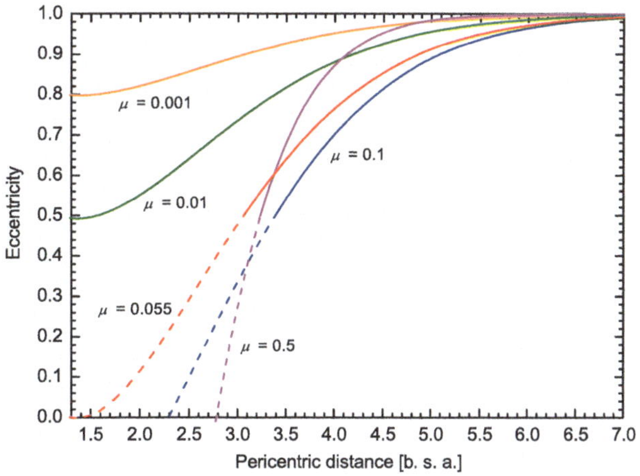

Fig. 14.4 Critical curves, given by Eq. (11.94), for several values of the mass parameter μ. The curves separate the chaotic and regular zones; the chaotic zone is on the left. Extrapolations are dashed. The pericentric distance is measured in the units of the central binary's semimajor axis, abbreviated as "b. s. a." at the horizontal axis caption (Figure 1 from Shevchenko 2015, by permission of IOP Publishing/AAS)

$\mu = 0.5$ hits the horizontal axis, corresponds to the ratio $\approx$4.7. Therefore, the extrapolation of the critical curve to the $e = 0$ axis gives q values that are rather insensitive to μ, if the latter's value is in the range 0.1–0.5, typical for stellar binaries. We see that if μ is in this range, then the border of the circumbinary chaotic zone corresponds to the circumbinary resonances from 7/2 to 5/1 with the central binary. In units of the central binary's semimajor axis, the chaotic zone radial size is $\approx$3 at moderate eccentricities, and $\approx$2.3–2.8 at zero eccentricities of the tertiaries; see Fig. 14.4.

To emphasize the extrapolative character of the critical curves at low eccentricities, in Fig. 14.4 they are dashed, if $e < 0.5$. The validity of the extrapolation is justified here post factum: the curves corresponding to $\mu \sim 0.1$–0.5 hit the $e = 0$ axis at high enough values of q, at which the higher-order harmonics in the Fourier expansion of the energy increment are relatively unimportant, because these harmonics are exponentially small with the harmonic order j, as they are proportional to $\exp[-4jq^{3/2}/(2^{1/2}3)]$ (Petrosky 1986; Petrosky and Broucke 1988).

At $e = 0$, the estimates for the chaotic zone size can be compared to that given by the numerical-experimental criterion of Holman and Wiegert (1999); see formula (11.103). Since formula (11.103) was proposed in Holman and Wiegert (1999) for the mass parameter interval $\mu \geq 0.1$, the comparisons can be made also solely at $\mu \geq 0.1$. Setting $e_b = 0$ in Eq. (11.103), at $\mu = 0.1$ and 0.5 one obtains $a_{cr}/a_b = 2.0$ and 2.4, respectively, whereas Fig. 14.4 gives $a_{cr}/a_b = 2.3$ and 2.8. Taking into account the "ragged" character of the global chaos border (see, for example, Fig. 14.3), one may state that the agreement is quite plausible.

The critical borders at $e = 0$ can be also compared to semi-analytical data of Szebehely (1980) and Szebehely and McKenzie (1981), who employed computations of the topology of the zero-velocity curves in the circular restricted three-body problem. At $\mu = 0.1,\ 0.24$, and 0.5, Szebehely (1980) and Szebehely and McKenzie (1981) obtained $a_{cr}/a_b \approx 2.24,\ 2.4$, and 2.17, whereas, from the presented above theoretical formulas, one has $a_{cr}/a_b \approx 2.29,\ 2.91$, and 2.79, respectively. At $\mu = 0.1$ the agreement is perfect, and it is even better than that with the data of Holman and Wiegert (1999). A comparison of the data of Szebehely and McKenzie (1981) with formula (11.103) is discussed in Holman and Wiegert (1999).

On increasing μ, the critical curves in Fig. 14.4 start to hit the $e = 0$ axis when $\mu \approx 0.0547 \sim 0.05$. This μ value thus represents the threshold at which the central continuous chaotic zone starts to be present at all eccentricities of the tertiary.

This μ threshold has a clear-cut dynamical sense: above it, the tertiary's orbit, even starting from a small eccentricity, can diffuse, following the staircase of the overlapping $p/1$ resonances with the central binary, up to $e \sim 1$, i.e., up to ejection from the system; close encounters with other bodies are not required for the escape.

Let us see how Holman–Wiegert's and Wisdom's μ dependences, Eqs. (11.103) and (11.5), can be related. If one extrapolates the polynomial fit (11.103) to zero μ, in the circular problem ($e_b = 0$), then one gets $a_{cr}/a_b = 1.6$, whereas in reality in this limit there is no circumbinary chaotic zone at all, and, as follows from Eq. (11.5), the width of the planetary chaotic band is zero. Therefore, the transition between Holman–Wiegert's and Wisdom's dependences takes place somewhere in the interval $0.001 \lesssim \mu \lesssim 0.1$. Indeed, Wisdom's law was verified in numerical experiments at least up to $\mu = 0.001$ (Murray and Dermott 1999; Quillen and Faber 2006); and Holman and Wiegert (1999) obtained fit (11.103) at $\mu \geq 0.1$.

The junction of the two relationships is illustrated in Fig. 14.5, where the "mass parameter–critical semimajor axis" analytical relationships are presented

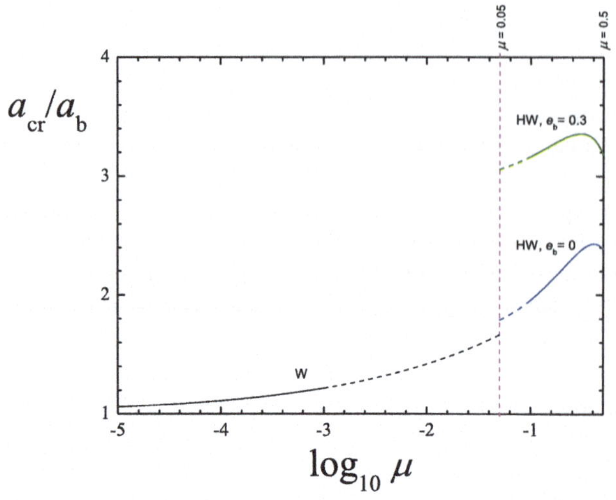

Fig. 14.5 The "mass parameter—critical semimajor axis" analytical relationships. Curve "W": Wisdom's law, given by Eq. (11.5). Curves "HW": Holman–Wiegert's empirical relation, given by Eq. (11.103), at two values of the central binary eccentricity. Extrapolations are dashed. The vertical dashed (magenta) line shows the theoretical threshold $\mu = 0.05$ (Figure 2 from Shevchenko 2015, by permission of IOP Publishing/AAS)

graphically. Wisdom's law is valid, as originally derived, at $e_b = 0$; Holman–Wiegert's curve, drawn for $e_b = 0$, joins Wisdom's curve at $\mu \approx 0.05$ rather smoothly.

As pointed out in Quillen and Faber (2006), at $\mu \leq 0.001$ and $e_b < 0.3$ the chaotic zone size is virtually independent of e_b and is described by Wisdom's law. Therefore, it is adequate to compare how Wisdom's and Holman–Wiegert's relations join, if $e_b = 0.3$. From Fig. 14.5, it is evident that a jump or a sharp rise should be definitely present in the "uncertainty interval" around $\mu \approx 0.05$, i.e., the chaotic zone size rises sharply somewhere in the interval. Therefore, in the eccentric case, the transition from the chaos originating due to the overlap of $(p+1)/p$ resonances to the chaos originating due to the overlap of $p/1$ resonances results not only in the change of the diffusion character, but also in the sharp change of the chaotic zone size.

Tantalizingly, the $\mu \approx 0.05$ threshold roughly corresponds to the μ value at which the loss of stability of the triangular Lagrangian points L_4 and L_5 takes place; the latter value is ≈ 0.04, see Szebehely (1967). Dynamically, this coincidence looks natural, because the transition to global chaos, originating due to the overlap of $p/1$ resonances, leaves no place for regular islands in the motion phase space around the triangular libration points.

The μ threshold existence seems to explain an old numerical-experimental result by Nacozy (1976) that the Sun–Jupiter–Saturn system becomes unstable if μ is increased ≈ 29 times, i.e., up to ~ 0.03. This is rather close to 0.05, taking into account that the problem differs from the restricted one. Kholshevnikov and Kuznetsov (2011) obtained an even smaller numerical-experimental threshold μ value for the system gross instability upsurge, namely, $\mu \sim 0.02$.

Further on, in Sect. 14.2.5, the theoretical prediction for the μ threshold is compared directly with relevant observational data on exoplanetary systems.

14.2.4 Structure of Chaos Border

Analytical approaches, proposed by Mudryk and Wu (2006), Mardling (2008a), Mardling (2008b) in the three-body problem framework, allows one to describe borders of chaotic domains in dynamical charts of various kinds by considering separatrices of individual mean-motion resonances, whose overlap forms the chaos global border.

Within the framework of the hierarchical three-body problem in the circumbinary case, Mardling (2008a,b) analytically represented the chaos global border as a combination of the borders (separatrices) of separate integer mean motion resonance cells. The circumbinary resonances form the sequence $n_b/n_{pl} = 1, 2, 3, \ldots$, where n_b are n_{pl} are the mean motions of the inner stellar binary and the planet, respectively.

According to Mardling (2008a,b), the function describing the separatrices of any resonance with $n_b/n_{pl} \gtrsim 3$ from this sequence is given by

$$\sigma(e_{out}) = k \pm 2|A_k(e_{out})|^{1/2}, \tag{14.8}$$

in units of the ratio of the mean motions of the central binary and the tertiary, $k \gtrsim 3$ are integers, and

$$A_k(e_{out}) = -\frac{9}{2} S(e_{in}) F_k(e_{out}) \left(M_{in} + M_{out} k^{2/3} \right), \tag{14.9}$$

$$S(e_{in}) \approx -3e_{in} + \frac{13}{8} e_{in}^3 + \frac{5}{192} e_{in}^5 - \frac{227}{3072} e_{in}^7, \tag{14.10}$$

$$F_k(e_{out}) = \frac{4}{3(2\pi)^{1/2}} \cdot \frac{(1 - e_{out}^2)^{3/4} k^{3/2}}{e_{out}^2 \exp\left[k\xi(e_{out})\right]}, \tag{14.11}$$

$$M_{in} = \frac{m_3}{m_1 + m_2 + m_3}, \tag{14.12}$$

$$M_{out} = \frac{m_1 m_2}{(m_1 + m_2)^2} \left(\frac{m_1 + m_2}{m_1 + m_2 + m_3} \right)^{2/3}, \tag{14.13}$$

$$\xi(e_{out}) = \cosh^{-1}\left(e_{out}^{-1}\right) - \left(1 - e_{out}^2\right)^{1/2}, \tag{14.14}$$

where $m_1 \geq m_2$ are the masses of the binary's components, $m_3 \ll m_2$ is the mass of the planet, e_{in} and e_{out} are the eccentricities of the central binary's orbit and the planet's orbit, respectively.

Series of analytical separatrices $\sigma(e_{out})$ are presented, as black thin curves, in Figs. 14.6 and 14.7 for the *Kepler*-16 and *Kepler*-47 systems. The nominal positions of integer mean motion resonances are shown by ref dashed lines. One may see that the theoretical borders, formed by the separatrices in total, are in an adequate agreement with numerical-experimental data also presented in the diagrams.

Note that the separatrices of any individual mean motion resonance are actually split due to interaction with neighbouring mean motion resonances, that is why the theoretical unperturbed separatrices describe the local chaotic borders in the stability charts.

The thick green solid curve describes the location of the averaged chaos border, as predicted by the Kepler map theory, Eq. (11.94). The thick black dotted curve is described by the same Equation, but for the border defined at the approximating standard map stochasticity parameter $K = 4$ instead of $K = K_G$. The latter curve, therefore, gives the border to the left of which any large regular islands are predicted to be absent, because the approximating standard map lacks any significant regular

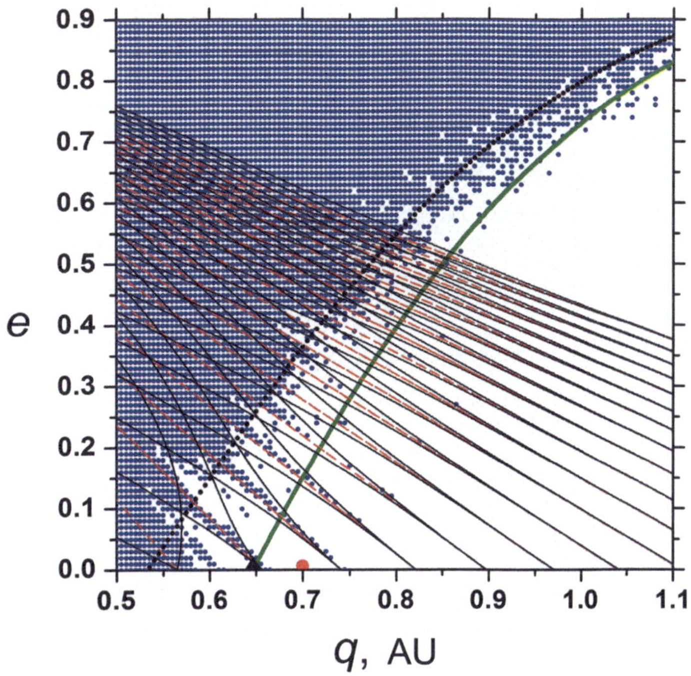

Fig. 14.6 The stability chart for *Kepler*-16. Blue shaded area: the chaos domain, as identified numerically in Popova and Shevchenko (2016b). Red dot: actual position of the planet. Black triangle: the critical semimajor axis, given by formula (11.103). For an explanation of the theoretical (green solid and black dotted) curves see the text (Figure 2 from Popova and Shevchenko 2016b. With permission from Pleiades Publishing Inc.)

component at $K > 4$, as the integer resonances bifurcate at this value of the stochasticity parameter.

Why the chaotic resonant bands in the stability charts for *Kepler*-16 (Fig. 14.6) look prominent, whereas for *Kepler*-47 (Fig. 14.7) the chaos border is not so ragged? The chaos inside the resonant bands is caused by interaction and overlap of subresonances of mean motion resonances, see Eqs. (13.14) and (13.17). However, as follows from these Equations, the splitting into subresonances occurs only when the central binary in non-circular. Indeed, only one term in the subresonance multiplet has a non-zero coefficient if $e_b = 0$. Therefore, any pronounced chaotic resonant bands in the stability diagrams are expected only if the central binary is eccentric.

For circular binaries, the separatrices of any individual mean motion resonance are split due to interaction with neighbouring mean motion resonances, and, in the stability diagrams, they look chaotic mostly at locations not far from the global chaos averaged border.

These qualitative considerations agree well with numerical data; indeed, for the systems with close-to-circular host stellar binaries, such as, e.g., *Kepler*-47 with

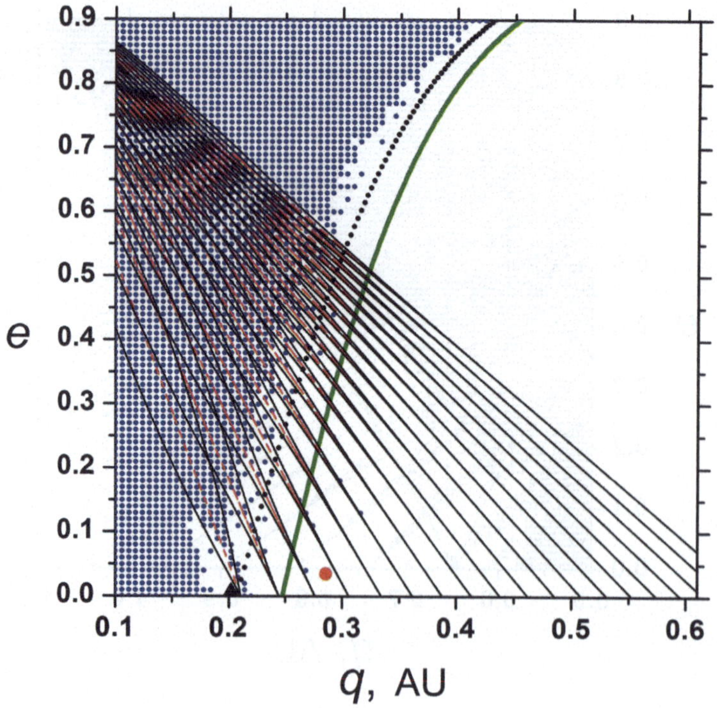

Fig. 14.7 Same as Fig. 14.6, but for *Kepler*-47 (Figure 4 from Popova and Shevchenko 2016b. With permission from Pleiades Publishing Inc.)

$e_b = 0.02$ or *Kepler*-413 with $e_b = 0.04$ (see Table 14.1), the "raggedness" of the global chaos border in the stability charts is significantly suppressed; see Figs. 14.7 and 12.6.

14.2.5 The Mass Parameter Threshold

The mass parameter threshold $\mu \approx 0.05$, introduced above in Sect. 14.2.3, essentially manifests itself in the diversity of the observed orbital configurations of exoplanet systems. To demonstrate this, let us consider an empirical relationship between the mass parameter μ of the primary binary and the ratio of the tertiary's and secondary's orbital periods, T_{out}/T_{in} (Shevchenko 2015).

Two classes of exoplanet systems are directly relevant: two-planet (a single star with two planets) and circumbinary (a binary star plus a planet in an orbit around it). In the first case, it is required that the planet in the outermost orbit have the smallest mass in the system.

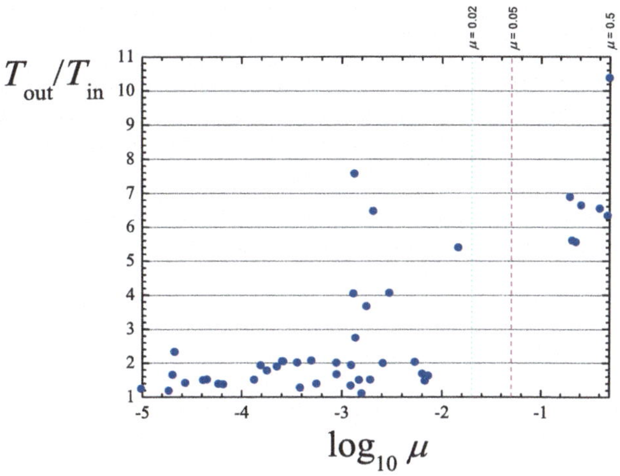

Fig. 14.8 The "mass parameter—orbital period ratio" relationship for two-planet systems of single stars and one-planet circumbinary systems. Dots: observational data from Exoplanet Encyclopedia (exoplanet.eu). The vertical dashed (magenta) line marks the theoretical threshold $\mu = 0.05$. The vertical dotted (cyan) line corresponds to $\mu = 0.02$, see the text (Figure 4 from Shevchenko 2015, by permission of IOP Publishing/AAS)

The resulting diagram is shown in Fig. 14.8. The actually observed two-planet configurations all turn out to reside to the left of two theoretical (vertical dashed and vertical dotted) lines, whereas the circumbinary systems are all situated to the right of them. The vertical dashed (magenta) line indicates the theoretical threshold $\mu = 0.05$ for the emergence of the circumbinary chaotic zone. The dotted (cyan) line at $\mu = 0.02$ marks the numerical-experimental threshold that signals the upsurge of instability of the Sun–Jupiter–Saturn system on raising the system mass parameter, as estimated in Nacozy (1976) and Kholshevnikov and Kuznetsov (2011).

In Fig. 14.8, the total absence of exoplanet systems with $T_{\text{out}}/T_{\text{in}} < 5$ at $\mu > 0.05$ is evident, in agreement with the theoretical prediction that, at $\mu > 0.05$, there exists the circumbinary continuous chaotic zone, in which all particles are eventually ejected from the system.

Two comments are in order (Shevchenko 2015). First, a certain gap in μ exists between the two classes of exoplanet systems comprising the sample. Indeed, in case of multiplanet systems of single stars, the mass ratio of a "Solar–Jovian" star–planet binary is $\sim$0.001; and, in case of planetary systems of binary stars, the mass parameter μ of the typical main-sequence binary is $\sim$1/2. The gap can be filled in the future, when more relevant exoplanet systems are identified, such as systems in which the central star–planet pair is composed of a brown dwarf and a Jovian-type planet, or such as systems in which the host stellar binaries are composed of a main-sequence star and a brown dwarf. A prototype of the latter system is the HD 202206 system (with $\mu \approx 0.014$), in which the star-orbiting inner planet has mass $\approx$17.4 in Jovian units; thus, it is most likely not a planet but a brown dwarf (Correia et al.

2005). In Fig. 14.8, the dot marking this system is the closest one to the vertical cyan line corresponding to $\mu = 0.02$.

Second, as follows from Fig. 14.8, many exoplanet systems with $\mu < 0.01$ are clustered close to the 2/1 orbital resonance; whereas the systems with $\mu > 0.1$ do not seem to prefer any integer resonance, but rather half-integer ones. This in accord with an observation that the known circumbinary planets survive, although they are at the brink of circumbinary chaos, because they are safe inside resonance cells whose borders are formed by unstable integer resonances; see Sect. 14.2.1.

Chapter 15
The Lidov–Kozai Effect: Chaotic Implications

Une telle orbite sera sous certaines conditions bouleversée par la planète perturbatrice.

Hugo von Zeipel (1910)

(Astron. Nachr., Bd. 183, Nr. 4390–92, 22.)

In this chapter, we consider the potential role of the Lidov–Kozai effect in evoking dynamical chaos in multiplanet systems and planetary systems of binary stars. The Lidov–Kozai effect reveals itself in coupled periodic variations (which can be very large) of the inclination and eccentricity of an orbiting body in presence of an inclined perturber. The effect is known to be important in the motion of many asteroids and planetary satellites. What is more, now it attracts more and more interest in the astronomical and astrophysical community due to its relevance to dynamics of many exoplanetary systems. The presentation in this chapter is partially based on the material of Section 8.6 in Shevchenko (2017a), by permission from Springer Nature, © 2017. The effect itself and its modern applications are described in detail in Shevchenko (2017a).

Highly-eccentric circumstellar orbits are observed in planetary systems of some stellar binaries, in particular, in the 16 Cygni and γ Cephei systems; this can be explained in the Lidov–Kozai effect framework (Holman et al. 1997; Innanen et al. 1997; Mazeh et al. 1997; Takeda and Rasio 2005; Haghighipour et al. 2010). The Lidov–Kozai effect can be active in planetary systems of binary stars, as well as in multiplanet systems of single stars, or acted in previous epochs of their long-term dynamical evolution, imprinting modern orbital configurations. Free-floating planets can be produced partly due to the Lidov–Kozai effect, because the effect is able to push planets to extreme apocentric distances, thus providing conditions for the planetary escape.

The Lidov–Kozai effect may have profound importance for theories of planet formation. In young binary stars that contain planetesimal discs around one of the stellar companions, the effect may excite orbital eccentricities of the planetesimals. Therefore, if the perturber's orbit is inclined enough with respect to the disc, the planet formation can be hindered (Marzari and Barbieri 2007; Fragner et al. 2011).

I. I. Shevchenko, *Dynamical Chaos in Planetary Systems*, Astrophysics and Space Science Library 463, https://doi.org/10.1007/978-3-030-52144-8_15

The discovery of a qualitative multitude of orbital behaviours in exoplanetary systems boosted celestial-mechanical studies of their secular dynamics (see Lee and Peale (2003), Greenberg and Van Laerhoven (2012) and references therein), based on earlier theoretical works on secular dynamics of triple stars (Marchal 1990; Ford et al. 2000). As an essential constituent of the dynamical exoplanetary studies, the Lidov–Kozai effect theory was extended and refined (Libert and Henrard 2007; Lithwick and Naoz 2011).

As we have already seen in Sect. 14.1.4 on the example of the 16 Cyg system, the Lidov–Kozai effect may cause dynamical chaos, if the planetary orbit is close enough to the separatrix of the Lidov–Kozai resonance. In this chapter, we consider the dynamical chaos due to the Lidov–Kozai effect in a general framework.

15.1 Multiplanet Systems

Already in the nineties, Innanen et al. (1997) showed that, in the presence of a distant stellar companion, the Lidov–Kozai oscillations of the eccentricity and inclination in exoplanetary systems may evoke planet-planet scattering. A planet orbiting one of the binary star components may attain, during the Lidov–Kozai oscillations, an eccentricity high enough for its encounters with other planets to become possible. If so, the planetary system of the component is eventually disrupted. In clusters of young stars, newly-born planets can be thus torn away from their host stars (Malmberg et al. 2007), to enrich the population of the *free-floating planets*. The same mechanism may work in multiplanet systems of single stars.

The Lidov–Kozai effect manifestations in systems of quadruple stars and triple stars with a planet can be even more pronounced (Beust and Dutrey 2006; Marzari and Barbieri 2007).

The Lidov–Kozai effect can be active in configuring binary planets in exoplanetary systems. Indeed, there exists an analogy with binary asteroids and binary TNOs in our Solar system. For the binary minor bodies, Perets and Naoz (2009) and Fang and Margot (2012) showed that the Lidov–Kozai effect may control the long-term inner dynamical evolution of such objects, as they are perturbed by the distant Sun.

Such binary systems as the binary asteroids and TNOs secularly evolve due to encounters with other bodies, mutual tides, and the Sun-induced Lidov–Kozai effect; this evolution has resulted in the observed almost isotropic distribution of the inclinations of the binary systems' planes with respect to the planes of their orbits around the Sun; the inclinations are therefore typically high (Naoz et al. 2010). For close enough binaries, the Lidov–Kozai effect in concert with the tidal friction may produce circularized short-period binaries and may force the components in the binaries to coalescence (Perets and Naoz 2009). These inferences on the binary minor bodies, after rescaling, might be applicable to assess statistics of orbital properties of binary planets in exoplanet systems.

If the system is in the Lidov–Kozai resonance, the secular variations in the eccentricity e and inclination i are coupled: they are in antiphase, if $i < \pi/2$, and in phase, if $i > \pi/2$ (see Shevchenko 2017a). If the initial inclination i_0 (of the inner binary with respect to the outer perturber's orbital plane) is greater than a critical value, then the maximum eccentricity achieved by the inner binary during the Lidov–Kozai cycle is insensitive to e_0 (if $e_0 \lesssim 0.1$) and is given by

$$e_{\max} \approx \left(1 - \frac{5}{3}\cos^2 i_0\right)^{1/2} \quad (15.1)$$

(Holman et al. 1997; Innanen et al. 1997). At the centre of the Lidov–Kozai resonance, the period of oscillations in the eccentricity and inclination is

$$P_{\mathrm{LK}} \approx P_1 \frac{(m_0 + m_1)}{m_2} \left(\frac{a_2}{a_1}\right)^3 (1 - e_2^2)^{3/2} \quad (15.2)$$

(Mazeh and Shaham 1979; Holman et al. 1997), where P_1, m_0, m_1, and a_1 are, respectively, the orbital period, masses and semimajor axis of the inner binary; P_2, m_2, and e_2 are, respectively, the orbital period, mass, and eccentricity of the outer perturber.

If the amplitude of ω-libration (libration of the argument of pericentre) is not too large, i.e., the system is not too close to the separatrix of the Lidov–Kozai resonance, then the libration period can be estimated by the same formula (15.2). In a slightly different version derived by Kiseleva et al. (1998), the formula reads

$$P_{\mathrm{LK}} \approx \frac{2}{3\pi} \frac{P_{\mathrm{pert}}^2}{P_1} \frac{(m_0 + m_1 + m_{\mathrm{pert}})}{m_{\mathrm{pert}}} (1 - e_{\mathrm{pert}}^2)^{3/2}. \quad (15.3)$$

If the Lidov–Kozai effect generic conditions are satisfied, then the number of the Lidov–Kozai oscillations in any typical exoplanetary system may exceed hundreds or even thousands, while the parent star stays on the main sequence (Kiseleva et al. 1998; Ford et al. 2000). For the Solar-like stars, the latter timescale is ~10 Gyr.

15.2 Planetary Systems of Binary Stars

As follows from formula (15.3), the Lidov–Kozai oscillation period sharply rises with increasing the relative size of the outer perturber orbit. Therefore, the Lidov–Kozai effect in wide stellar binaries is much less significant than in close binaries. However, even in the wide binaries, the Lidov–Kozai oscillation timescale can be sufficiently small to make the planetary orbit highly eccentric in realistic times.

The presence of the Lidov–Kozai oscillations in a number of planetary systems of binary stars is confirmed by observational data statistics, which show that planets with extreme orbital eccentricities mostly orbit around stars with distant stellar companions (Naoz et al. 2011).

In exoplanetary studies, the Lidov–Kozai effect was first invoked to explain a highly-eccentric orbit of an exoplanet in the 16 Cyg B system, as already discussed above in Sect. 14.1.4.

The 16 Cyg A–B system is a binary consisting of two Solar-like (G1.5V and G2.5V) stars, separated from each other by $\sim 10^3$ AU. The 16 Cyg A has a minor stellar companion, an M-dwarf 16 Cyg C, whose orbit around 16 Cyg A has radius $\approx$73 AU. Radial-velocity measurements showed that the binary component 16 Cyg B has a Jovian-mass planet, now designated 16 Cyg Bb (Cochran et al. 1997). According to Plávalová and Solovaya (2013), the planet's mass, semimajor axis, orbital period, and eccentricity are 2.38 M_J, 1.693 AU, 799.5 d, and 0.689, respectively. Such a large eccentricity, $e \approx 0.7$, looks rather unexpected for a giant planet, judging from the behaviour of giant planets in our Solar system; besides, the planets are believed to form in protoplanetary discs in which the planetesimal motions are circularized.

The planet's orbit around 16 Cyg B is subject to perturbations from 16 Cyg A, the distant companion of the host star. Holman et al. (1997) and Mazeh et al. (1997) put forward a hypothesis that the planet's large eccentricity is excited by an inclined outer stellar perturber, by the Lidov–Kozai mechanism. In this system, the perturber is 16 Cyg A. However, as pointed out by Holman et al. (1997), the general relativity may suppress the Lidov–Kozai effect. For the suppression to happen, the inequality $a_2 q_2 \lesssim 3 \cdot 10^5$ must hold, where the stellar binary's semimajor axis a_2 and pericentric distance q_2 are measured in AU. This inequality is derived by combining analytical expressions for the Lidov–Kozai and relativistic timescales; see Holman et al. (1997).

Holman et al. (1997) showed that, given a small initial eccentricity and the initial inclination in the range 45–135°, the planet may suffer secular eccentricity variations up to 0.8, being in the high-eccentricity ($e > 0.6$) state up to 35% of its lifetime. The amplitudes of changes in the eccentricity and inclination are independent of the stellar binary's a_2 and q_2; these two quantities only affect the Lidov–Kozai oscillation timescale. On the other hand, the relativistic apsidal precession of 16 Cyg Bb may have the period as low as 70 mln years.

In Fig. 14.2, the contour plot is shown as built by Holman et al. (1997) to describe the secular dynamics of 16 Cyg Bb. This is a vivid example, though a strictly model one, demonstrating the pronounced Lidov–Kozai resonant pattern in the (ω, x) plane. Here ω is the argument of pericentre, $x = 1 - e^2$, e is eccentricity; the motion constant

$$c_1 = (1 - e^2) \cos^2 i,$$

where i is inclination, is set to 0.25.

In the diagram, results of a direct numerical integration of the non-averaged equations of motion are superimposed on the contour plot. The initial conditions for the integration are chosen to be near the separatrix of the Lidov–Kozai resonance;

therefore, the model motion is chaotic and the resulting curve on the graph is irregular.

15.3 Chaos Due to the Lidov–Kozai Effect

The double-averaged Hamiltonian of the elliptic restricted three-body problem can be written explicitly in terms of canonical variables. Namely, the expansion of the Hamiltonian in a power series of a_1/a_{pert} up to the second (quadrupole) and third (octupole) orders is given by

$$\mathcal{H} = \mathcal{H}_{\mathrm{quad}} + \epsilon_{\mathrm{oct}} \mathcal{H}_{\mathrm{oct}}, \tag{15.4}$$

where

$$\mathcal{H}_{\mathrm{quad}} = \frac{1}{2}(1 - J^2) - \frac{J_z^2}{J^2} - \frac{3(1 - J^2)J_z^2}{2J^2} - \frac{5}{2}\frac{(1 - J^2)J^2}{(J^2 - J_z^2)} \cos 2\omega, \tag{15.5}$$

$$\begin{aligned}
\mathcal{H}_{\mathrm{oct}} = & -\frac{5}{16}\left[(1 - J^2)^{1/2} + \frac{3}{4}(1 - J^2)^{3/2}\right] \\
& \times \left[\left(1 - \frac{11J_z}{J} - \frac{5J_z^2}{J^2} + \frac{15J_z^3}{J^3}\right) \cos(\omega - \Omega) \right. \\
& \left. + \left(1 + \frac{11J_z}{J} - \frac{5J_z^2}{J^2} - \frac{15J_z^3}{J^3}\right) \cos(\omega + \Omega)\right] \\
& + \frac{175}{64}(1 - J^2)^{3/2} \times \left[\left(1 - \frac{J_z}{J} - \frac{J_z^2}{J^2} + \frac{J_z^3}{J^3}\right) \cos(3\omega - \Omega) \right. \\
& \left. + \left(1 + \frac{J_z}{J} - \frac{J_z^2}{J^2} - \frac{J_z^3}{J^3}\right) \cos(3\omega + \Omega)\right]
\end{aligned} \tag{15.6}$$

(Li et al. 2014b; see also Lithwick and Naoz 2011, and Li et al. 2014a). Here $(J, \ \omega)$ and $(J_z, \ \Omega)$ are pairs of canonically conjugated variables, and

$$J = (1 - e_1^2)^{1/2}$$

is the unitless (scaled) angular momentum of the inner orbit,

$$J_z = (1 - e_1^2)^{1/2} \cos i_1$$

is the vertical component of the angular momentum; e_1 and i_1 are the eccentricity and inclination of the inner orbit; ω and Ω are the argument of pericentre and the longitude of ascending node of the inner orbit, respectively; $\epsilon_{\rm oct}$, by definition, is

$$\epsilon_{\rm oct} = \frac{a_1}{a_{\rm pert}} \cdot \frac{e_{\rm pert}}{(1 - e_{\rm pert}^2)}. \tag{15.7}$$

In the general three-body problem, the role of the octupole terms can be estimated by the value of the parameter

$$\epsilon_{\rm oct}^{\rm gen} = \frac{(m_0 - m_1)}{(m_0 + m_1)} \cdot \frac{a_1}{a_{\rm pert}} \cdot \frac{e_{\rm pert}}{(1 - e_{\rm pert}^2)}, \tag{15.8}$$

where $m_0 > m_1$ are the masses of the star and its planet, a_1 is the semimajor axis of the planet's orbit, $a_{\rm pert}$ and $e_{\rm pert}$ are the semimajor axis and eccentricity of the outer perturber's orbit (Shappee and Thompson 2013).

The time variable in the equations of motion can be made unitless by scaling (dividing) it by the constant

$$t_{\rm LK} = \frac{8}{3} \cdot \frac{m_0}{m_{\rm pert}} \left(\frac{a_{\rm pert}}{a_1}\right)^3 \cdot (1 - e_{\rm pert}^2)^{3/2}, \tag{15.9}$$

where m_0 and $m_{\rm pert}$ are the masses of the host star and the outer perturber, respectively. The Hamiltonian (15.6) is scaled accordingly.

Numerically built sections of the phase space of the Hamiltonian (15.4) at several representative values of $\mathcal{H}$ and $\epsilon_{\rm oct}$ are given in Li et al. (2014b). They are reproduced here in Figs. 15.1 and 15.2. The chaotic layers around the separatrices of

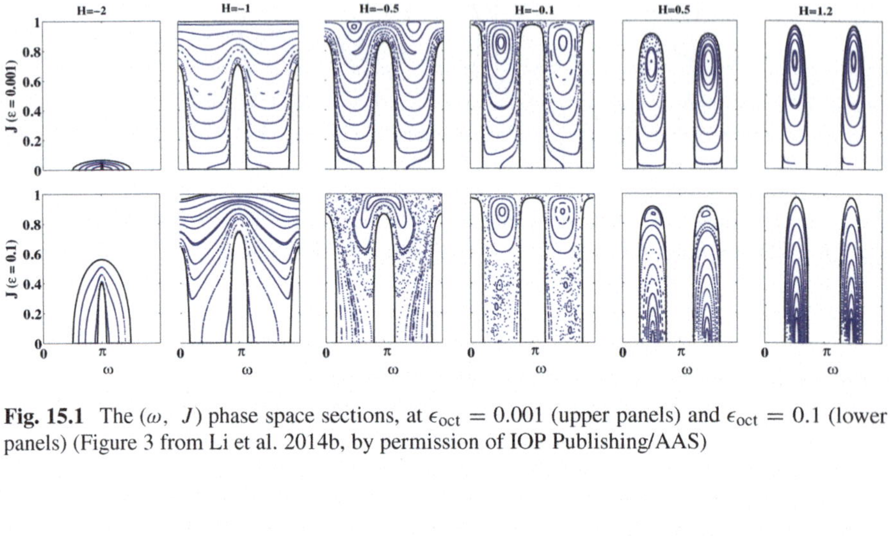

Fig. 15.1 The $(\omega,\ J)$ phase space sections, at $\epsilon_{\rm oct} = 0.001$ (upper panels) and $\epsilon_{\rm oct} = 0.1$ (lower panels) (Figure 3 from Li et al. 2014b, by permission of IOP Publishing/AAS)

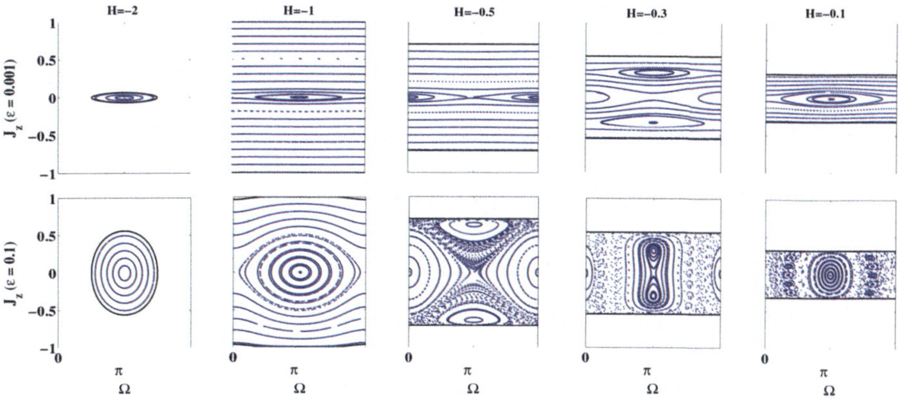

Fig. 15.2 The (Ω, J_z) phase space sections, at the same values of ϵ_{oct} as in Fig. 15.1 (Figure 4 from Li et al. 2014b, by permission of IOP Publishing/AAS)

resonances are clearly identifiable in the both Figs. 15.1 and 15.2, namely, in their panels corresponding to $-0.5 \le \mathcal{H} \le -0.1$.

Analytical descriptions of the properties of motion in the chaotic layers of the Lidov–Kozai resonance (widths of the layers, the Lyapunov and diffusion timescales, etc.) represent a problem for future studies.

Chapter 16
Epilogue

Worlds scoop their Arcs –
And Firmaments – row –

Emily Dickinson, Poem 216 (1861)

"We are living in the new age of discovery"—by these words Murray and Dermott (1999) start their book *Solar System Dynamics*. Indeed, in the last three decades of the twentieth century, space probes and space telescopes made the Solar system planets, comets, asteroids, and planetary satellites as just familiar to us, as, some five hundred years ago, other continents became familiar to mankind thanks to seafarers. In this sense, our times are in no way less important and impressive than the epoch of great geographical discoveries.

In view that thousands of planetary systems, far away from our Solar system, have been discovered and explored in astronomical observations during the last three decades, nowadays this citation acquires a novel broader meaning, and in 2020 one may repeat it with even greater surety and emphasis. Ages of novel geographical or planetary discoveries are rare in the history of mankind, and we are lucky to live in such times.

Exoplanet discoveries trigger studies in relevant areas of fundamental science, including non-linear dynamics and dynamical chaos theory. In the field of numerical-experimental studies, fast and effective numerical tools, allowing one to visualize global charts of chaotic transport, are being developed and improved. Current and perspective studies of chaotic behaviour in exoplanetary systems comprise investigations of various clearing effects and mechanisms, explaining how the low-mass material in planetary systems is steadily removed from chaotic zones due to chaotic diffusion in orbital elements. Developments of analytical criteria of stability of planetary systems, in broad ranges of parameter values and initial conditions, are warranted.

A lot of dynamical and statistical phenomena, newly found in observations of exoplanet systems, are inherently related to various dynamical chaos effects. These phenomena still await rigorous and complete theoretical explanation and interpretation, especially in conjunction with planetary formation theories. Among the actual

I. I. Shevchenko, *Dynamical Chaos in Planetary Systems*, Astrophysics and Space Science Library 463, https://doi.org/10.1007/978-3-030-52144-8_16

subjects of studies in exoplanetary dynamics, there are: resonant and near-resonant architectures of circumbinary systems, peculiarities in statistical distributions of planet period ratios, prevalence of particular resonances in exoplanetary systems, resonant architectures of closely packed exoplanet systems, circumbinary planetary dynamics at the "edge of chaos," "rapid dynamical chaos" and "bounded dynamical chaos" in planetary systems, long-lived planetary systems with relatively short Lyapunov times, and many others. After all, these studies help to understand our own Solar system.

Appendix A
Basic Notations

In this Appendix, mathematical, astronomical and physical notations used in the book are listed. Note that, in the book, some variations in symbol definitions, as well as deviating notations, can be sometimes encountered. They were introduced for convenience of reading, where appropriate.

Mathematical and Physical Quantities

- i is the imaginary unit
- $\psi(z) = \frac{\Gamma'(z)}{\Gamma(z)}$ is the digamma-function
- $K(m)$ or $K(k)$ (where $m = k^2$) is the complete elliptic integral of the first kind with modulus k
- $E(m)$ or $E(k)$ is the complete elliptic integral of the second kind
- x_0 is the initial value of a variable x

Coordinates and Frames

- x, y, z are the Cartesian (orthogonal) coordinates
- r, ϕ, α are the spherical coordinates (radial distance, longitude, and latitude)

Orbital Elements and Corresponding Quantities

- $a, e, i, \omega, \varpi, \Omega, M, l$ are, respectively the semimajor axis, eccentricity, inclination, argument of pericentre, longitude of pericentre, longitude of ascending node, mean anomaly, mean longitude of a test body
- $q = a(1 - e)$ is the pericentric distance (the letter q is also used to designate resonance order)
- $Q = a(1 + e)$ is the apocentric distance

I. I. Shevchenko, *Dynamical Chaos in Planetary Systems*, Astrophysics and Space Science Library 463, https://doi.org/10.1007/978-3-030-52144-8

- $a_{\rm pert}$, $e_{\rm pert}$, $i_{\rm pert}$, $\omega_{\rm pert}$, $\varpi_{\rm pert}$, $\Omega_{\rm pert}$, $M_{\rm pert}$, $l_{\rm pert}$ are, respectively the semimajor axis, eccentricity, inclination, argument of pericentre, longitude of pericentre, longitude of ascending node, mean anomaly, mean longitude of a perturber's orbit
- $a_{\rm pl}$, $e_{\rm pl}$ are, respectively the semimajor axis and eccentricity of a planetary orbit
- $a_{\rm b}$, $e_{\rm b}$ are, respectively the semimajor axis and eccentricity of a binary
- a_1, a_2 are the semimajor axes of the inner and outer binaries, respectively
- e_1, e_2 are the eccentricities of the inner and outer binaries, respectively
- $\alpha = a_1/a_2$ is the ratio of semimajor axes of the inner and outer binaries
- $e_{\rm f}$ is the forced eccentricity
- e_i, ϖ_i, i_i, and Ω_i are, respectively, the eccentricity, longitude of pericentre, inclination, and longitude of ascending node of planet i
- n is the mean motion of a test body
- $n_{\rm pert}$ is the mean motion of a perturber
- P_1 is the orbital period of the inner binary
- P_2 is the orbital period of the outer binary
- τ is the time of a pericentre transit

Dynamical Definitions

- n is the number of degrees of freedom
- $\mathcal{R}$ is the perturbing function
- $\mathcal{H}$ is a Hamiltonian
- $\mathbf{q}$ is the vector of canonical coordinates
- $\mathbf{p}$ is the vector of conjugate canonical momenta
- l, g, h are the Delaunay canonical angles, corresponding to the mean anomaly M, argument of pericentre ω, and longitude of ascending node Ω, respectively
- L, G, H are the Delaunay canonical momenta, conjugate to the Delaunay canonical angles
- $\omega \equiv \varpi - \Omega$ is the argument of pericentre, the critical angle of the Lidov–Kozai resonance
- g_k and s_k are the planetary frequencies, $k = 1, 2, \ldots, 8$
- ν_5 is a secular resonance
- ν_6 is a secular resonance
- ν_{16} is a secular resonance
- K is the stochasticity parameter
- $K = K_{\rm G} = 0.971635406\ldots$ is the critical (Greene's) value of the stochasticity parameter
- λ is the adiabaticity parameter

Physical Quantities

- m_1 is the mass of the primary, the most massive body in a system
- m is the mass of a test body (an asteroid, a planet)
- $m_{\rm pert}$ is the mass of a perturber
- $\mu = m_2/(m_1 + m_2)$ is the mass parameter of a binary
- $M_{\rm Sun}$ is the mass of the Sun

- m_{pl} is the mass of a planet
- $\mathcal{G}$ is the gravitational constant
- c is the speed of light
- R_{pl} is the mean radius of a planet
- R_{Sun} is the radius of the Sun
- R_{Earth} is the mean radius of the Earth
- R_{Moon} is the mean radius of the Moon
- R_{H} is the Hill radius
- G is the module of the angular momentum
- H is the angular momentum vector's vertical component
- Ω is the perturbation frequency
- τ is the phase angle of perturbation
- L is the maximum Lyapunov exponent
- $T_{\text{L}} \equiv L^{-1}$ is the Lyapunov time
- ω_0 is the frequency of small-amplitude oscillations on a resonance
- C_{J} is the Jacobi integral, also called the Jacobi constant

List of Abbreviations

CBP	Circumbinary planet
FA	Frequency analysis
FLI	Fast Lyapunov indicators
GJ	Gliese–Jahreiss catalogue
HD	Henry Draper catalogue
HQR	"Householder transformation—QR-decomposition" method
IAU	International Astronomical Union
KAM	Kolmogorov–Arnold–Moser theory
KBO	Kuiper belt object
LCE	Lyapunov characteristic exponent
LF1	Lévy flights of the first kind
LF2	Lévy flights of the second kind
LKE	Lidov–Kozai effect
MA	Melnikov–Arnold integrals
MEGNO	Mean exponential growth factor of nearby orbits
MEM	Maximum eccentricity method
MS	Main sequence
NEA	Near-Earth asteroids
QR	QR-decomposition, i.e., a decomposition of a matrix into a product of an orthogonal matrix (Q) and an upper triangular matrix (R)
RV	"Radial velocity" method
TNO	Trans-Neptunian object
TTV	"Transit timing variations" method

Appendix B
Astronomical Constants and Parameters

The astronomical constants and parameters are given in Table B.1 as adopted by the 16th General Assembly of the International Astronomical Union in 1976. Numerical estimates presented in this book are usually based on these quantities.

Table B.1 Astronomical constants and parameters

Constant	Value
Gravitational constant $\mathcal{G}$	$6.672 \cdot 10^{-8}$ g^{-1}cm^{3}s^{-2}
Speed of light c	$2.997925 \cdot 10^{10}$ cm/s
Astronomical unit (AU)	$1.49600 \cdot 10^{13}$ cm
Parsec	$3.0857 \cdot 10^{18}$ cm = 206265 AU
Solar mass $M_{\rm Sun}$	$1.989 \cdot 10^{33}$ g
Solar radius $R_{\rm Sun}$	$6.960 \cdot 10^{10}$ cm
Earth's mass $M_{\rm Earth}$	$5.977 \cdot 10^{27}$ g
Mass ratio of Earth and Moon	81.30
Mass ratio of Sun and Earth	332958
Earth's equatorial radius $R_{\rm Earth}$	$6.378140 \cdot 10^8$ cm
Earth's mean radius $\langle R_{\rm Earth} \rangle$	$6.371032 \cdot 10^8$ cm
Jovian mass $M_{\rm J}$	$1.898 \cdot 10^{30}$ g = $1/1047\ M_{\rm Sun}$ = $317.8\ M_{\rm Earth}$
Neptune's mass $M_{\rm N}$	$1.030 \cdot 10^{29}$ g =$1/19314\ M_{\rm Sun}$ = $17.25\ M_{\rm Earth}$
Jovian mean radius $\langle R_{\rm J} \rangle$	$0.70 \cdot 10^{10}$ cm = $11.0\ \langle R_{\rm Earth} \rangle$
Neptune's mean radius $\langle R_{\rm N} \rangle$	$2.45 \cdot 10^9$ cm = $3.85\ \langle R_{\rm Earth} \rangle$

(continued)

I. I. Shevchenko, *Dynamical Chaos in Planetary Systems*, Astrophysics and Space Science Library 463, https://doi.org/10.1007/978-3-030-52144-8

Table B.1 (continued)

Constant	Value
Semimajor axes of planetary orbits in the Solar system	
Mercury	0.3871 AU
Venus	0.7233 AU
Earth	1 AU
Mars	1.524 AU
Jupiter	5.204 AU
Saturn	9.583 AU
Uranus	19.22 AU
Neptune	30.11 AU
Pluto	39.48 AU
Planetary orbital periods in the Solar system	
Mercury	0.2408 yr
Venus	0.6153 yr
Earth	1 yr
Mars	1.881 yr
Jupiter	11.86 yr
Saturn	29.46 yr
Uranus	84.02 yr
Neptune	164.8 yr
Pluto	248.0 yr

Appendix C
The Melnikov–Arnold Integrals

The Melnikov–Arnold integrals, as defined in (Chirikov 1977, 1979), are given by

$$A_n(\lambda) = \int_{-\infty}^{\infty} \cos\left(\frac{n}{2}\phi(t) - \lambda t\right) dt, \tag{C.1}$$

where λ is any real, and $n \geq 0$ is generally also real, but hereafter we assume n to be integer; and

$$\phi(t) = 4 \arctan \exp(t) - \pi. \tag{C.2}$$

The functions can be recast as

$$A_n(\lambda) = 2 \int_{0}^{\infty} \cos(n \arctan \sinh(t) - \lambda t)\, dt. \tag{C.3}$$

Related functions, given by

$$B_n(\lambda) = \int_{0}^{\infty} \sin(n \arctan \sinh(t) - \lambda t)\, dt, \tag{C.4}$$

are hereafter called the Melnikov–Arnold integrals of the second kind, whereby A_n are called the Melnikov–Arnold integrals of the first kind. We also introduce

$$\alpha_n^c(\lambda) = A_n(\lambda) + A_n(-\lambda),$$
$$\alpha_n^s(\lambda) = A_n(\lambda) - A_n(-\lambda),$$

I. I. Shevchenko, *Dynamical Chaos in Planetary Systems*, Astrophysics and Space Science Library 463, https://doi.org/10.1007/978-3-030-52144-8

$$\beta_n^c(\lambda) = B_n(\lambda) + B_n(-\lambda),$$
$$\beta_n^s(\lambda) = B_n(\lambda) - B_n(-\lambda). \tag{C.5}$$

Equivalently

$$\alpha_n^c(\lambda) = 4\int_0^\infty \cos(n \arctan \sinh(t)) \cos(\lambda t)\, dt, \tag{C.6}$$

$$\alpha_n^s(\lambda) = 4\int_0^\infty \sin(n \arctan \sinh(t)) \sin(\lambda t)\, dt, \tag{C.7}$$

$$\beta_n^c(\lambda) = 2\int_0^\infty \sin(n \arctan \sinh(t)) \cos(\lambda t)\, dt, \tag{C.8}$$

$$\beta_n^s(\lambda) = -2\int_0^\infty \cos(n \arctan \sinh(t)) \sin(\lambda t)\, dt. \tag{C.9}$$

The upper indices c and s thus indicate the presence of the cosine or sine, respectively, in the integrand.

As functions of t, the primitives of the integrand expressions in α_n^c, α_n^s, β_n^c, β_n^s, A_n, and B_n oscillate, and, depending on the parity of n, in the limit $t \to +\infty$ the oscillation amplitude can be nonzero. As in Chirikov (1979), to determine the Melnikov–Arnold integrals of both kinds, we average over these oscillations. Therefore, in particular, $A_0 = 0$.

In Chirikov (1979), A_n were estimated analytically at any real n by means of the residue theory. For n natural, the both integrals A_n and B_n can be calculated by induction.

Zhirov's recurrent relation for A_n in terms of A_{n-1} and A_{n-2}, where $n \geq 2$, is presented in (Chirikov 1979, Equation A.8). It specifies that, as soon as A_0 and A_1 are known, any A_n can be found by induction:

$$A_0(\lambda) = 0,$$
$$A_1(\lambda) = \frac{\alpha_1^c(\lambda) + \alpha_1^s(\lambda)}{2},$$
$$\ldots$$
$$A_n(\lambda) = \frac{2\lambda}{n-1} A_{n-1}(\lambda) - A_{n-2}(\lambda). \tag{C.10}$$

For B_n, a recurrent relation can be derived analogously (see Shevchenko 2000a), and it is given by

$$
\begin{aligned}
B_0(\lambda) &= -\frac{1}{\lambda}, \\
B_1(\lambda) &= \frac{\beta_1^c(\lambda) + \beta_1^s(\lambda)}{2}, \\
&\dots \\
B_n(\lambda) &= \frac{2\lambda}{n-1} B_{n-1}(\lambda) - B_{n-2}(\lambda) + \frac{2}{n-1}.
\end{aligned}
\tag{C.11}
$$

In the induction bases, A_1 and B_1 are expressed through α_1^c, α_1^s, β_1^c, and β_1^s, which themselves can be expressed through known integrals contained in Gradshtein and Ryzhik (1962) or Beitmen and Erdeii (1969); one has

$$\alpha_1^c(\lambda) = \frac{2\pi}{\cosh \frac{\pi\lambda}{2}}, \tag{C.12}$$

$$\alpha_1^s(\lambda) = \frac{2\pi}{\sinh \frac{\pi\lambda}{2}}, \tag{C.13}$$

$$\beta_1^c(\lambda) = 2\,\mathrm{Re}\left(\psi\left(i\frac{\lambda}{2}\right) - \psi\left(i\frac{\lambda}{4}\right)\right) - 2\ln 2, \tag{C.14}$$

$$\beta_1^s(\lambda) = \pi \tanh\frac{\pi\lambda}{2} - 2\,\mathrm{Im}\,\psi\left(\frac{1+i\lambda}{4}\right). \tag{C.15}$$

Here i is the imaginary unit, and $\psi(z) = \frac{\Gamma'(z)}{\Gamma(z)}$ is the digamma-function. The real part of the digamma-function of an imaginary argument is given by

$$\mathrm{Re}\,\psi(iy) = -C + y^2 \sum_{n=1}^{\infty} \frac{1}{n(n^2+y^2)} \tag{C.16}$$

(Abramowitz and Stegun 1970), where y is any real, $C \approx 0.577216$ is Euler's constant.

Equations (C.11), (C.14) and (C.15) allows one to calculate the Melnikov–Arnold integrals of the second kind of any order n in terms of special functions. For algorithms of effective numerical estimating the Melnikov–Arnold integrals see Shevchenko (2000a).

Bibliography

Abdullaev, S. S. (2006) *Construction of Mappings for Hamiltonian Systems and Their Applications* (Springer, Berlin, Heidelberg)
Abdullaev, S. S., & Zaslavsky, G. M. (1995) "Self-similarity of stochastic magnetic field lines near the X-point." Phys. Plasmas, **2**, 4533–4541
Abdullaev, S. S., & Zaslavsky, G. M. (1996) "Application of the separatrix map to study perturbed magnetic field line near the separatrix." Phys. Plasmas, **3**, 516–528
Abramowitz, M., & Stegun, I. A. (1970) *Handbook of Mathematical Functions* (Dover, New York)
Adrianova, L. Ya. (1995) *Introduction to Linear Systems of Differential Equations* (Translations of Mathematical Monographs **146**) (American Mathematical Society, Providence)
Agekian, T. A., Anosova, Zh. P., & Orlov, V. V. (1983) "Decay time of triple systems." Astrophysics, **19**, 66–70
Agol, E., Steffen, J., Sari, R., & Clarkson, W. (2005) "On detecting terrestrial planets with timing of giant planet transits." Mon. Not. R. Astron. Soc., **359**, 567–579
Ahn, T., Kim, G., & Kim, S. (1996) "Analysis of the separatrix map in Hamiltonian systems." Physica D, **89**, 315–328
Anglada-Escudé, G., Amado, P. J., Barnes, J., Berdiñas, Z. M., Butler, R. P., et al. (2016) "A terrestrial planet candidate in a temperate orbit around Proxima Centauri." Nature, **536**, 437–440
Anosova, J. P., Orlov, V. V., & Pavlova, N. A. (1994) "Dynamics of nearby multiple stars. The α Centauri system." Astron. Astrophys., **292**, 115–118
Arnold, V. I. (1963a) "Proof of Kolmogorov's theorem on the preservation of quasi-periodic motions under small perturbations of the Hamiltonian." Russian Math. Surveys, **18**, 6, 9–36
Arnold, V. I. (1963b) "Small denominators and problems of stability of motion in classical celestial mechanics." Russian Math. Surveys, **18**, 6, 85–193
Arnold, V. I. (1989) *Mathematical Methods of Classical Mechanics* (Springer-Verlag, New York)
Arnold, V. I., Kozlov, V. V., & Neishtadt, A. I. (2006) *Mathematical Aspects of Classical and Celestial Mechanics*, 3rd Edition (Springer, New York)
Artymowicz, P., & Lubow, S. H. (1994) "Dynamics of binary-disk interaction. I. Resonances and disk gap sizes." Astrophys. J., **421**, 651–667
Artymowicz, P., & Lubow, S. H. (1996) "Interaction of young binaries with protostellar disks." In: *Disks and Outflows around Young Stars*, ed. by Beckwith, S. V. W., et al. (Springer, New York) pp. 115–131
AstDyS website: http://hamilton.dm.unipi.it/cgi-bin/astdys/ (Maintained by A. Milani, Z. Knežević and coworkers)

I. I. Shevchenko, *Dynamical Chaos in Planetary Systems*, Astrophysics and Space Science Library 463, https://doi.org/10.1007/978-3-030-52144-8

Aubry, S. (1983) "The twist map, the extended Frenkel–Kontorova model and the devil's staircase." Physica D, **7**, 240–258

A'Hearn, M. F., Belton, M. J. S., Delamere, W. A., Feaga, L. M., Hampton, D., et al. (2011) "EPOXI at Comet Hartley 2." Science, **332**, 1396–1400

Balescu, R. (1998) "Hamiltonian non-twist map for magnetic field lines with locally reversed shear in a toroidal geometry." Phys. Rev. E, **58**, 3781–3792

Balescu, R., Vlad, M., & Spineanu, F. (1998) "Tokamap: A Hamiltonian twist map for magnetic field lines in a toroidal geometry." Phys. Rev. E, **58**, 951–964

Baluev, R. V. (2011) "Orbital structure of the GJ876 extrasolar planetary system based on the latest Keck and HARPS radial velocity data." Celest. Mech. Dyn. Astron., **111**, 235–266

Barclay, T., Rowe J. F., Lissauer, J. J., Huber D., Fressin F., et al. (2013) "A sub-Mercury-sized exoplanet." Nature, **494**, 452–454

Bardin, B. S., & Markeev, A. P. (1995) "The stability of the equilibrium of a pendulum for vertical oscillations of the point of suspension." Journal of Applied Mathematics and Mechanics, **59**, 879–886

Barnes, R., & Greenberg, R. (2006) "Extrasolar planetary systems near a secular separatrix." Astrophys. J., **638**, 478–487

Bartczak, P., & Breiter, S. (2003) "Double material segment as the model of irregular bodies." Celest. Mech. Dyn. Astron., **86**, 131–141

Batygin, K., Morbidelli, A., & Tsiganis, K. (2011) "Formation and evolution of planetary systems in presence of highly inclined stellar perturbers." Astron. Astrophys., **533**, A7 (8pp)

Batygin, K., & Morbidelli, A. (2013a) "Analytical treatment of planetary resonances." Astron. Astrophys., **556**, A28 (20pp)

Batygin, K., & Morbidelli, A. (2013b) "Dissipative divergence of resonant orbits." Astron. J., **145**, 1 (10pp)

Batygin, K., Morbidelli, A., & Holman, M. J. (2015) "Chaotic disintegration of the inner solar system." Astrophys. J., **799**, 120 (16pp)

Batygin, K., Laughlin, G., Meschiari, S., Rivera, E., Vogt, S., et al. (2009) "A quasi-stationary solution to Gliese 436b's eccentricity." Astrophys. J., **699**, 23–30

Beitmen, G., & Erdeii, A. (1969) *Tables of integral transforms. V. 1* (Nauka Publishers, Moscow) (In Russian)

Belbruno, E. A. (2004) *Capture Dynamics and Chaotic Motions in Celestial Mechanics* (Princeton University Press, Princeton)

Beletsky, V. V. (1965) *The Motion of an Artificial Satellite about its Mass Center* (Nauka Publishers, Moscow) (in Russian)

Belton, M. J. S., Chapman, C. R., Thomas, P. C, Davies, M. E., Greenberg, R., et al. (1995) "Bulk density of asteroid 243-Ida from the orbit of its satellite Dactyl." Nature, **374**, 785–788

Benest, D. (1988a) "Planetary orbits in the elliptic restricted problem I. The α Centauri system." Astron. Astrophys., **206**, 143–146

Benest, D. (1988b) "Stable planetary orbits around one component in nearby binary stars." Celest. Mech., **43**, 47–53

Benest, D. (1989) "Planetary orbits in the elliptic restricted problem II. The Sirius system." Astron. Astrophys., **223**, 361–364

Benettin, G., & Galgani, L. (1979) "Lyapunov characteristic exponents and stochasticity." In: *Intrinsic Stochasticity in Plasmas*, ed. by Laval, G., & Gresillon, D. (Les Editions de Physique, Courtaboeuf, Orsay) pp. 93–114

Benettin, G., Galgani, L., & Strelcyn, J.-M. (1976) "Kolmogorov entropy and numerical experiments." Phys. Rev. A, **14**, 2338–2345

Benettin, G., Galgani, L., Giorgilli, A., & Strelcyn, J.-M. (1980) "Lyapunov characteristic exponents for smooth dynamical systems and for Hamiltonian systems – A method for computing all of them. I — Theory. II – Numerical application." Meccanica, **15**, 9–30

Bertini, I., Gutiérrez, P. J., Lara, L. M., Marzari, F., Moreno, F., et al. (2015) "Search for satellites near comet 67P/Churyumov–Gerasimenko using Rosetta/OSIRIS images." Astron. Astrophys., **583**, A19

Beust, H., & Dutrey, A. (2006) "Dynamics of the young multiple system GG Tauri. II. Relation between the stellar system and the circumbinary disk." Astron. Astrophys., **446**, 137–154

Beutler, G. (2005) *Methods of Celestial Mechanics*. Vols. 1–2 (Springer, Berlin)

Black, G. J., Nicholson, P. D., & Thomas, P. C. (1995) "Hyperion: Rotational dynamics." Icarus, **117**, 149–161

Boccaletti, D., & Pucacco, G. (1996) *Theory of Orbits. Vol. 1: Integrable Systems and Non-perturbative Methods* (Springer, Berlin Heidelberg)

Boccaletti, D., & Pucacco, G. (1999) *Theory of Orbits. Vol. 2. Perturbative and Geometrical Methods* (Springer, Berlin Heidelberg)

Bodenheimer, P., Hubickyj, O., & Lissauer, J. J. (2000) "Models of the in situ formation of detected extrasolar giant planets." Icarus, **143**, 2–14

Boekholt, T. C. N., Pelupessy, F. I., Heggie, D. C., & Portegies Zwart, S. F. (2016) "The origin of chaos in the orbit of comet 1P/Halley." Mon. Not. R. Astron. Soc., **461**, 3576–3584

Bogomolov, A. V., Pavluchenko, S. A., & Toporensky, A. V. (2011) "Power-law tails in triple system decay statistics." Eprint arXiv: 1101.0399 (4pp)

Boime, A. (1990) *Van Gogh: La Nuit étoilée. L'Histoire de la Matière et la Matière de L'Histoire* (Adam Biro, Paris)

Borderies, N., Goldreich, P., & Tremaine, S. (1984) "Unsolved problems in planetary ring dynamics." In: *Planetary Rings*, ed. by Greenberg, R., & Brahic, A. (University of Arizona Press, Tucson) pp. 713–734

Borucki, W. J., Koch, D. G., Basri, G., Batalha, N., & Brown, T. M. (2011) "Characteristics of planetary candidates observed by Kepler. II. Analysis of the first four months of data." Astrophys. J., **736**, 19 (22pp)

Bosh, A. S., & Rivkin, A. S. (1996) "Observations of Saturn's inner satellites during the May 1995 ring-plane crossing." Science, **272**, 518–521

Boss, A. P. (2006) "Rapid formation of super-Earths around M dwarf stars." Astrophys. J., **644**, L79–L82

Von Bremen, H. F., Udwadia, F. E., & Proskurowski, W. (1997) "An efficient QR based method for the computation of Lyapunov exponents." Physica D, **101**, 1–16

Borgonovi, F., Guarneri, I., & Sempio, P. (1988) "Long-time decay properties of the Kepler map." Nuovo Cimento Soc. Ital. Fis. B, **102**, 151–158

Boué, G., Laskar, J., & Farago, F. (2012) "A simple model of the chaotic eccentricity of Mercury." Astron. Astrophys., **548**, A43 (11pp)

Brasser, R. (2002) "Hill stability of a triple system with an inner binary of large mass ratio." Mon. Not. R. Astron. Soc., **332**, 723–728

Breiter, S. (2003) "Extended fundamental model of resonance." Celest. Mech. Dyn. Astron., **85**, 209–218

Bretagnon, P. (1990) "An iterative method for the construction of a general planetary theory." Astron. Astrophys., **231**, 561–570

Brouwer D., & Clemence, G. M. (1961) *Methods of Celestial Mechanics* (Academic Press, New York)

Bruhwiler, D. L., & Cary, J. R. (1989) "Diffusion of particles in a slowly modulated wave." Physica D, **40**, 265–282

Burns, J. A., & Safronov, V. S. (1973) "Asteroid nutation angles." Mon. Not. R. Astron. Soc., **165**, 403–411

Butler, R. P., Marcy, G. W., Fischer, D. A., et al. (1999) "Evidence for multiple companions to υ Andromedae." Astrophys. J., **526**, 916–927

Cachucho, F., Cincotta, P. M., & Ferraz-Mello, S. (2010) "Chirikov diffusion in the asteroidal three-body resonance (5,-2,-2)." Celest. Mech. Dyn. Astron., **108**, 35–58

Caldas, I. L., Bartoloni, B. F., Ciro, D., et al. (2017) "Symplectic maps for diverted plasmas." Eprint arXiv: 1710.10127 (15pp)

Callegari, N. Jr., & Yokoyama, T. (2007) "Dynamics of two satellites in the 2/1 mean motion resonance: Application to the case of Enceladus and Dione." Celest. Mech. Dyn. Astron., **98**, 5–30

Carrasco-González, C., Henning, T., Chandler, C. J., et al. (2016) "The VLA view of the HL Tau disk: Disk mass, grain evolution, and early planet formation." Astrophys. J., **821**, L16 (6pp)

Carpintero, D. D., Maffione, N., & Darriba, L. (2014) "LP-VIcode: A program to compute a suite of variational chaos indicators." Astron. and Comput., **5**, 19–27

Carter, J. A., Agol, E., Chaplin, W. J., Basu, S., & Bedding, T. R. (2012) "Kepler-36: A pair of planets with neighboring orbits and dissimilar densities." Science, **337**, 556–559

Cary, J. R., Meiss, J. D., & Bhattacharjee, A. (1981) "Statistical characterization of periodic, area-preserving mappings." Phys. Rev. A, **23**, 2744–2746

Casati, G., Guarneri, I., & Shepelyansky, D. L. (1987) "Exponential photonic localization for the hydrogen atom in a monochromatic field." Phys. Rev. A, **36**, 3501–3504

Casati, G., Guarneri, I., & Shepelyansky, D. L. (1988) "Hydrogen atom in monochromatic field: Chaos and dynamical photonic localization." IEEE J. Quantum Electron., **24**, 1420–1444

Celletti, A. Z. (1990) "Analysis of resonances in the spin-orbit problem in Celestial Mechanics: The synchronous resonance (Part I)." ZAMP Zeitschrift fur angewandte Mathematik und Physik, **41**, 174–204

Chambers J. (1999) "A hybrid symplectic integrator that permits close encounters between massive bodies." Mon. Not. R. Astron. Soc., **304**, 793–799

Champenois, S., & Vienne, A. (1999a) "The role of secondary resonances in the evolution of the Mimas–Tethys system." Icarus, **140**, 106–121

Champenois, S., & Vienne, A. (1999b) "Chaos and secondary resonances in the Mimas–Tethys system." Celest. Mech. Dyn. Astron., **74**, 111–146

Char, B. W., Geddes, K. O., Gonnet, G. H., Leong, B. L., Monagan, M. B., & Watt, S. M. (1993) *Maple V Library Reference Manual* (Springer, New York)

Charbonneau, D., Brown, T. M., Latham, D.W., & Mayor, M. (2000) "Detection of planetary transits across a Sun-like star." Astrophys. J., **529**, L45–L48

Chauvin, G., Lagrange, A. M., Dumas, C., Zuckerman, B., Mouillet, D., et al. (2004) "A giant planet candidate near a young brown dwarf. Direct VLT/NACO observations using IR wavefront sensing." Astron. Astrophys., **425**, L29–L32

Chauvin, G., Lagrange, A. M., Dumas, C., Zuckerman, B., Mouillet, D., et al. (2005) "Giant planet companion to 2MASSW J1207334–393254." Astron. Astrophys., **438**, L25–L28

Chauvineau, B., Farinella, P., & Mignard, F. (1993) "Planar orbits about a triaxial body: application to asteroidal satellites." Icarus, **105**, 370–384

Chirikov, B. V. (1959) "Resonance processes in magnetic traps." Atomnaya Energiya, **6**, 630–638 (in Russian) [Chirikov, B. V. (1960) "Resonance processes in magnetic traps." J. Nucl. Energy, Part C: Plasma Physics, **1**, 253–260]

Chirikov, B. V. (1977) *Nonlinear Resonance* (Novosib. Gos. Univ., Novosibirsk) (in Russian)

Chirikov, B. V. (1978) *Interaction of Nonlinear Resonances* (Novosib. Gos. Univ., Novosibirsk) (in Russian)

Chirikov, B. V. (1979) "A universal instability of many-dimensional oscillator systems." Phys. Rep., **52**, 263–379

Chirikov, B. V. (1982) "Nonlinear resonances and dynamical stochasticity." Priroda, No. 7, 15–25 (in Russian)

Chirikov, B. V. (1990) "Patterns in Chaos." INP Preprint 90–109 (Institute of Nuclear Physics, Novosibirsk)

Chirikov, B. V. (1991) "Patterns in Chaos." Chaos, Solitons and Fractals, **1**, 79–103

Chirikov, B. V. (1996) "Anomalous diffusion in a microtron and critical structure at the chaos boundary." J. Exp. Theor. Phys., **83**, 646–652

Chirikov, B. V. (2000) " Poincaré recurrences in microtron and the global critical structure." INP Preprint 99–7 (Institute of Nuclear Physics, Novosibirsk) Eprint arXiv:nlin/0006013 (31pp)

Chirikov, B. V., & Shepelyansky, D. L. (1981) "Statistics of Poincaré recurrences and the structure of the stochastic layer of the non-linear resonance." INP Preprint 81–69 (Institute of Nuclear Physics, Novosibirsk) (in Russian)

Chirikov, B. V., & Shepelyansky, D. L. (1984) "Correlation properties of dynamical chaos in Hamiltonian systems." Physica D, **13**, 395–400

Chirikov, B. V., & Shepelyansky, D. L. (1986) "Chaos border and statistical anomalies." INP Preprint 86–174 (Institute of Nuclear Physics, Novosibirsk) (in Russian)

Chirikov, B. V., & Vecheslavov, V. V. (1986) "Chaotic dynamics of comet Halley." INP Preprint 86–184 (Institute of Nuclear Physics, Novosibirsk)

Chirikov, B. V., & Vecheslavov, V. V. (1989) "Chaotic dynamics of comet Halley." Astron. Astrophys., **221**, 146–154

Chirikov, B. V., & Vecheslavov, V. V. (2000a) "Adiabatic invariance and separatrix: Single separatrix crossing." J. Exp. Theor. Phys., **90**, 562–569

Chirikov, B. V., & Vecheslavov, V. V. (2000b) "Multiple separatrix crossing: A chaos structure." J. Exp. Theor. Phys., **90**, 897–904

Cincotta, P. M. (2002) "Arnold diffusion: An overview through dynamical astronomy." New Astronomy Reviews, **46**, 13–39

Cincotta, P., & Simó, C. (2000) "Simple tools to study global dynamics in non-axisymmetric galactic potentials – I." Astron. Astrophys. Suppl., **147**, 205–228

Cincotta, P. M., Giordano, C. M., & Simó, C. (2003) "Phase space structure of multi-dimensional systems by means of the mean exponential growth factor of nearby orbits." Physica D, **182**, 151–178

Clemence, G. (1947) "The relativity effect in planetary motions." Rev. Mod. Phys., **19**, 361–364

Cochran, W. D., Hatzes, A., Butler, R. P., & Marcy, G. W. (1997) "The discovery of a planetary companion to 16 Cygni B." Astrophys. J., **483**, 457–463

Cooper, N. J., & Murray, C. D. (2004) "Dynamical influences on the orbits of Prometheus and Pandora." Astron. J., **127**, 1204–1217

Correia, A. C. M., Udry, S., Mayor, M., et al. (2005) "The CORALIE survey for southern extra-solar planets. XIII. A pair of planets around HD 202206 or a circumbinary planet?" Astron. Astrophys., **440**, 751–758

Correia, A. C. M., Udry, S., Mayor, M., et al. (2009) "The HARPS search for southern extra-solar planets. XVI. HD 45364, a pair of planets in a 3:2 mean motion resonance." Astron. Astrophys., **496**, 521–526

Correia, A. C. M., Boué, G., Laskar, J., & Morais, M. H. M. (2013) "Tidal damping of the mutual inclination in hierarchical systems." Astron. Astrophys., **553**, A39 (15pp)

Correia, A. C. M., Leleu, A., Rambaux, N., & Robutel, P. (2015) "Spin-orbit coupling and chaotic rotation for circumbinary bodies. Application to the small satellites of the Pluto–Charon system." Astron. Astrophys., **580**, L14 (7pp)

Cristadoro, G., & Ketzmerick, R. (2008) "Universality of algebraic decays in hamiltonian systems." Phys. Rev. Lett., **100**, 184101 (4pp)

Danby, J. M. A. (1988) *Fundamentals of Celestial Mechanics*. 2nd ed. (Willmann-Bell, Richmond)

Davies, M. B., Adams, F. C., Armitage, P., Chambers, J., Ford, E., Morbidelli, A., Raymond, S. N., & Veras, D. (2014) "The long-term dynamical evolution of planetary systems." In: *Protostars and Planets VI*, ed. by Beuther, H., Klessen, R. S., Dullemond, C. P., & Henning, T. (University of Arizona Press, Tucson) pp. 787–808

Deck, K. M., Holman, M. J., Agol, E., et al. (2012) "Rapid dynamical chaos in an exoplanetary system." Astrophys. J., 755, L21 (7pp) [Erratum: (2013) Astrophys. J., **774**, L15 (3pp)]

Deck, K. M., Payne, M., & Holman, M. J. (2013) "First-order resonance overlap and the stability of close two-planet systems." Astrophys. J., **774**, 129–141

Deeg, H. J., & Belmonte, J. A., eds. (2018) *Handbook of Exoplanets* (Springer International Publishing, Switzerland)

Deller, A. T., & Maddison, S. T. (2005) "Numerical modeling of dusty debris disks." Astrophys. J., **625**, 398–413

Demidova, T. V., & Shevchenko, I. I. (2015) "Spiral patterns in planetesimal circumbinary disks." Astrophys. J., **805**, 38 (8pp)

Demidova, T. V., & Shevchenko, I. I. (2016) "Three-lane and multi-lane signatures of planets in planetesimal disks." Mon. Not. R. Astron. Soc., **463**, L22–L26

Demidova, T. V., & Shevchenko, I. I. (2018) "Simulations of the dynamics of the debris disks in the systems Kepler-16, Kepler-34, and Kepler-35." Astron. Lett., **44**, 119–125

Denisov, S., Klafter, J., & Urbakh, M. (2002) "Ballistic flights and random diffusion as building blocks for Hamiltonian kinetics." Phys. Rev. E, **66**, 046217 (6pp)

Dermott, S. F., Malhotra, R., & Murray, C. D. (1988) "Dynamics of the Uranian and the Saturnian satellite systems: A chaotic route to melting Miranda?" Icarus, **76**, 295–334

Dermott, F. S., & Murray, C. D. (1983) "Nature of the Kirkwood gaps in the asteroid belt." Nature, **301**, 201–205

Devaney, R. (1986) *An Introduction to Chaotic Dynamical Systems* (Benjamin/Cummings, Menlo Park)

Devyatkin, A. V., Gorshanov, D. L., Gritsuk, A. N., et al. (2002) "Observations and theoretical analysis of lightcurves of natural satellites of planets." Sol. Sys. Res., **36**, 248–259

Dobrovolskis, A. R. (1995) "Chaotic rotation of Nereid?" Icarus, **118**, 181–198

Dones, L., Levison, H. F., & Duncan, M. (1996) "On the dynamical lifetimes of planet-crossing objects." In: *Completing the Inventory of the Solar System* (ASP Conf. Ser., v. 107), ed. by Rettig, T. W., & Hahn, J. M. (Astron. Soc. Pacific, San Francisco) pp. 233–244

Donnison, J. R. (2006) "The Hill stability of a binary or planetary system during encounters with a third inclined body." Mon. Not. R. Astron. Soc., **369**, 1267–1280

Doyle, L. R., Carter, J. A., & Fabrycky, D. C., et al. (2011) "Kepler-16: A transiting circumbinary planet." Science, **333**, 1602–1606

Dumusque, X., Pepe, F., Lovis, C., et al. (2012) "An Earth-mass planet orbiting α Centauri B." Nature, **491**, 207–211

Duncan, M., Quinn, T., & Tremaine, S. (1989) "The long-term evolution of orbits in the solar system. A mapping approach." Icarus, **82**, 402–418

Duquennoy, A., & Mayor, M. (1991) "Multiplicity among solar-type stars in the solar neighbourhood. II – Distribution of the orbital elements in an unbiased sample." Astron. Astrophys., **248**, 485–524

Dvorak, R., Pilat-Lohinger, E., Schwarz, R., & Freistetter, F. (2004) "Extrasolar Trojan planets close to habitable zones." Astron. Astrophys., **426**, L37–L40

Elskens, Y., & Eskande, D. F. (1991) "Slowly pulsating separatrices sweep homoclinic tangles where islands must be small: An extension of classical adiabatic theory." Nonlinearity, **4**, 615–667

Elskens, Y., & Eskande, D. F. (1993) "Infinite resonance overlap: A natural limit for Hamiltonian chaos." Physica D, **62**, 66–74

Éphémérides Astronomiques (1999) *Annuaire du Bureau des Longitudes* (Masson, Paris)

Emelyanenko, V. V. (1990) "Motion of nearly parabolic comets under the action of weak planetary perturbations." Sov. Astron. Lett., **16**, 318–321

Emelyanenko, V. V. (1992) "Dynamics of periodic comets and meteor streams." Celest. Mech. Dyn. Astron., **54**, 91–110

Escande, D. F. (1985) "Stochasticity in classical Hamiltonian systems: Universal aspects." Phys. Rep., **121**, 165–261

Escande, D. F., & Doveil, F. (1981) "Renormalization method for computing the threshold of the large-scale stochastic instability in two degrees of freedom Hamiltonian systems." J. Stat. Phys., **26**, 257–284

Fabrycky, D. C., Ford, E. B., Steffen, J. H., Rowe, J. F., Carter, J. A., et al. (2012) "Transit timing observations from Kepler. IV. Confirmation of four multiple-planet systems by simple physical models." Astrophys. J., **750**, 114 (17pp)

Fabrycky, D. C., Lissauer, J. J., Ragozzine, D., et al. (2014) "Architecture of Kepler's multi-transiting systems: II. New investigations with twice as many candidates." Astrophys. J., **790**, 146 (12pp)

Fang, J., & Margot, J.-L. (2012) "The role of Kozai cycles in near-Earth binary asteroids." Astron. J., **143**, 59 (8pp)

Farinella, P., Froeschlé, Ch., Froeschlé, C., Gonczi, R., Hahn, G., Morbidelli, A., & Valsecchi, G. B. (1994) "Asteroids falling onto the Sun." Nature, **371**, 314–317

Farmer, A. J., & Goldreich, P. (2006) "Understanding the behavior of Prometheus and Pandora." Icarus, **180**, 403–411

Feng, J., Noomen, R., Visser, P., & Yuan, J. (2016) "Numerical analysis of orbital motion around a contact binary asteroid system." Advances in Space Research, **58**, 387–401

Ferraz-Mello, S. (1997) "A symplectic mapping approach to the study of the stochasticity in asteroidal resonances." Celest. Mech. Dyn. Astron., **65**, 421–437

Ferraz-Mello, S. (2007) *Canonical Perturbation Theories* (Springer, Berlin)

Ferraz-Mello, S., & Dvorak, R. (1987) "Chaos and secular variations of planar orbits in 2:1 resonance with Dione." Astron. Astrophys., **179**, 304–310

Fischer, D. A., & Valenti, J. (2005) "The planet–metallicity correlation." Astrophys. J., **622**, 1102–1117

Ford, E. B., Kozinsky, B., & Rasio, F. A. (2000) "Secular evolution of hierarchical triple star systems." Astrophys. J., **535**, 385–401

Fragner, M. M., Nelson, R. P., & Kley, W. (2011) "On the dynamics and collisional growth of planetesimals in misaligned binary systems." Astron. Astrophys., **528**, A40 (21pp)

French, R. G., McGhee, C. A., Dones, L., & Lissauer, J. J. (2003) "Saturn's wayward shepherds: The peregrinations of Prometheus and Pandora." Icarus, **162**, 143–170

Froeschlé, Cl. (1984) "The Lyapunov characteristic exponents – applications to celestial mechanics." Celest. Mech., **34**, 95–115

Froeschlé, Ch., & Morbidelli, A. (1994) "The secular resonances in the solar system." In: *Asteroids, Comets, Meteors 1993*, ed. by Milani, A., di Martino, M., & Cellino, A. (Kluwer, Dordrecht) pp. 189–204

Froeschlé, C., Lega, E., & Gonczi, R. (1997) "Fast Lyapunov indicators. Application to asteroidal motion." Celest. Mech. Dyn. Astron., **67**, 41–62

Froeschlé, Ch., Morbidelli, A., & Scholl, H. (1991) "Complex dynamical behaviour of the asteroid 2335 James associated with the secular resonances ν_5 and ν_{16}: numerical studies and theoretical interpretation." Astron. Astrophys., **249**, 553–562

Fujiwara, A., Kawaguchi, J., Yeomans, D. K., et al. (2006) "The rubble-pile asteroid Itokawa as observed by Hayabusa." Science, **312**, 1330–1334

Fulle, M. (1997) "Injection of large grains into orbits around comet nuclei." Astron. Astrophys., **325**, 1237–1248

Gallardo, T. (2006) "Atlas of the mean motion resonances in the Solar System." Icarus **184**, 29–38

Gallardo, T. (2014) "Atlas of three body mean motion resonances in the Solar System." Icarus, **231**, 273–286

Gaudi, B. S., Seager, S., & Mallen-Ornelas, G. (2005) "On the period distribution of close-in extrasolar giant planets." Astrophys. J., **623**, 472–481

Gayon, J., Marzari, F., & Scholl, H. (2008) "Stable chaos in the 55 Cnc exoplanetary system?" Mon. Not. R. Astron. Soc., **389**, L1–L3

Geist, K., Parlitz, U., & Lauterborn, W. (1990) "Comparison of different methods for computing Lyapunov exponents." Progr. Theor. Phys., **83**, 875–893

Gelfreich, V. G. (1997) "Reference systems for splitting of separatrices." Nonlinearity, **10**, 175–193

Gelfreich, V. G. (1999) "A proof of the exponentially small transversality of the separatrices for the standard map." Commun. Math. Phys., **201**, 155–216

Giorgilli, A., Locatelli, U., & Sansottera, M. (2017) "Secular dynamics of a planar model of the Sun–Jupiter– Saturn–Uranus system; effective stability in the light of Kolmogorov and Nekhoroshev theories." Regular and Chaotic Dynamics, **22**, 54–77

Gladman, B. (1993) "Dynamics of systems of two close planets." Icarus, **106**, 247–263

Gladman, B., Lawler, S. M., Petit, J.-M., et al. (2012) "The resonant trans-Neptunian populations." Astron. J., **144**, 23 (24pp)

Goldreich, P., & Rappaport, N. (2003a) "Chaotic motions of Prometheus and Pandora." Icarus, **162**, 391–399

Goldreich, P., & Rappaport, N. (2003b) "Origin of chaos in the Prometheus–Pandora system." Icarus, **166**, 320–327

Gontis, V., & Kaulakys, B. (1987) "Stochastic dynamics of hydrogenic atoms in the microwave field: Modelling by maps and quantum description." Journal of Physics B: Atomic, Molecular, and Optical Physics, **20**, 5051–5064

Goździewski, K. (1997) "Rotational dynamics of Janus and Epimetheus." In: *Dynamics and Astrometry of Natural and Artificial Celestial Bodies*, ed. by Wytrzyszczak, I. M., et al. (Kluwer Academic Publishers, Dordrecht) pp. 269–274

Goździewski, K. (2003) "Stability of the HD 12661 Planetary System." Astron. Astrophys., **398**, 1151–1161

Goździewski, K., & Maciejewski, A. J. (1995) "On the gravitational fields of Pandora and Prometheus." Earth, Moon and Planets, **69**, 25–50

Goździewski, K., Migaszewski, C., Panichi, F., & Szuszkiewicz, E. (2016) "The Laplace resonance in the Kepler-60 planetary system." Mon. Not. R. Astron. Soc., **455**, L104–L108

Gradshtein, I. S., & Ryzhik, I. M. (1962) *Tables of Integrals, Sums, Series, and Products* (Fizmatlit, Moscow) (in Russian)

Grav, T., Holman, M.J., & Kavelaars, J.J. (2003) "The short rotation period of Nereid." Astrophys. J., **591**, L71–L74

Greaves, J. S., Fischer, D. A., Wyatt, M. C., Beichman, C. A., & Bryden, G. (2007) "Predicting the frequencies of diverse exo-planetary systems." Mon. Not. R. Astron. Soc., **378**, L1–L5

Greene, G. M. (1968) "Two-dimensional measure-preserving mappings." J. Math. Phys., **9**, 760–768

Greene, G. M. (1979) "A method for determining a stochastic transition." J. Math. Phys., **20**, 1183–1201

Greenberg, R., & Van Laerhoven, C. (2012) "Aligned major axes in a planetary system without tidal evolution: The 61 Virginis example." Mon. Not. R. Astron. Soc., **419**, 429–435

Grether, D., & Lineweaver, C. H. (2004) "At least a quarter of Sun-like stars have planets." In: *Stars as Suns: Activity, Evolution, and Planets* (Proc. IAU Symp. 219), ed. by Dupree, A. K., & Benz, A. O. (Cambridge Univ. Press, Cambridge) pp. 798–802

Guzzo, M. (2006) "The web of three-planet resonances in the outer Solar System. II. A source of orbital instability for Uranus and Neptune." Icarus, **181**, 475–485

Haghighipour, N., Dvorak, R., & Pilat-Lohinger, E. (2010) "Planetary dynamics and habitable planet formation in binary star systems." In: *Planets in Binary Star Systems*, ed. by Haghighipour, N. (Springer, Dordrecht) pp. 285–327

Hairer, E., Nørsett, S.P., & Wanner, G. (1987) *Solving Ordinary Differential Equations I. Nonstiff Problems* (Springer-Verlag, Berlin)

Hamilton, D. P., & Burns, J. A. (1992) "Orbital stability zones about asteroids II. The destabilizing effects of eccentric orbits and of solar radiation." Icarus, **96**, 43–64

Harbison, R. A., Thomas, P. C., & Nicholson, P. C. (2011) "Rotational modeling of Hyperion." Celest. Mech. Dyn. Astron., **110**, 1–16

Hastings, D. M., Ragozzine, D., Fabrycky, D. C., Burkhart, L. D., Fuentes, C., et al. (2016) "The short rotation period of Hi'iaka, Haumea's largest satellite." Astron. J., **152**, 195 (12pp)

Hayes, W. B. (2007) "Is the outer Solar System chaotic?" Nature Phys., **3**, 689–691

Hayes, W. B., Malykh, A. V., & Danforth, C. M. (2010) "The interplay of chaos between the terrestrial and giant planets." Mon. Not. R. Astron. Soc., **407**, 1859–1865

Heggie, D. C. (1975) "Binary evolution in stellar dynamics." Mon. Not. R. Astron. Soc., **173**, 729–787

Henrard, J. (1982) "Capture into resonance – an extension of the use of adiabatic invariants." Celest. Mech., **27**, 3–22

Henrard, J. (1983) "Orbital evolution of the Galilean satellites: Capture into resonance." Icarus, **53**, 55–67

Henrard, J., & Lemaître, A. (1983) "A second fundamental model for resonance." Celest. Mech., **30**, 197–218

Henrard, J., & Lemaître, A. (1983) "A mechanism of formation for the Kirkwood gaps." Icarus, **55**, 482–494

Henrard, J., & Moons, M. (1992) "Capture probabilities in secondary resonances." Icarus, **95**, 244–252

Heppenheimer, T. A. (1978) "On the formation of planets in binary star systems." Astron. Astropys., **65**, 421–426

Hesselbrock, A. J., Alexander, S. G., Harp, T. W., & Abel, N. P. (2013) "An investigation of the relationship between shape and rotation to explain the light curve of Nereid." Astron. J., **145**, 144 (11pp)

Hénon, M. & Heiles, C. (1964) "The applicability of the third integral of motion: Some numerical experiments." Astron. J., **69**, 73–79

Hénon, M. & Petit, J.-M. (1986) "Series expansions for encounter-type solutions of Hill's problem." Celest. Mech., **38**, 67–100

Hill, G. W. (1878) "Researches in the lunar theory." Am. J. Math., **1**, 5–26, 129–147, 245–261

Holman, M. J., & Murray, N. W. (1996) "Chaos in high order mean motion resonances in the outer asteroid belt." Astron. J., **112**, 1278–1293

Holman, M. J., & Murray, N. W. (2005) "The use of transit timing to detect terrestrial-mass extrasolar planets." Science, **307**, 1288–1291

Holman, M. J., & Wiegert, P. A. (1999) "Long-term stability of planets in binary systems." Astron. J., **117**, 621–628

Holman, M., Touma, J., & Tremaine, S. (1997) "Chaotic variations in the eccentricity of the planet orbiting 16 Cygni B." Nature, **386**, 254–256

Hu, W., & Scheeres, D. J. (2004) "Numerical determination of stability regions for orbital motion in uniformly rotating second degree and order gravity fields." Planet. Space Sci., **52**, 685–692

Huang, S.-S. (1960) "Life-supporting regions in the vicinity of binary systems." Publications of the Astronomical Society of the Pacific, **72**, 106–114

Hut, P. (1993) "Binary–single-star scattering. III. Numerical experiments for equal-mass hard binaries." Astrophys. J., **403**, 256–270

IAU, 2006, IAU 2006 General Assembly: Result of the IAU Resolution votes (iau0603 – Press Release). Available at: http://www.iau.org/news/pressreleases/detail/iau0603/

Innanen, K. A., Zheng, J. Q., Mikkola, S., & Valtonen, M. J. (1997) "The Kozai mechanism and the stability of planetary orbits in binary star systems." Astron. J., **113**, 1915–1919

Jacobson, R. A., & French, R. G. (2004) "Orbits and masses of Saturn's coorbital and F-ring shepherding satellites." Icarus, **172**, 382–387

Jensen, R. V., Leopold, J. G., & Richards, D. (1988) "High-frequency microwave ionisation of excited hydrogen atoms." J. Phys. B, **21**, L527–L531

Jensen, R. V., Susskind, M. M., & Sanders, M. M. (1991) "Chaotic ionization of highly excited hydrogen atoms: Comparison of classical and quantum theory with experiment." Phys. Rep., **201**, 1–56

Jeon, D., Bai, M., Chu, C. M., et al. (1996) "Role of parametric resonances in global chaos." Phys. Rev. E, **54**, 4192–4201

Jorda, L., Gaskell, R., Capanna, C., et al. (2016) "The global shape, density and rotation of Comet 67P/Churyumov–Gerasimenko from preperihelion Rosetta/OSIRIS observations." Icarus, **277**, 257–278

Kandrup, H. E., Sideris, I. V., & Bohn, C. L. (2002) "Chaos, ergodicity, and the thermodynamics of lower-dimensional time-independent Hamiltonian systems." Phys. Rev. E, **65**, 016214

Kane, T. R. (1995) "Attitude stability of Earth-pointing satellites." AIAA Journal, **3**, 726–731

Karch, M., & Dvorak, R. (1988) "New results on the possible chaotic motion of Enceladus." Celest. Mech. Dyn. Astron., **43**, 361–369

Kapitsa, P. L. (1951) "Dynamic stability of a pendulum with a weight oscillating about a point." Zh. Exp. Teor. Fiz., **21**, 588–607 (in Russian)

Kapitsa, P. L. (1954) "Pendulum with vibrating weight." Uspekhi Fiz. Nauk., **44**, 7–20 (in Russian)

Karney, C. F. F., Rechester, A. B., & White, R. B. (1982) "Effect of noise on the standard mapping." Physica D, **4**, 425–438

Kaulakys, B., & Vilutis, G. (1999) "Kepler map." Physica Scripta, **59**, 251–256

Keller, H. U., Mottola, S., Hviid, S. F.,et al. (2017) "Seasonal mass transfer on the nucleus of comet 67P/Churyumov-Gerasimenko." Mon. Not. R. Astron. Soc., **469**, S357–S371

Kepler, J. (1618) *Epitome astronomiae Copernicanae* (Linz)

Kepler, J. (1619) *Harmonices Mundi Libri V* (Linz)

Kholshevnikov, K. V. (1997) "D'Alembertian functions in celestial mechanics." Astron. Rep., **41**, 135–142

Kholshevnikov, K. V. (2001) "The Hamiltonian in the planetary or satellite problem as a D'Alembertian function." Astron. Rep., **41**, 135–142 , **45**, 577–579

Kholshevnikov, K. V., & Kuznetsov, E. D. (2011) "Stability of planetary systems with respect to masses." Celest. Mech. Dyn. Astron., **109**, 201–210

Kirkwood, D. (1867) *Meteoric Astronomy* (Lippincott, Philadelphia)

Kiseleva, L. G., Eggleton, P. P., & Mikkola, S. (1998) "Tidal friction in triple stars." Mon. Not. R. Astron. Soc., **300**, 292–392

Klavetter J. J. (1989a) "Rotation of Hyperion. 1. Observations." Astron. J., **97**, 570–579

Klavetter, J. J. (1989b) "Rotation of Hyperion. II — Dynamics." Astron. J., **98**, 1855–1874, 1946–1947

Knežević, Z., & Milani, A. (1994) "Asteroid proper elements: the big picture." In: *Asteroids, Comets, Meteors 1993*, ed. by Milani, A., di Martino, M., & Cellino, A. (Kluwer, Dordrecht) pp. 143–158

Kolmogorov, A. N. (1954) "On the conservation of conditionally periodic motions under small perturbation of the Hamiltonian." Dokl. Akad. Nauk. SSSR, **98**, 527–530

Koren, T., Klafter, J., & Magdziarz, M. (2007) "First passage times of Lévy flights coexisting with subdiffusion." Phys. Rev. E, **76**, 031129

Koren, T., Lomholt, M. A., Chechkin, A. V., Klafter, J., & Metzler, R. (2007) "Leapover lengths and first passage time statistics for Lévy flights." Phys. Rev. Lett., **99**, 160602

Kostov, V. B., McCullough, P. R., Hinse, T. C., et al. (2013) "A gas giant circumbinary planet transiting the F star primary of the eclipsing binary star KIC 4862625 and the independent discovery and characterization of the two transiting planets in the Kepler-47 system." Astrophys. J., **770**, 52 (19pp)

Kostov, V. B., McCullough, P. R., Carter, J. A., et al. (2014) "Kepler-413b: A slightly misaligned, Neptune-size transiting circumbinary planet." Astrophys. J., **784**, 14 (18pp) [Erratum: **787**, 93 (1pp)]

Kostov, V. B., Orosz, J. A., Welsh, W. F., et al. (2016) "Kepler-1647b: The largest and longest-period Kepler transiting circumbinary planet." Astrophys. J., **827**, 86 (26pp)

Kouprianov, V. V., & Shevchenko, I. I. (2003) "On the chaotic rotation of planetary satellites: The Lyapunov exponents and the energy." Astron. Astrophys., **410**, 749–757

Kouprianov, V. V., & Shevchenko, I. I. (2005) "Rotational dynamics of planetary satellites: A survey of regular and chaotic behavior." Icarus, **176**, 224–234

Kozai, Y. (1962) "Secular perturbations of asteroids with high inclination and eccentricity." Astron. J., **67**, 591–598

Küchner, M. J. (2003) "Volatile-rich Earth-mass planets in the habitable zone." Astrophys. J., **596**, L105–L108

Küchner, M. J., & Holman, M. J. (2003) "The geometry of resonant signatures in debris disks with planets." Astrophys. J., **588**, 1110–1120

Van Laerhoven, C., & Greenberg, R. (2012) "Characterizing multi-planet systems with classical secular theory." Celest. Mech. Dyn. Astron., **113**, 215–234

Lages, J., & Shevchenko, I. I., (2020) "Dynamical environments of MU69: a state of chaotic clearing." In: *Origins: From the Protosun to the First Steps of Life* (Proc. IAU Symp. 345), ed. by Elmegreen, B. G., Viktor Tóth, L., & Güdel, M. (Cambridge Univ. Press, Cambridge) pp. 227–229

Lages, J., Shepelyansky, D. L., & Shevchenko, I. I. (2017) "Chaotic zones around rotating small bodies." Astron. J., **153**, 272 (10pp)

Lages, J., Shepelyansky, D. L., & Shevchenko, I. I. (2018) "Kepler map." Scholarpedia, 13(2):33238

Lages, J., Shevchenko, I. I., & Rollin, G. (2018) "Chaotic dynamics around cometary nuclei." Icarus, **307**, 391–399

Lagrange, J. L. (1892) *Oeuvres* (Gauthier-Villars, Paris, 1867–1892)

Laplace, P. S. (1895) *Oeuvres complètes* (Gauthier-Villars, Paris, 1895)

Laskar, J. (1985) "Accurate methods in general planetary theory." Astron. Astrophys., **144**, 133–146

Laskar, J. (1989) "A numerical experiment on the chaotic behaviour of the Solar system." Nature, **338**, 237–238

Laskar, J. (1990) "The chaotic motion of the solar system: a numerical estimate of the size of the chaotic zones." Icarus, **88**, 266–291

Laskar, J. (1993) "Frequency analysis for multi-dimensional systems. Global dynamics and diffusion." Physica D, **67**, 257–281

Laskar, J. (1994) "Large-scale chaos in the solar system." Astron. Astrophys., **287**, L9–L12

Laskar, J. (2008) "Chaotic diffusion in the Solar System." Icarus, **196**, 1–15

Laskar, J. (2014) "Michel Hénon and the stability of the Solar system." Eprint arXiv: 1411.4930 (13pp)

Laskar, J. (2015) "Stability of the solar system." Scholarpedia, www.scholarpedia.org/article/Stability_of_the_solar_system

Laskar, J., & Correia, A. C. M. (2009) "HD 60532, a planetary system in a 3:1 mean motion resonance." Astron. Astrophys., **496**, L5–L8

Laskar, J., & Gastineau, M. (2009) "Existence of collisional trajectories of Mercury, Mars and Venus with the Earth." Nature, **459**, 817–819

Laskar, J., & Robutel, Ph. (1993) "The chaotic obliquity of the planets." Nature, **361**, 608–612

Laskar, J., Froeschlé, C., & Celletti, A. (1992) "The measure of chaos by the numerical analysis of the fundamental frequencies." Physica D, **56**, 253–269

Laskar, J., Joutel, F., & Robutel, Ph. (1993) "Stabilization of the Earth's obliquity by the Moon." Nature, **361**, 615–617

Lazutkin, V. F. (1991) "On the width of the instability zone near the separatrices of a standard mapping." Sov. Math. Dokl., **42**, 5–9

Lazutkin, V. F. (2005) "Splitting of separatrices for the Chirikov standard map." J. Math. Sci., **128**, 2687–2705

Lecar, M., Franklin, F., & Murison, M. (1992) "On Predicting long-term orbital instability: A relation between the Lyapunov time and sudden orbital transitions." Astron. J., **104**, 1230–1236

Lee, M. H., & Peale, S. J. (2002) "Dynamics and origin of the 2:1 orbital resonances of the GJ 876 planets." Astrophys. J., **567**, 596–609

Lee, M. H., & Peale, S. (2003) "Secular evolution of hierarchical planetary systems." Astrophys. J., **592**, 1201–1216 [Erratum: (2003) Astrophys. J., **597**, 644 (1p)]

Lemaître, A. (1984) "High-order resonances in the restricted three-body problem." Celest. Mech. Dyn. Astron., **32**, 109–126

Levison, H. F., & Duncan, M. J. (1993) "The gravitational sculpting of the Kuiper belt." Astrophys. J., **406**, L35–L38

Le Verrier, U.-J. (1844) "Théorie de la comète périodique de 1770." Comptes rendues de l'Academie des sciences, **19**, 982–994

Le Verrier, U.-J. (1848) "Mémoire sur la comète périodique de 1770." Comptes rendues de l'Academie des sciences, **26**, 465–469

Le Verrier, U.-J. (1857) "'Théorie de la comète périodique de 1770." Annales de l'Observatoire impérial de Paris, **3**, 203–270

Lexell, A. J. (1777a) "Tentamen astronomicum de temporibus periodicis cometarum et speciatim de tempore revolutionis cometae, aA. 1770 observati." (An astronomical exam on periodic comets, and in particular about the revolving comets observed in 1770.) Acta Academiae scientiarum imperialis Petropolitanae, **I**, 332–369

Lexell, A. J. (1777b) "Coniectura de locis coeli, in quibus cometa anni 1770, in proximo suo ad perihelium reditu, e tellure nostra conspici debet." (A conjecture on the places in the sky, in which, comet year 1770 in its next return to perihelion from our world should be observed.) Acta Academiae scientiarum imperialis Petropolitanae, **II**, 328–342

Lexell, A. J. (1778a) "Ulteriores disquisitiones de tempore periodico cometae anno 1770 observati." (Further studies of the periodic comet observed in 1770.) Acta Academiae scientiarum imperialis Petropolitanae, **I**, 317–352

Lexell, A. J. (1778b) "Réflexions sur le temps périodique des Comètes en général, & principalement sur celui de la Comète observée en 1770." Acta Academiae scientiarum imperialis Petropolitanae, **II**, 12–34

Libert, A.-S., & Henrard, J. (2007) "Exoplanetary systems: The role of an equilibrium at high mutual inclination in shaping the global behavior of the 3-D secular planetary three-body problem." Icarus, **191**, 469–485

Li, G. & Batygin, K. (2014) "On the spin-axis dynamics of a Moonless Earth." Astrophys. J., **790**, 69 (7pp)

Li, G., Naoz, S., Kocsis, B., & Loeb, A. (2014a) "Eccentricity growth and orbit flip in near-coplanar hierarchical three-body systems." Astrophys. J., **785**, 116 (8pp)

Li, G., Naoz, S., Holman, M., & Loeb, A. (2014b) "Chaos in the test particle eccentric Kozai–Lidov mechanism." Astrophys. J., **791**, 86 (10pp) [Erratum: (2015) Astrophys. J., **802**, 71 (1p)]

Lichtenberg, A. J., & Lieberman, M. A. (1992) *Regular and Chaotic Dynamics*. 2nd ed. (Springer-Verlag, New York)

Lidov, M. L. (1961) "Evolution of artificial planetary satellites under the action of gravitational perturbations due to external bodies." Iskusstviennye Sputniki Zemli (Artificial Satellites of the Earth), **8**, 5–45 (in Russian)

Lidov, M. L. (1962) "The evolution of orbits of artificial satellites of planets under the action of gravitational perturbations of external bodies." Planet. Space Sci., **9**, 719–759 (An English translation of Lidov's (1961) article.)

Ligi, R., et al. (2012) "A new interferometric study of four exoplanet host stars: θ Cygni, 14 Andromedae, υ Andromedae and 42 Draconis." Astron. Astrophys., **545**, A5

Lissauer, J. J. (1999) "Chaotic motion in the Solar System." Rev. Mod. Phys., **71**, 835–845

Lissauer, J. J., Ragozzine, D., Fabrycky, D. C., et al. (2011a) "Architecture and dynamics of *Kepler*'s candidate multiple transiting planet systems." Astrophys. J. Suppl. Ser., **197**, 8 (26pp)

Lissauer, J. J., Fabrycky, D. C., Ford, E. B., Borucki, W. J., Fressin, F., Marcy, G. W., Orosz, J. A., & Rowe, J. F. (2011b) "A closely packed system of low-mass, low-density planets transiting Kepler-11." Nature, **470**, 53–58

Lissauer, J. J., Barnes, J. W., & Chambers, J. E. (2012) "Obliquity variations of a moonless Earth." Icarus, **217**, 77–87

Lissauer, J. J., Jontof-Hutter, D., Rowe, J. F., Fabrycky, D. C., & Lopez, E. D., (2013) "All six planets known to orbit Kepler-11 have low densities." Astrophys. J., **770**, 131 (15pp)

Lithwick, Y., & Naoz, S. (2011) "The eccentric Kozai mechanism for a test particle." Astrophys. J., **742**, 94 (8pp)

Lithwick, Y., & Wu, Y. (2011) "Theory of secular chaos and Mercury's orbit." Astrophys. J., **739**, 31

Lithwick, Y., & Wu, Y. (2012) "Resonant repulsion of Kepler planet pairs." Astrophys. J., **756**, L11 (5pp)

Liu, J., & Sun, Y. S. (1994) "Chaotic motion of comets in near-parabolic orbit: Mapping approaches." Celest. Mech. Dyn. Astron., **60**, 3–28

Locatelli, U., & Giorgilli, A. (2000) "Invariant tori in the secular motions of the three-body planetary systems." Celest. Mech. Dyn. Astron., **78**, 47–74

Lovis, C., Ségransan, D., Mayor, M., Udry, S., Benz, W., et al. (2011) "The HARPS search for southern extra-solar planets. XXVIII. Up to seven planets orbiting HD 10180: Probing the architecture of low-mass planetary systems." Astron. Astrophys., **528**, A112 (16pp)

Maffione, N. P., Darriba, L. A., Cincotta, P. M., & Giordano, C. M. (2013) "Chaos detection tools: Application to a self-consistent triaxial model." Mon. Not. R. Astron. Soc., **429**, 2700–2717

Malhotra, R. (1990) "Capture probabilities for secondary resonances." Icarus, **87**, 249–264

Malhotra, R. (1994) "Nonlinear resonances in the solar system." Physica D, **77**, 289–304

Malhotra, R. (1998) "Orbital resonances and chaos in the Solar system." In: *Solar System Formation and Evolution* (ASP Conf. Ser., v. 149), ed. by Lazzaro, D., Vieira Martins, R., Ferraz-Mello, S., Fernández, J., & Beaugé, C. (Astron. Soc. of the Pacific, San Francisco) pp. 37–63

Malhotra, R. (2012) "Orbital resonances in planetary systems." In: *Encyclopedia of Life Support Systems by UNESCO*. Volume 6.119.55 *Celestial Mechanics*. Preprint: September 2012. 31 pp.

Malhotra, R., & Dermott, S. F. (1990) "The role of secondary resonances in the orbital history of Miranda." Icarus, **85**, 444–480

Malmberg, D., Davies, M. B., & Chambers, J. E. (2007) "The instability of planetary systems in binaries: how the Kozai mechanism leads to strong planet-planet interactions." Mon. Not. R. Astron. Soc., **377**, L1–L4

Malyshkin, L., & Tremaine, S. (1999) "The Keplerian map for the planar restricted three-body problem as a model of comet evolution." Icarus, **142**, 341–353

Mandushev, G., O'Donovan, F. T., Charbonneau, D., et al. (2007) "TrES-4: A transiting hot jupiter of very low density." Astrophys. J., **667**, L195–L198

Marchal, C. (1990) *The Three-Body Problem* (Elsevier, Amsterdam)

Marchis, F., Durech, J., Castillo-Rogez, J., et al. (2014) "The puzzling mutual orbit of the binary Trojan asteroid (624) Hektor." Astrophys. J., **783**, L37 (6pp)

Mardling, R. A. (2008a) "Resonance, chaos and stability: The three-body problem in astrophysics." In: *The Cambridge N-Body Lectures*, Lecture Notes in Physics, **760**, ed. by Aarseth, S. J., Tout, C. A., & Mardling, R. A. (Springer, Berlin) pp. 59–96

Mardling, R. (2008b) "Resonance, chaos and stability in the general three-body problem." In: *Dynamical Evolution of Dense Stellar Systems* (Proc. IAU Symp. 246), ed. by Vesperini, E., Giersz, M., & Sills, A. (Cambridge Univ. Press, Cambridge) pp. 199–208

Marcy, G., Butler, R. P., Fischer, D., Vogt, S., Wright, J. T., Tinney, C. G., & Jones, H. R. A. (2005) "Observed properties of exoplanets: Masses, orbits, and metallicities." Progr. Theor. Phys. Suppl., **158**, 24–42

Markeev, A. P. (1994) "Third-order resonance in a Hamiltonian system with one degree of freedom." Journal of Applied Mathematics and Mechanics, **58**, 793–804

Markeev, A. P. (1995) "The behaviour of a non-linear Hamiltonian system with one degree of freedom at the boundary of a parametric resonance domain." Journal of Applied Mathematics and Mechanics, **59**, 541–551

Martin, D. V. (2018) "Populations of planets in multiple star systems." In:*Handbook of Exoplanets*, ed. by Deeg, H., & Belmonte, J. A. (Springer Nature) pp. 2035–2060

Martí, J. G., Giuppone, C. A., & Beaugé, C. (2013) "Dynamical analysis of the Gliese-876 Laplace resonance." Mon. Not. R. Astron. Soc., **433**, 928–934

Marzari, F., & Barbieri, M. (2007) "Planet dispersal in binary systems during transient multiple star phases." Astron. Astrophys., **472**, 643–647

Mathieu, R. D., Ghez, A. M., Jensen, E. L. N., & Simon, M. (2000) "Young binary stars and associated disks." In: *Protostars and Planets IV*, ed. by Mannings, V., Boss, A. P., & Russell, S. S. (University of Arizona Press, Tucson) pp. 703–730

Mayor, M., & Queloz, D. (1995) "A Jupiter-mass companion to a solar-type star." Nature, **378**, 355–359

Mazeh, T., & Shaham, J. (1979) "The orbital evolution of close triple systems — the binary eccentricity." Astron. Astrophys., **77**, 145–151

Mazeh, T., Krymolowski, Y., & Rosenfeld, G. (1997) "The high eccentricity of the planet orbiting 16 Cygni B." Astrophys. J., **477**, L103–L106

McArthur, B. E., Benedict, G. F., Barnes, R., et al. (2010) "New observational constraints on the υ Andromedae system with data from the Hubble Space Telescope and Hobby-Eberly Telescope." Astrophys. J., **715**, 1203–1220

McGhee, C. A., Nicholson, P. D., French, R. G., & Hall, K. J. (2001) "HST observations of saturnian satellites during the 1995 ring plane crossings." Icarus, **152**, 282–315

Meiss, J. D. (1992) "Symplectic maps, variational principles, and transport." Rev. Mod. Phys., **64**, 795–848

Meiss, J. D. (2007) *Differential Dynamical Systems* (SIAM, Philadelphia)

Meiss, J. D. (2008) "Visual explorations of dynamics: The standard map." Pramana – Journal of Physics, **70**, 965–988

Meiss, J. D., Cary, J. R., Grebogi, C., Crawford, J. D., Kaufman, A. N., & Abarbanel, H. D. I. (1983) "Correlations of periodic, area-preserving maps." Physica D, **6**, 375–384

Melnikov, V. K. (1963) "On the stability of the center for time-periodic perturbations." Trans. Moscow Math. Soc., **12**, 1–56

Melnikov, A. V. (2001) "Bifurcation regime of a synchronous resonance in the translational-rotational motion of non-spherical natural satellites of planets." Cosmic Res., **39**, 68–77

Melnikov, A. V. (2002) "Modelling of lightcurves of minor planetary satellites." IAA Transactions, No. 8, pp. 131–132

Melnikov, A. V. (2016) "On the chaotic orbital dynamics of the planet in the system 16 Cyg." Astron. Lett., **42**, 115–125

Melnikov, A. V., & Shevchenko, I. I. (1998) "The stability of the rotational motion of nonspherical natural satellites, with respect to tilting the axis of rotation." Sol. Sys. Res., **32**, 480–490

Melnikov, A. V., & Shevchenko, I. I. (2000) "On the stability of the rotational motion of nonspherical natural satellites in synchronous resonance." Sol. Sys. Res., **34**, 434–442

Melnikov, A. V., & Shevchenko, I. I. (2005) "Chaotic dynamics of satellite systems." Sol. Sys. Res., **39**, 322–332

Melnikov, A. V., & Shevchenko, I. I. (2008) "On the rotational dynamics of Prometheus and Pandora." Celest. Mech. Dyn. Astron., **101**, 31–47

Melnikov A. V., & Shevchenko I. I. (2010) "The rotation states predominant among the planetary satellites." Icarus, **209**, 786–794

Mendonça, J. T. (1991) "Diffusion of magnetic field lines in a toroidal geometry." Phys. Fluids B, **3**, 87–94

Meschiari, S. (2012) "Circumbinary planet formation in the Kepler-16 system. I. N-body simulations." Astrophys. J., **752**, 71 (6pp)

Meschiari, S. (2014) "Circumbinary planet formation in the Kepler-16 system. II. A toy model for in situ planet formation within a debris belt." Astrophys. J., **790**, 41 (14pp)

Michtchenko, T. A., & Malhotra, R. (2004) "Secular dynamics of the three-body problem: Application to the υ Andromedae planetary system." Icarus, **168**, 237–248

Migaszewski, C., Słonina, M., & Goździewski, K. (2012) "A dynamical analysis of the Kepler-11 planetary system." Mon. Not. R. Astron. Soc., **427**, 770–789

Mikkola, S. (2008) "Dynamics and stability of triple stars." In: *Multiple stars across the H–R diagram*, ed. by Hubrig, S., Petr-Gotzens, M., & Tokovinin, A. (Springer, Berlin) pp. 11–20

Mikkola, S., & Tanikawa, K. (2007) "Correlation of macroscopic instability and Lyapunov times in the general three-body problem." Mon. Not. R. Astron. Soc., **379**, L21–L24

Milani, A., & Knežević, Z. (1990) "Secular perturbation theory and computation of asteroid proper elements." Celest. Mech., **49**, 347–411

Milani, A., & Knežević, Z. (1994) "Asteroid proper elements and the dynamical structure of the asteroid main belt." Icarus, **107**, 219–254

Milani, A., & Nobili, A. M. (1992) "An example of stable chaos in the Solar System." Nature, **357**, 569–571

Milani, A., & Nobili, A. M. (1993) "Asteroid 522 Helga is chaotic and stable." Celest. Mech. Dyn. Astron., **56**, 323–324

Milani, A., Nobili, A. M., & Knežević, Z. (1997) "Stable chaos in the asteroid belt." Icarus, **125**, 13–31

Mills, S. M., Fabrycky, D. C., Migaszewski, C., et al. (2016) "A resonant chain of four transiting, sub-Neptune planets." Nature. **533**, 509–512

Moons, M., & Henrard, J. (1994) "Surfaces of section in the Miranda–Umbriel 1:3 inclination problem." Celest. Mech. Dyn. Astron., **59**, 129–148

Moons, M., & Morbidelli, A. (1995) "Secular resonances inside mean-motion commensurabilities: the 4/1, 3/1, 5/2 and 7/3 cases." Icarus, **114**, 33–50

Morbidelli, A. (2002) *Modern Celestial Mechanics. Aspects of Solar System Dynamics* (Taylor and Francis, Padstow)

Morbidelli, A. (2014) Private communication

Morbidelli, A., & Froeschlé, C. (1996) "On the relationship between lyapunov times and macroscopic instability times." Celest. Mech. Dyn. Astron., **63**, 227–239

Morbidelli, A., & Guzzo, M. (1997) "The Nekhoroshev theorem and the asteroid belt dynamical system." Celest. Mech. Dyn. Astron., **65**, 107–136

Morbidelli, A., & Nesvorný, D. (1999) "Numerous weak resonances drive asteroids toward terrestrial planets orbits." Icarus, **139**, 295–308

Moriwaki, K., & Nakagawa, Y. (2004) "A planetesimal accretion zone in a circumbinary disk." Astrophys. J., **609**, 1065–1070

Morrison, S., & Malhotra, R. (2015) "Planetary chaotic zone clearing: Destinations and timescales." Astrophys. J., **799**, 41 (8pp)

Moser, J. (1962) "On invariant curves of area-preserving mappings of an annulus." Nachr. Akad. Wiss. Goettingen, Math.-Phys. Kl. IIa, No. 1, 1–20

Mudryk, L. R., & Wu, Y. (2006) "Resonance overlap is responsible for ejecting planets in binary systems." Astrophys. J., **639**, 423–431

Muñoz-Gutiérrez, M. A., Reyes-Ruiz, M., & Pichardo, B. (2015) "Chaotic dynamics of Comet 1P/Halley: Lyapunov exponent and survival time expectancy." Mon. Not. R. Astron. Soc., **447**, 3775–3784

Murison, M., Lecar, M., & Franklin, F. (1994) "Chaotic motion in the outer asteroid belt and its relation to the age of the Solar System." Astron. J., **108**, 2323–2329

Murray, C. D., & Dermott, S. F. (1999) *Solar System Dynamics* (Cambridge Univ. Press, Cambridge)

Murray, C. D., & Fox, K. (1984) "Structure of the 3/1 Jovian Resonance: A Comparison of Numerical Methods." Icarus, **59**, 221–233

Murray, N. W., & Holman, M. J. (1997) "Diffusive chaos in the outer asteroid belt." Astron. J., **114**, 1246–1259

Murray, N., & Holman, M. (1999) "The origin of chaos in the outer Solar system." Science, **283**, 1877–1881

Murray, N., & Holman, M. (2001) "The role of chaotic resonances in the Solar System." Nature, **410**, 773–779

Murray, N., Holman, M., & Potter, M. (1998) "On the origin of chaos in the asteroid belt." Astron. J., **116**, 2583–2589

Mustill, A. J., & Wyatt, M. C. (2012) "Dependence of a planet's chaotic zone on particle eccentricity: The shape of debris disc inner edges." Mon. Not. R. Astron. Soc., **419**, 3074–3080

Mysen, E., & Aksnes, K. (2007) "On the dynamical stability of the Rosetta orbiter. II." Astron. Astrophys., **470**, 1193–1199

Mysen, E., Olsen, O., & Aksnes, K. (2006) "Chaotic gravitational zones around a regularly shaped complex rotating body." Planet. Space Sci., **54**, 750–760

Nacozy, P. E. (1976) "On the stability of the solar system." Astron. J., **81**, 787–791

Nagasawa, M., Ida, S., & Bessho, T. (2008) "Formation of hot planets by a combination of planet scattering, tidal circularization, and the Kozai mechanism." Astrophys. J., **678**, 498–508

Namouni, E., Luciani, J. F., Tabachnik, S., & Pellat, R. (1996) "A mapping approach to Hill's distant encounters." Astron. Astrophys., **313**, 979–992

Naoz, S., Perets, H. B., & Ragozzine, D. (2010) "The observed orbital properties of binary minor planets." Astrophys. J., **719**, 1775–1783

Naoz, S., Farr, W. M., Lithwick, Y., Rasio, F. A., & Teyssandier, J. (2011) "Hot Jupiters from secular planet-planet interactions." Nature, **473**, 187–189

Nauenberg, M. (1990) "Canonical Kepler map." Europhys. Lett., **13**, 611–616

Nayakshin, S. (2011) "Hot Super Earths: disrupted young Jupiters?" Mon. Not. R. Astron. Soc., **416**, 2974–2980

Neishtadt, A. I. (1986) "Change in adiabatic invariant at a separatrix." Sov. J. Plasma Phys., **12**, 568–573

Neishtadt, A. I., Sidorenko, V. V., & Treschev, D. V. (1997) "Stable periodic motions in the problem on passage through a separatrix." Chaos, **7**, 2–11

Nekhoroshev, N. N. (1977) "Exponential estimates of the stability time of near-integrable Hamiltonian systems." Russian Math. Surveys, **32**, 6, 1–65

Néron de Surgy, O., & Laskar, J. (1997) "On the long term evolution of the spin of the Earth." Astron. Astrophys., **318**, 975–989

Nesvorný, D., & Morbidelli, A. (1998) "Three-body mean motion resonances and the chaotic structure of the asteroid belt." Astron. J., **116**, 3029–3037

Nesvorný, D., & Morbidelli, A. (1999) "An analytic model of three-body mean motion resonances." Celest. Mech. Dyn. Astron., **71**, 243–271

Nesvorný, D., & Roig, F. (2001) "Mean motion resonances in the transneptunian region. Part II. The 1:2, 3:4, and weaker resonances." Icarus, **150**, 104–123

Nesvorný, D., Kipping, D. M., Buchhave, L. A., Bakos, G. A., Hartman, J., & Schmitt, A. R. (2012) "The detection and characterization of a nontransiting planet by transit timing variations." Science, **336**, 1133–1136

Nicholson, P. D., Showalter, M. R., Dones, L., et al. (1996) "Observations of Saturn's ring-plane crossings in August and November 1995." Science, **272**, 509–515

Noyelles, B., & Vienne, A. (2005) "The k:k+4 resonances in planetary systems." In: *Dynamics of Populations of Planetary Systems* (Proc. IAU Coll. 197), ed. by Knežević, Z., & Milani, A. (Cambridge, Cambridge Univ. Press) pp. 453–458

Noyelles, B., & Vienne, A. (2007) "Chaos induced by De Haerdtl inequality in the Galilean system." Icarus, **190**, 594–607

Ollivier, M., Roques, F., Casoli, F., Encrenaz, T., & Selsis, F. (2009) *Planetary systems: Detection, formation and habitability of extrasolar planets* (Springer, Berlin Heidelberg)

Olsen, O. (2006) "Orbital resonance widths in a uniformly rotating second degree and order gravity field." Astron. Astrophys., **449**, 821–826

Oort, J. H. (1950) "The structure of the cloud of comets surrounding the Solar System and a hypothesis concerning its origin." Bull. Astron. Inst. Neth., **11**, 91–110

OrbFit software and orbit9 integrator website: http://adams.dm.unipi.it/~orbmaint/orbfit/ (Maintained by OrbFit consortium)

Orlov, V. V., Rubinov, A. V., & Shevchenko, I. I. (2010) "The disruption of three-body gravitational systems: Lifetime statistics." Mon. Not. R. Astron. Soc., **408**, 1623–1627

Orosz, J., Welsh, W. F., Carter, J. A., et al. (2012a) "The Neptune-sized circumbinary planet Kepler-38b." Astrophys. J., **758**, 87 (14pp)

Orosz, J., Welsh, W. F., Carter, J. A., et al. (2012b) "Kepler-47: A transiting circumbinary multiplanet system." Science, **337**, 1511–1514

Oseledets, V. I. (1968) "A multiplicative ergodic theorem. Characteristic Laypunov exponents of dynamical systems." Trudy Moscow Math. Soc., **19**, 179–210 (in Russian)

Ozernoy, L. M., Gorkavyi, N. N., Mather, J. C., & Taidakova, T. A., (2000) "Signatures of exosolar planets in dust debris disks." Astrophys. J., **537**, L147–L151

Paardekooper, S.-J., Leinhardt, Z. M., Thébault, T., & Baruteau, C. (2012) "How not to build Tatooine: the difficulty of in situ formation of circumbinary planets Kepler 16b, Kepler 34b, and Kepler 35b." Astrophys. J., **754**, L16 (5pp)

Pakoński, P., & Zakrzewski, J. (2001) "Kepler map for H atom driven by microwaves with arbitrary polarization." Acta Physica Polonica B, **32**, 2801–2812

Pan, M., & Sari, R. (2004) "A generalization of the Lagrangian points: Studies of resonance for highly eccentric orbits." Astron. J., **128**, 1418–1429

Pannekoek, A. (1961) *A History of Astronomy.* (George Allen and Unwin Ltd., London)

Peale, S. J. (1976) "Orbital resonances in the Solar system." Ann. Rev. Astron. Astrophys., **14**, 215–246

Peale, S. J. (1977) "Rotation histories of the natural satellites." In: *Planetary Satellites*, ed. by Burns, J. A. (University of Arizona Press, Tucson) pp. 87–111

Peale, S. J. (1986) "Orbital resonances, unusual configurations and exotic rotation states among planetary satellites." In: *Satellites*, ed. by Burns, J. A., & Matthews, M. S. (University of Arizona Press, Tucson) pp. 159–223

Peale, S. J. (1999) "Origin and evolution of the natural satellites." Ann. Rev. Astron. Astrophys., **37**, 533–602

Perets, H. B., & Naoz, S. (2009) "Kozai cycles, tidal friction, and the dynamical evolution of binary minor planets." Astrophys. J. Lett., **699**, L17–L21

Pérez-Hernández, J. A., & Benet, L. (2019) "On the dynamics of Comet 1P/Halley: Lyapunov and power spectra." Mon. Not. R. Astron. Soc., **487**, 296–303

Perryman, M. (2018) *The Exoplanet Handbook.* 2nd edition (Cambridge Univ. Press, Cambridge)

Petrovich, C., Malhotra, R., & Tremaine, S. (2013) "Planets near mean-motion resonances." Astrophys. J., **770**, 24 (16pp)

Pesin, Ya. B. (1977) "Characteristic Lyapunov exponents and smooth ergodic theory." Russian Math. Surveys, **32**, 55–114

Petit, J. M., & Hénon, M. (1986) "Satellite encounters." Icarus, **66**, 536–555

Petit, J.-M., Durda, D. D., Greenberg, R., Hurford, T. A., & Geissler, P. E. (1997) "The long-term dynamics of Dactyl's orbit." Icarus, **130**, 177–197

Petit, A. C., Laskar, J., & Boué, G. (2017) "AMD-stability in the presence of first-order mean motion resonances." Astron. Astrophys., **607**, A35 (17pp)

Petrosky, T. Y. (1986) "Chaos and cometary clouds in the solar system." Phys. Lett. A, **117**, 328–332

Petrosky, T. Y., & Broucke, R. (1988) "Area-preserving mappings and deterministic chaos for nearly parabolic motions." Celest. Mech. Dyn. Astron., **42**, 53–79

Pichierri, G., Batygin, K., & Morbidelli, A. (2019) "The role of dissipative evolution for three-planet, near-resonant extrasolar systems." Astron. Astrophys., **625**, A7 (14pp)

Pierens, A., & Nelson, R. P. (2007) "On the migration of protoplanets embedded in circumbinary disks." Astron. Astrophys., **472**, 993–1001

Pierens, A., & Nelson, R. P. (2008) "On the formation and migration of giant planets in circumbinary discs." Astron. Astrophys., **483**, 633–642

Piftankin, G. N., & Treshchev, D. V. (2007) "Separatrix maps in Hamiltonian systems." Russian Math. Surveys, **62**, 1–102

Pilat-Lohinger, E., & Dvorak, R. (2002) "Stability of S-type orbits in binaries." Celest. Mech. Dyn. Astron., **82**, 143–153

Pilat-Lohinger, E., Funk, B., & Dvorak, R. (2003) "Stability limits in double stars. A study of inclined planetary orbits." Astron. Astrophys., **400**, 1085–1094

Plávalová, E., & Solovaya, N. A. (2013) "Analysis of the motion of an extrasolar planet in a binary system." Astron. J., **146**, 108 (8pp)

Poincaré, H. (1890) "Sur le problème des trois corps et les équations de la dynamique." Acta Math., **13**, Nos. 1–2, 1–270

Poincaré, H. (1899) *Les Méthodes Nouvelles de la Mécanique Céleste III* (Gauthier–Villars, Paris)

Poincaré, H. (1905) *Leçons de Mécanique Céleste* (Gauthier–Villars, Paris)

Popova, E. A., & Shevchenko, I. I. (2012) "Planetary dynamics in the system α Centauri: The stability diagrams." Astron. Lett., **38**, 581–588

Popova, E. A., & Shevchenko, I. I. (2013) "Kepler-16b: Safe in a resonance cell." Astrophys. J., **769**, 152 (7pp)

Popova, E. A., & Shevchenko, I. I. (2014) "Orbital resonances in exoplanetary systems." Journal of Physics Conf. Series., **572**, 012006 (6pp)

Popova, E. A., & Shevchenko, I. I. (2016a) "On possible circumbinary configurations of the planetary systems of α Centauri and EZ Aquarii." Astron. Lett., **42**, 260–267

Popova, E. A., & Shevchenko, I. I. (2016b) "On the stability of circumbinary planetary systems." Astron. Lett., **42**, 474–481

Pravec, P., & Harris, A. W. (2000) "Fast and slow rotation of asteroids." Icarus, **148**, 12–20

Pu, B., & Wu, Y. (2015) "Spacing of *Kepler* planets: Sculpting by dynamical instability." Astrophys. J., **807**, 44 (10pp)

Quarles, B., & Lissauer, J. J. (2016) "Long-term stability of planets in the α Centauri system." Astron. J., **151**, 111 (12pp)

Quarles, B., Lissauer, J. J., & Kaib, N. (2018) "Long-term stability of planets in the α Centauri system. II. Forced eccentricities." Astron. J., **155**, 64 (11pp)

Quillen, A. C. (2006) "Reducing the probability of capture into resonance." Mon. Not. R. Astron. Soc., **365**, 1367–1382

Quillen, A. C. (2011) "Three-body resonance overlap in closely spaced multiple-planet systems." Mon. Not. R. Astron. Soc., **418**, 1043–1054

Quillen, A. C., & Faber, P. (2006) "Chaotic zone boundary for low free eccentricity particles near an eccentric planet." Mon. Not. R. Astron. Soc., **373**, 1245–1250

Quillen, A. C., & Thorndike, S. (2002) "Structure in the ϵ Eridani dusty disk caused by mean motion resonances with a 0.3 eccentricity planet at periastron." Astrophys. J., **578**, L149–L152

Quillen, A. C., Nichols-Fleming, F., Chen, Y.-Y., & Noyelles, B. (2017) "Obliquity evolution of the minor satellites of Pluto and Charon." Icarus, **293**, 94–113

Ramos, X. S., Correa-Otto, J. A., & Beaugé, C. (2015) "The resonance overlap and Hill stability criteria revisited." Celest. Mech. Dyn. Astron., **123**, 453–479

Raymond, S. N., Izidoro, A., & Morbidelli, A. (2018) "Solar system formation in the context of extra-solar planets." Eprint arXiv: 1812.01033 (35pp)

Rechester, A. B., Rosenbluth, M. N., & White, R. B. (1981) "Fourier-space paths applied to the calculation of diffusion for the Chirikov–Taylor model." Phys. Rev. A, **23**, 2664–2672

Rein, H. (2012) "Period ratios in multiplanetary systems discovered by Kepler are consistent with planet migration." Mon. Not. R. Astron. Soc., **427**, L21–L24

Renner, S., & Sicardy, B. (2003) "Consequences of the chaotic motions of Prometheus and Pandora." Bulletin of the American Astronomical Society, **35**, 1044 (abstract)

Renner, S., Sicardy, B., & French, R. G. (2005) "Prometheus and Pandora: Masses and orbital positions during the Cassini tour." Icarus, **174**, 230–240

Ribas, I. (2010) "The Sun and stars as the primary energy input in planetary atmospheres." In: *Solar and Stellar Variability – Impact on Earth and Planets* (Proc. IAU Symp. 264), ed. by Kosovichev, A. G., Andrei, A. H., & Rozelot, J.-P. (Cambridge Univ. Press, Cambridge) pp. 3–18

Rivera E. J., Laughlin G., Butler R. P., Vogt S. S., Haghighipour, N., & Meschiari S. (2010) "The Lick-Carnegie exoplanet survey: A Uranus-mass fourth planet for GJ 876 in an extrasolar Laplace configuration." Astrophys. J., **719**, 890–899

Roy, A., & Haddow, M. (2003) "Energy change in a hard binary due to distant encounters." Celest. Mech. Dyn. Astron., **87**, 411–435

Sagdeev, R. Z., & Zaslavsky, G. M. (1987) "Stochasticity in the Kepler problem and a model of possible dynamics of comets in the Oort cloud." Nuovo Cimento Soc. Ital. Fis. B, **97**, 119–130

Sagdeev, R. Z., Usikov, D. A., & Zaslavsky, G. M. (1991) *Nonlinear Physics. From the Pendulum to Turbulence and Chaos* (Harwood Academics, Chur, Switzerland)

Sahu, K. C., Casertano, S., Bond, H. E., Valenti, J., Ed Smith, T., et al. (2006) "Transiting extrasolar planetary candidates in the Galactic bulge." Nature, **443**, 534–540

Sahu, K. C., Casertano, S., Valenti, J., Bond, H. E., Brown, T. M., et al. (2008) "Planets in the Galactic Bulge: Results from the SWEEPS Project." In: *Extreme Solar Systems* (ASP Conf. Ser., v. 398), ed. by Fischer, D., Rasio, F. A., Thorsett, S. E., & Wolszczan, A. (Astron. Soc. Pacific, San Francisco) pp. 93–100

Saito, M. M., Tanikawa, K., & Orlov, V. V. (2012) "Disintegration process of hierarchical triple systems. I. Small-mass planet orbiting equal-mass binary." Celest. Mech. Dyn. Astron., **112**, 235–251

Saito, M. M., Tanikawa, K., & Orlov, V. V. (2013) "Disintegration process of hierarchical triple systems. II. Non-small mass third body orbiting equal-mass binary." Celest. Mech. Dyn. Astron., **116**, 1–10

Scheeres, D. J. (1994) "Dynamics about uniformly rotating triaxial ellipsoids: Application to asteroids." Icarus, **110**, 225–238

Scheeres, D. J. (2007) "Rotational fission of contact binary asteroids." Icarus, **189**, 370–385

Scheeres, D. J. (2012) "Orbital mechanics about small bodies." Acta Astronautica, **72**, 1–14

Scheeres, D. J., & Marzari, F. (2000) "Temporary orbital capture of ejecta from comets and asteroids: Application to the Deep Impact experiment." Astron. Astrophys., **356**, 747–756

Scheeres, D. J., Ostro, S. J., Hudson, R. S., & Werner, R. A. (1996) "Orbits close to asteroid 4769 Castalia." Icarus, **121**, 67–87

Scheeres, D. J., Ostro, S. J., Hudson, R. S., DeJong, E. M., & Suzuki, S. (1998) "Dynamics of orbits close to asteroid 4179 Toutatis." Icarus, **132**, 53–79

Scheeres, D. J., Williams, B. G., & Miller, J. K. (2000) "Evaluation of the dynamic environment of an asteroid: applications to 433 Eros." J. Guidance Control Dyn., **23**, 466–475

Schlagheck, P., & Buchleitner, A. (2001) "Algebraic decay of the survival probability in chaotic helium." Phys. Rev. A, **63**, 024701

Schneider, J. (2018) *The Extrasolar Planets Encyclopaedia*, http://exoplanet.eu

Schwamb, M., Orosz, J. A., Carter, J. A., et al. (2013) "Planet Hunters: A transiting circumbinary planet in a quadruple star system." Astrophys. J., **768**, 127 (21pp)

Schwarz, R., Haghighipour, N., Eggl, S., Pilat-Lohinger, E., & Funk, B. (2011) "Prospects of the detection of circumbinary planets with Kepler and CoRoT using the variations of eclipse timing." Mon. Not. R. Astron. Soc., **414**, 2763–2770

Seidelmann, P. K., Abalakin, V. K., Burša, M., Davies, M. E., de Bergh, C., Lieske, J. H., Oberst, J., Simon, J. L., Standish, E. M., Stooke, P., & Thomas, P. C. (2002) "Report of the IAU/IAG Working Group on cartographic coordinates and rotational elements of the planets and satellites: 2000." Celest. Mech. Dyn. Astron., **82**, 83–110

Shappee, B. J., & Thompson, T. A. (2013) "The mass-loss-induced eccentric Kozai mechanism: a new channel for the production of close compact object–stellar binaries." Astrophys. J., **766**, 64 (12pp)

Sheppard, S. (2006) "Small bodies in the outer solar system." In: *Frank N. Bash Symposium 2005: New Horizons in Astronomy* (ASP Conf. Ser., v. 352), ed. by Kannappan, S. J., et al. (Astron. Soc. Pacific, San Francisco) pp. 3–14

Shevchenko, I. I. (1996) "Spectra of winding numbers of chaotic asteroidal motion." In: *Chaos in Gravitational N-Body Systems*, ed. by Muzzio, J. C., et al. (Kluwer, Dordrecht) pp. 311–314

Shevchenko, I. I. (1998a) "On the recurrence and Lyapunov time scales of the motion near the chaos border." Phys. Lett. A, **241**, 53–60

Shevchenko, I. I. (1998b) "Marginal resonances and intermittent behavior in the motion in the vicinity of a separatrix." Physica Scripta, **57**, 185–191

Shevchenko, I. I. (1999a) "The separatrix algorithmic map: Application to the spin-orbit motion." Celest. Mech. Dyn. Astron, **73**, 259–268

Shevchenko, I. I. (1999b) "On the critical phenomena in the dynamics of asteroids." In: *Impact of modern dynamics in astronomy (IAU Colloquium 172)*, ed. by Henrard, J., & Ferraz-Mello, S. (Kluwer, Dordrecht) pp. 383–386

Shevchenko, I. I. (2000a) "Geometry of a chaotic layer." J. Exp. Theor. Phys., **91**, 615–625

Shevchenko, I. I. (2000b) "Orbital resonances and the separatrix algorithmic map." In: *The Chaotic Universe*, ed. by Gurzadyan, V. G., & Ruffini, R. (World Scientific, London) pp. 599–608

Shevchenko, I. I. (2002a) "On the maximum Lyapunov exponents of the chaotic rotation of natural planetary satellites." Cosmic Res., **40**, 296–304

Shevchenko, I. I. (2002b) "Quaternary cycles in the chaotic transport of NEAs." In: *Proceedings of "Asteroids, Comets, Meteors 2002"*, ed. by Warmbein, B. (ESA, Berlin) pp. 367–370

Shevchenko, I. I. (2004a) "On the maximum Lyapunov exponent of the motion in a chaotic layer." JETP Lett., **79**, 523–528

Shevchenko, I. I. (2004b) "Isentropic perturbations of a chaotic domain." Phys. Lett. A, **333**, 408–414

Shevchenko, I. I. (2007a) "On the Lyapunov exponents of the asteroidal motion subject to resonances and encounters." In: *Near Earth Objects, our Celestial Neighbors: Opportunity and Risk* (Proc. IAU Symp. 236), ed. by Milani, A., Valsecchi, G. B., & Vokrouhlický, D. (Cambridge Univ. Press, Cambridge) pp. 15–29

Shevchenko, I. I. (2007b) "The dynamical temperature and the standard map." Physica A, **386**, 85–91

Shevchenko, I. I. (2008a) "The width of a chaotic layer." Phys. Lett. A, **372**, 808–816

Shevchenko, I. I. (2008b) "Adiabatic chaos in the Prometheus–Pandora system." Mon. Not. R. Astron. Soc., **384**, 1211–1220; **407**, 704

Shevchenko, I. I. (2010) "Hamiltonian intermittency and Lévy flights in the three-body problem." Phys. Rev. E, **81**, 066216 (11pp)

Shevchenko, I. I. (2011a) "Lyapunov and diffusion timescales in the solar neighborhood." Astrophys. J., **733**, 39 (8pp)

Shevchenko, I. I. (2011b) "The Kepler map in the three-body problem." New Astronomy, **16**, 94–99

Shevchenko, I. I. (2011c) "Dynamical chaos in the Solar system." In: *Chaos, Diffusion and Non-Integrability in Hamiltonian Systems – Applications to Astronomy*, ed. by Cincotta, P., Giordano, C., & Efthymiopoulos, C. (AAA Workshop Series, Asociación Argentina de Astronomía, La Plata) pp. 217–246

Shevchenko, I. I. (2012) "Width of the chaotic layer: Maxima due to marginal resonances." Phys. Rev. E, **85**, 066202 (7pp)

Shevchenko, I. I. (2014) "Lyapunov exponents in resonance multiplets." Phys. Lett. A, **378**, 34–42

Shevchenko, I. I. (2015) "Chaotic zones around gravitating binaries." Astrophys. J., **799**, 8 (7pp)

Shevchenko, I. I. (2017a) *The Lidov–Kozai Effect – Applications in Exoplanet Research and Dynamical Astronomy* (Springer International Publishing Switzerland AG)

Shevchenko, I. I. (2017b) "Habitability properties of circumbinary planets." Astron. J., **153**, 273 (14pp)

Shevchenko, I. I. (2018) "Tidal decay of circumbinary planetary systems." Astron. J., **156**, 52 (8pp)

Shevchenko, I. I. (2020) "Lyapunov and clearing timescales in planetary chaotic zones." Astron. J., in press

Shevchenko, I. I., & Melnikov, A. V. (2003) "Lyapunov Exponents in the Hénon–Heiles Problem." JETP Lett., **77**, 642–646

Shevchenko, I. I., & Kouprianov, V. V. (2002) "On the chaotic rotation of planetary satellites: The Lyapunov spectra and the maximum Lyapunov exponents." Astron. Astrophys., **394**, 663–674

Shevchenko, I. I., & Scholl, H. (1996) "Chaotic Asteroidal Trajectories Exhibiting Multiple Bursts of Eccentricity: A Statistical Analysis." In: Dynamics, Ephemerides and Astrometry of the Solar System, eds. S. Ferraz-Mello et al. (Kluwer, Dordrecht) pp. 183–186

Shevchenko, I. I., & Scholl, H. (1997) "Intermittent trajectories in the 3/1 Jovian resonance." Celest. Mech. Dyn. Astron., **68**, 163–175

Shevchenko, I. I., Melnikov, A. V., Popova, E. A., Bobylev, V. V., & Karelin, G. M. (2019) "Circumbinary planetary systems in the Solar neighborhood: Stability and habitability." Astron. Lett., **45**, 620–626

Shevchenko, I. I., Rollin, G., Melnikov, A. V., & Lages, J. (2020) "Massive evaluation and analysis of Poincare recurrences on grids of initial data: a tool to map chaotic diffusion." Computer Phys. Commun., **246**, 106868 (8pp)

Shimada, I., & Nagashima, T. (1979) "A numerical approach to ergodic problem of dissipative dynamical systems." Progr. Theor. Phys. **61**, 1605–1616

Shinkin, V. N. (1995) "The integrable cases of the general spatial three-body problem at third-order resonance." Celest. Mech. Dyn. Astron., **62**, 323–334

Shklovsky, I. S. (1979) "On possible uniqueness of intellectual life in the Universe." In: *Astronomy. Methodology. World Outlook*, ed. by Kazyutinsky, V. V., et al. (Nauka Publishers, Moscow) pp. 252–274 (In Russian)

Showalter, M. R., & Hamilton, D. P. (2015) "Resonant interactions and chaotic rotation of Pluto's small moons." Nature, **522**, 45–49

Sidorenko, V. V. (2018) "The eccentric Kozai–Lidov effect as a resonance phenomenon." Celest. Mech. Dyn. Astron., **130**, 4 (23pp)

Sitnikov, K. A. (1960) "The existence of oscillatory motions in the three-body problems." Sov. Phys. Doklady **5**, 647–650

Skokos, Ch. (2010) "The Lyapunov characteristic exponents and their computation." Lect. Notes Phys. **790**, 63–135

Smirnov, E. A., & Shevchenko, I. I. (2013) "Massive identification of asteroids in three-body resonances." Icarus, **222**, 220–228

Snellgrove, M. D., Papaloizou J. C. B., & Nelson R. P. (2001) "On disc driven inward migration of resonantly coupled planets with application to the system around GJ 876." Astron. Astrophys., **374**, 1092–1099

Soper, M., Franclin, F., & Lecar, M. (1990) "On the original distribution of the asteroids." Icarus, **87**, 265–284

Soskin, S. M., & Mannella, R. (2009) "Maximal width of the separatrix chaotic layer." Phys. Rev. E, **80**, 066212

Soskin, S. M., Mannella, R., Yevtushenko, O. M., Khovanov, I. A., & McClintock, P. V. E. (2012) "A new approach to the treatment of separatrix chaos." Fluctuation and Noise Letters, **11**, 1240002

Soter, S. (2006) "What is a planet?" Astron. J., **132**, 2513–2519

Sotin, C., Grasset, O., & Mocquet, A. (2007) "Mass-radius curve for extrasolar Earth-like planets and ocean planets." Icarus, **191**, 337–351

Spencer, J. R., Stern, S. A., Moore, J. M., Weaver, H. A., Singer, K. N., et al. (2020) "The geology and geophysics of Kuiper Belt object (486958) Arrokoth." Science, 10.1126/science.aay3999 (19pp)

Stephenson, A. (1908) "New type of dynamical stability." Manchester University Memoirs, **LII**, No. 8, March 5th

Stern, S. A., & Levison, H. F. (2002) "Regarding the criteria for planethood and proposed planetary classification schemes." Highlights of Astronomy, **12**, 205–213

Stern, S. A., Weaver, H. A., Spencer, J. R., Olkin, C. B., Gladstone, G. R., et al. (2019) "Initial results from the New Horizons exploration of 2014 MU69, a small Kuiper Belt object." Science, **364**, eaaw9771 (12pp)

Stevens, D. J., & Gaudi, B. S. (2013) "A posteriori transit probabilities." Publications of the Astronomical Society of the Pacific, **125**, 933–950

Sussman, G.J., & Wisdom, J. (1988) "Numerical evidence that Pluto is chaotic." Science, **241**, 433–437

Sussman, G. J., & Wisdom, J. (1992) "Chaotic evolution of the solar system." Science, **257**, 56–62

Synnott, S. P., Terrile, R. J., Jacobson, R. A., & Smith, B. A. (1983) "Orbits of Saturn's F ring and its shepherding satellites." Icarus, **53**, 156–158

Szebehely, V. (1967) *The Theory of Orbits* (Academic Press, New York)

Szebehely, V. (1980) "Stability of planetary orbits in binary systems." Celest. Mech., **22**, 7–12

Szebehely, V., & McKenzie, R. (1981) "Stability of outer planetary systems." Celest. Mech., **23**, 3–7

Takeda, G., & Rasio, F. A. (2005) "High orbital eccentricities of extrasolar planets induced by the Kozai mechanism." Astrophys. J., **627**, 1001–1010

Tamayo, D., Burns, J. A., Hamilton, D. P., & Nicholson, P. D. (2013) "Dynamical instabilities in high-obliquity systems." Astron. J., **145**, 54 (12pp)

Tancredi, G. (1995) "The dynamical memory of Jupiter family comets." Astron. Astrophys., **299**, 288–292

Tancredi, G. (1999) "Chaotic dynamics of planet-encountering bodies." Celest. Mech. Dyn. Astron., **70**, 181–200

Tarnopolski, M. (2015) "Nonlinear time-series analysis of Hyperion's lightcurves." Astrophys. Space Sci., **357**, 160 (12pp)

Terai, T., & Itoh, Y. (2013) "High-precision measurements of the brightness variation of Nereid." Publ. Astron. Soc. Japan, **65**, Id. 46 (7pp)

Terzić, B., & Kandrup, H. E. (2003) "Semi-analytic estimates of Lyapunov exponents in lower-dimensional systems." Phys. Lett. A, **311**, 165–171

Thébault, P., Marzari, F., & Scholl, H. (2006) "Relative velocities among accreting planetesimals in binary systems: The circumprimary case." Icarus, **183**, 193–206

Thirouin, A., Noll, K. S., Ortiz, J. L., & Morales, N. (2014) "Rotational properties of the binary and non-binary populations in the trans-Neptunian belt." Astron. Astrophys., **569**, A3 (20pp)

Thomas P. C., Black G. J., & Nicholson P. D. (1995) "Hyperion: Rotation, shape and geology from Voyager images." Icarus, **117**, 128–148

Thomas, P., Veverka, J., & Dermott, S. (1986) "Small satellites." In: *Satellites*, ed. by Burns, J. A., & Matthews, M. S. (University of Arizona Press, Tucson) pp. 802–835

Thomas, P. C., Armstrong, J. W., Asmar, S. W., Burns, J. A., Denk, T., et al. (2007) "Hyperion's sponge-like appearance." Nature, **448**, 50–56

Timofeev, A. V. (1978) "On the constancy of an adiabatic invariant when the nature of the motion changes." Sov. Phys. JETP, **48**, 656–659

Tisserand, F.-F. (1896) *Traité de Mécanique Céleste IV* (Gauthier-Villars, Paris)

Tittemore, W. C. (1990) "Chaotic motion of Europa and Ganymede and the Ganymede–Callisto dichotomy." Science, **250**, 263–267

Tittemore, W., & Wisdom, J. (1988) "Tidal evolution of the Uranian satellites. I. Passage of Ariel and Umbriel through the 5:3 mean-motion commensurability." Icarus, **74**, 172–230

Tittemore, W., & Wisdom, J. (1989) "Tidal evolution of the Uranian satellites. II. An explanation for the anomalously high orbital inclination of Miranda." Icarus, **78**, 63–89

Tittemore, W., & Wisdom, J. (1990) "Tidal evolution of the Uranian satellites. III. Evolution through the Miranda–Umbriel 3:1, Miranda–Ariel 5:3, and Ariel–Umbriel 2:1 mean-motion commensurabilities." Icarus, **85**, 394–443

Tokovinin, A. A. (1997) "On the multiplicity of spectroscopic binary stars." Astron. Lett., **23**, 727–730

Torzhevsky, A. P. (1964) "Periodic solutions of the equation of planar librations of a satellite in an elliptic orbit." Kosmich. Issled., **2**, 667–678 (in Russian)

Treshchev, D. V. (1998) *An Introduction to the Perturbation Theory of Hamiltonian Systems* (FAZIS, Moscow) (in Russian)

Tsiganis, K., Varvoglis, H., & Dvorak, R. (2005) "Chaotic diffusion and effective stability of Jupiter Trojans." Celest. Mech. Dyn. Astron., **92**, 71–87

Udry, S., Fischer, D., & Queloz, D. (2007) "A decade of radial-velocity discoveries in the exoplanet domain." In: *Protostars and Planets V*, ed. by Reipurth, B., Jewitt, D., & Keil, K. (University of Arizona Press, Tucson) pp. 685–699

Umberger, D. K., & Farmer, J. D. (1985) "Fat fractals on the energy surface." Phys. Rev. Lett., **55**, 661–664

Urminsky, D. J., & Heggie, D. C. (2008) "On the relationship between instability and Lyapunov times for the three-body problem." Mon. Not. R. Astron. Soc., **392**, 1051–1059

Valluri, M., Debattista, V. P., Quinn, T. R., Roškar, R., & Wadsley, J. (2012) "Probing the shape and history of the Milky Way halo with orbital spectral analysis." Mon. Not. R. Astron. Soc., **419**, 1951–1969

Valsecchi, G. B. (2007) In: *Near Earth Objects, our Celestial Neighbors: Opportunity and Risk* (Proc. IAU Symp. 236), ed. by Milani, A., Valsecchi, G. B., & Vokrouhlický, D. (Cambridge Univ. Press, Cambridge) pp. xvii–xx

Valtonen, M., Mylläri, A., Orlov, V., et al. (2008) "The problem of three stars: Stability limit." In: *Dynamical Evolution of Dense Stellar Systems* (Proc. IAU Symp. 246), ed. by Vesperini, E., Giersz, M., & Sills, A. (Cambridge Univ. Press, Cambridge) pp. 209–217

Vecheslavov, V. V. (1999) "Fine structure of splitting of the separatrix of a nonlinear resonance." J. Exp. Theor. Phys., **89**, 182–187

Vecheslavov, V. V. (2002) "The chaotic layer of a nonlinear resonance under low-frequency perturbation." Tech. Phys., **47**, 160–167

Vecheslavov, V. V. (2004) "Chaotic layer of a pendulum under low and medium-frequency perturbations." Tech. Phys., **49**, 521–525

Vecheslavov, V. V., & Chirikov, B. V. (1988) "Chaotic dynamics of Comet Halley." Sov. Astron. Lett., **14**, 151–153

Vecheslavov, V. V., & Chirikov, B. V. (1998) "High-precision measurement of separatrix splitting in a nonlinear resonance." Sov. Phys. JETP, **87**, 823–831

Veerman, P. & Holmes, P. (1986) "Resonance bands in a two degree of freedom Hamiltonian system." Physica D, **20**, 413–422

Venegeroles, R. (2009) "Universality of algebraic laws in hamiltonian systems." Phys. Rev. Lett. **102**, 064101

Veras, D., & Armitage, P.J. (2004) "The dynamics of two massive planets on inclined orbits." Icarus, **172**, 349–371

Vokrouhlický, D., Nesvorný, D., & Bottke, W. F. (2003) "The vector alignments of asteroid spins by thermal torques." Nature, **425**, 147–151

Wang S., Ji J., & Zhou J.-L. (2012) "Predicting the configuration of a planetary system: KOI-152 observed by Kepler." Astrophys. J., **753**, 170

Weaver, H. A. (2016) "The small satellites of Pluto as observed by New Horizons." Science, **351**, 1281

Weiss, L. M., Marcy, G. W., Rowe, J. F., Howard, A. W., Isaacson, H., et al. (2013) "The mass of KOI-94d and a relation for planet radius, mass, and incident flux." Astrophys. J., **768**, 14 (19pp)

Welsh, W. F., Orosz, J. A., Carter, J. A., et al. (2012) "Transiting circumbinary planets Kepler-34b and Kepler-35b." Nature, **481**, 475–479

Welsh, W. F., Orosz, J. A., Carter, J. A., et al. (2014) "Recent Kepler results on circumbinary planets." In: *Formation, Detection, and Characterization of Extrasolar Habitable Planets* (Proc. IAU Symp. 293), ed. by Haghighipour, N. (Cambridge Univ. Press, Cambridge) pp. 125–132

Welsh, W., et al. (2015) "Kepler 453 b — The 10th Kepler transiting circumbinary planet." Astrophys. J., **809**, 26 (17pp)

Werner, R. A., & Scheeres, D. J. (1996) "Exterior gravitation of a polyhedron derived and compared with harmonic and mascon gravitation representations of asteroid 4769 Castalia." Celest. Mech., **65**, 313–344

Whipple, A. L. (1995) "Lyapunov times of the inner asteroids." Icarus, **115**, 347–353

Whitmire, D., Matese, J., Criswell, L., & Mikkola, S. (1998) "Habitable planet formation in binary star systems." Icarus, **132**, 196–203

Whittaker, E. T. (1927) *A Treatise on the Analytical Dynamics*, 3rd edition (Cambridge Univ. Press, Cambridge)

Williams, G. E. (1993) "History of the Earth's obliquity." Earth Sci. Rev., **34**, 1–45

Williams, J. G. (1969) "Secular perturbations in the solar system." Ph. D. thesis. (Univ. of California, Los Angeles)

Williams, J. G., & Faulkner, J. (1981) "The positions of secular resonance surfaces." *Icarus*, **46**, 390–399

Wisdom, J. (1980) "The resonance overlap criterion and the onset of stochastic behavior in the restricted three-body problem." Astron. J., **85**, 1122–1133

Wisdom, J. (1982) "The origin of the Kirkwood gaps: a mapping for asteroidal motion near the 3/1 commensurability." Astron. J., **87**, 577–593

Wisdom, J. (1983) "Chaotic behavior and the origin of the 3/1 Kirkwood gap." Icarus, **56**, 51–74

Wisdom, J. (1985) "A perturbative treatment of motion near the 3/1 commensurability." Icarus, **63**, 272–289

Wisdom, J. (1987a) "Rotational dynamics of irregularly shaped natural satellites." Astron. J., **94**, 1350–1360

Wisdom, J. (1987b) "Urey prize lecture: Chaotic dynamics in the Solar system." Icarus, **72**, 241–275

Wisdom, J. (2018) "The dynamical systems approach to numerical integration." Mon. Not. R. Astron. Soc., **474**, 3273–3279

Wisdom, J., Peale, S.J., & Mignard, F. (1984) "The chaotic rotation of Hyperion." Icarus, **58**, 137–152

Wobig, H. (1987) "Magnetic surfaces and localized perturbations in the Weldelstein VII-A stellarator." Z. Naturforsch., **42a**, 1054–1066

Wolszczan A., & Frail D.A. (1992) "A planetary system around the millisecond pulsar PSR 1257+12." Nature, **355**, 145–147

Wright, J. T., Upadhyay, S., Marcy, G. W., et al. (2009) "Ten new and updated multi-planet systems, and a survey of exoplanetary systems." Astrophys. J., **693**, 1084–1099

Wright, J. T., Veras, D., Ford, E. B., et al. (2011) "The California Planet Survey. III. A possible 2:1 resonance in the exoplanetary triple system HD 37124." Astrophys. J., **730**, 93 (9pp)

Wu, Y., & Lithwick, Y. (2011) "Secular chaos and the production of hot Jupiters." Astrophys. J., **735**, 109 (12pp)

Yamagishi, T. (1995) "Evaluation of the stochastic layer near separatrix in toroidal divertor." Fusion Technology, **27**, 505–508

Yu, Y., & Baoyin, H. X. (2012) "Orbital dynamics in the vicinity of asteroid 216 Kleopatra." Astron. J., **143**, 62–71

Zapatero Osorio, M. R., Béjar, V. J. S., Martín, E. L., Rebolo, R., Barrado y Navascués, D., Bailer-Jones, C. A. L., & Mundt, R. (2000) "Discovery of young, isolated planetary mass objects in the σ Orionis star cluster." Science, **290**, 103–107

Zappalà, V., Bendjoya, Ph., Cellino, A., Farinella, P., & Froeschlé, C. (1995) "Asteroid families: Search of a 12,487-asteroid sample using two different clustering techniques." Icarus, **116**, 291–314

Zaslavsky, G.M. (1984) *Stochasticity of Dynamical Systems* (Nauka, Moscow) (In Russian. English translation: Harwood Academics, Chur, Switzerland, 1985)

Zaslavsky, G. M. (1998) *Physics of Chaos in Hamiltonian Systems* (Imperial College Press, London)

Zaslavsky, G. M. (2005) *Hamiltonian Chaos and Fractional Dynamics* (Oxford University Press, New York)

Zaslavsky, G. M., & Abdullaev, S. S. (1995) "Scaling properties and anomalous transport of particles inside the stochastic layer." Phys. Rev. E, **51**, 3901–3910

Zaslavsky, G. M., & Chirikov, B. V. (1965) "Fermi acceleration mechanism in the one-dimensional case." Sov. Phys. Doklady, **9**, 989–992

Zaslavsky, G. M., & Chirikov, B. V. (1972) "Stochastic instability of non-linear oscillations." Sov. Phys. Uspekhi, **14**, 549–568

Zaslavsky, G. M., & Filonenko, N. N. (1968) "Stochastic instability of trapped particles and conditions of applicability of the quasi-linear approximation." Sov. Phys. JETP, **25**, 851–857

Zaslavsky, G. M., & Sagdeev, R. Z. *Introduction to Nonlinear Physics. From the Pendulum to Turbulence and Chaos* (Nauka, Moscow, 1988) (in Russian)

Zaslavsky, G. M., Sagdeev, R. Z., Usikov, D. A., & Chernikov, A. A. (1991) *Weak Chaos and Quasi-Regular Patterns* (Cambridge University Press, Cambridge)

Zhou, J.-L., & Sun, Y.-S. (2001) "Lévy flights in comet motion and related chaotic systems." Phys. Lett. A, **287**, 217–222

Zhou, J.-L., Sun, Y.-S., Zheng, J.-Q., & Valtonen, M. J. (2000) "The transfer of comets from near-parabolic to short-period orbits: Map approach." Astron. Astrophys., **364**, 887–893

Zhou, J.-L., Sun, Y.-S., & Zhou, L.-Y. (2002) "Evidence for Lévy random walks in the evolution of comets from the Oort Cloud." Celest. Mech. Dyn. Astron., **84**, 409–427

Zinnecker, H. (2001) "A free-floating planet population in the Galaxy?" In: *Microlensing 2000: A New Era of Microlensing Astrophysics* (ASP Conf. Ser., v. 239), ed. by Menzies, J., & Sackett, P. D. (Astron. Soc. Pacific, San Francisco) pp. 223–227

Index

I. I. Shevchenko, *Dynamical Chaos in Planetary Systems*, Astrophysics and Space Science Library 463, https://doi.org/10.1007/978-3-030-52144-8

The manufacturer's authorised representative in the EU is Springer
Nature Customer Service Centre GmbH, Europaplatz 3, 69115 Heidelberg,
Germany. If you have any concerns regarding our products, please
contact ProductSafety@springernature.com

Printed and bound by CPI Group (UK) Ltd, Croydon, CR0 4YY
30/09/2026
02225689-0014